W9-AFJ-190

Universitext

*Editorial Board
(North America):*

S. Axler
F.W. Gehring
P.R. Halmos

Springer
*New York
Berlin
Heidelberg
Barcelona
Budapest
Hong Kong
London
Milan
Paris
Santa Clara
Singapore
Tokyo*

Universitext

Editors (North America): S. Axler, F.W. Gehring, and P.R. Halmos

(continued after index)

Paul A. Fuhrmann

A Polynomial Approach
to Linear Algebra

 Springer

Paul A. Fuhrmann
Department of Mathematics
Ben-Gurion University of the Negev
Beer Sheva
Israel

Editorial Board
(North America):

S. Axler
Department of
    Mathematics
Michigan State University
East Lansing, MI 48824
USA

F.W. Gehring
Department of
    Mathematics
University of Michigan
Ann Arbor, MI 48109
USA

P.R. Halmos
Department of
    Mathematics
Santa Clara University
Santa Clara, CA 95053
USA

Mathematics Subject Classification (1991): 15-02, 15A04, 15A63, 93Axx

Library of Congress Cataloging-in-Publication Data
Fuhrmann, Paul Abraham.
    A polynomial approach to linear algebra/P.A. Fuhrmann.
        p.   cm. — (Universitext)
    Includes bibliographical references and index.
    ISBN 0-387-94643-8 (softcover: alk. paper)
    1. Algebras, Linear.   I. Title.
    QA184.F8   1996
    512′.5 – dc20   95-49237

Printed on acid-free paper.

© 1996 Springer-Verlag New York, Inc.
All rights reserved. This work may not be translated or copied in whole or in part without the
written permission of the publisher (Springer-Verlag New York, Inc., 175 Fifth Avenue, New
York, NY 10010, USA), except for brief excerpts in connection with reviews or scholarly
analysis. Use in connection with any form of information storage and retrieval, electronic
adaptation, computer software, or by similar or dissimilar methodology now known or here-
after developed is forbidden.
The use of general descriptive names, trade names, trademarks, etc., in this publication, even
if the former are not especially identified, is not to be taken as a sign that such names, as
understood by the Trade Marks and Merchandise Marks Act, may accordingly be used freely
by anyone.

Production managed by Francine McNeill; manufacturing supervised by Jacqui Ashri.
Photocomposed copy prepared using Springer's svsing.sty macro.
Printed and bound by R.R. Donnelley and Sons, Harrisonburg, VA.
Printed in the United States of America.

9 8 7 6 5 4 3 2 1

ISBN 0-387-94643-8 Springer-Verlag New York Berlin Heidelberg   SPIN 10490948

To Nilly

Preface

Linear algebra is a well-entrenched mathematical subject that is taught in virtually every undergraduate program in both the sciences and engineering. Over the years, many texts have been written on linear algebra; therefore, it is up to the author to justify the presentation of another book in this area to the public.

I feel that my justification for the writing of this book is based on a different choice of material and a different approach to the classical core of linear algebra. The main innovation in it is the emphasis placed on functional models and polynomial algebra as the best vehicle for the analysis of linear transformations and quadratic forms. In pursuing this innovation, a long-lasting trend in mathematics is being reversed. Modern algebra went from the specific to the general, abstracting the underlying unifying concepts and structures. The epitome of this trend was represented by the Bourbaki school. No doubt this was an important part in the development of modern mathematics, but it had its faults, too. It led to several generations of students who could not compute, nor could they give interesting examples of theorems they proved. Even worse, it increased the gap between pure mathematics and the general user of mathematics. It is the last group, which is made up of engineers and applied mathematicians, that is interested not only in understanding a problem, but also in its computational aspects. A very similar development occurred in functional analysis and operator theory. Initially, the axiomatization of Banach and Hilbert spaces led to a search for general methods and results. Although there were some significant successes in these directions, it soon became apparent, especially when trying to understand the structure of bounded operators, that one has to be much more specific. In particular, the introduction of functional models, through the work of Livsic, De Branges, Sz.-Nagy, and Foias, provided a new approach to structure theory. It is these ideas that I have taken as my motivation in the writing of this book.

In the present book, at least where the structure theory is concerned, we look at a special class of shift operators. These are defined by using polynomial modular arithmetic. The interesting fact about this class is its property of universality, in the sense that every cyclic operator is similar to a shift and every linear operator on a finite-dimensional vector space is similar to a direct sum of shifts. Thus, the shifts are the building blocks of an arbitrary linear operator.

Basically, the approach taken in this book is a variation on the study of a linear transformation via the study of the module structure induced by it over the ring of polynomials. While module theory provides great elegance, it is also difficult to grasp by students. Furthermore, it seems too far removed from computation. Matrix theory seems to be at the other extreme; it is concerned too much with computation and not enough with structure. Functional models, especially the polynomial models, lie on an intermediate level of abstraction between module theory and matrix theory.

The book includes specific chapters devoted to quadratic forms and the establishments of algebraic stability criteria. The emphasis is shared between the general theory and the specific examples, which are in this case the study of the Hankel and Bezout forms. This general area, via the work of Hermite, is one of the roots of the theory of Hilbert spaces. I feel that it is most illuminating to see the Euclidean algorithm and the associated Bezout identity not as isolated results, but as an extremely effective tool in the development of fast inversion algorithms for structured matrices.

Another innovation in this book is the inclusion of basic system-theoretic ideas. It is my conviction that it no longer is possible to separate in a natural way the study of linear algebra from the study of linear systems. The two topics have benefited greatly from cross-fertilization. In particular, the theory of finite-dimensional linear systems seems to provide an unending flow of problems, ideas, and concepts that are quickly assimilated in linear algebra. Realization theory is as much a part of linear algebra as is the long familiar companion matrix.

The inclusion of a whole chapter on Hankel norm approximation theory, or AAK theory as it is commonly known, is also a new addition as far as linear algebra books are concerned. This part requires very little mathematical knowledge not covered in the book, but a certain mathematical maturity is assumed. I believe that it is very much within the grasp of a well-motivated undergraduate. In this part, several results from early chapters are reconstructed in a context where stability is central. Thus, the rational Hardy spaces enter, and we have analytic models and shifts. Lagrange and Hermite interpolations are replaced by the Nevanlinna–Pick interpolation. Finally, coprimeness and the Bezout identity reappear, but over a different ring. I believe that the study of these analogies goes a long way toward demonstrating to the student the underlying unity of mathematics.

Let me explain the philosophy that underlies the writing of this book. In a way I share the aim of Halmos [1958] in trying to treat linear transformations on finite-dimensional vector spaces by methods of more general theories. These theories were functional analysis and operator theory in Hilbert space; this is still the case in this book. However, in the intervening years, operator theory has changed remarkably. The emphasis has moved from the study of self-adjoint and normal operators to the study of non-self-adjoint operators. The hope that a general structure theory for linear operators might be developed seems to be too naive. The methods utilizing Riesz–Dunford integrals proved to be too restrictive. On the other hand, a whole new area centering around the theory of invariant subspaces, and the construction and study of functional models, was developed. This new development had its roots not only in pure mathematics, but also in many applied areas, notably scattering, network, control theories, and some areas of stochastic processes as estimation and prediction theories.

I hope that this book will show how linear algebra is related to other, more advanced areas of mathematics. Polynomial models have their root in operator theory, especially that part of operator theory that centered around invariant subspace theory and Hardy spaces. Thus, the point of view adopted here provides a natural link with that area of mathematics, as well as those application areas I have already mentioned.

In writing this book, I chose to work almost exclusively with scalar polynomials, the one exception being the invariant factor algorithm and its application to structure theory. My choice was influenced by the desire to have the book accessible to most undergraduates. Virtually all results about scalar polynomial models have polynomial matrix generalizations, and some of the appropriate references are pointed out in the "Notes and Remarks" sections.

The exercises at the end of chapters have been chosen partly to indicate directions not covered in the book. I have refrained from including routine computational problems. This does not indicate a negative attitude toward computation. Quite to the contrary, I am a great believer in the exercise of computation, and I suggest that readers choose, and work out, their own problems. This is the best way to get a better grasp of the presented material.

I usually use the first seven chapters for a one-year course on linear algebra at the Ben-Gurion University. If the group is a bit more advanced, one can supplement this by more material on quadratic forms. The material on quadratic forms and stability can be used as a one-semester course of special topics in linear algebra. Also, the material on linear systems and Hankel norm approximations can be used as a basis for either a one-term course or a seminar.

Beer Sheva, Israel   Paul A. Fuhrmann

Contents

1
Preliminaries

1.1 Maps

Let S be a set. If between elements of the set a relation $a \simeq b$ is defined, so that either $a \simeq b$ holds or not, then we say that we have a **binary relation**. If a binary relation in S satisfies the following conditions:

1. $a \simeq a$ holds for all $a \in S$,

2. $a \simeq b \Leftarrow b \simeq a$,

3. $a \simeq b$ and $b \simeq c \Leftarrow a \simeq c$,

then we say that we have an **equivalence relation** in S. The three conditions are referred to as **reflexivity**, **symmetry**, and **transitivity**, respectively.

For each $a \in S$ we define its equivalence class S_a by $S_a = \{x \in S | x \simeq a\}$. Clearly, $S_a \subset S$ and $S_a \neq \emptyset$.

An equivalence relation leads to a "partition" of the set S. By a **partition** of S we mean a representation of S as the disjoint union of subsets. Since, clearly, using transitivity, either $S_a \cap S_b = \emptyset$ or $S_a = S_b$, and $S = \cup_{a \in S} S_a$, the set of equivalence classes is a partition of S.

Similarly, any partition $S = \cup_\alpha S_\alpha$ defines an equivalence relation by letting $a \simeq b$ if for some α we have $a, b \in S_\alpha$.

A rule that assigns to each member $a \in A$ a unique member $b \in B$ is called a **map** or a **function** from A into B. We will denote this by $f : A \longrightarrow B$ or $A \overset{f}{\longrightarrow} B$.

We denote by $f(A)$ the image of the set A defined by $f(A) = \{y|y \in B,$ there exists an $x \in A$ s.t. $y = f(x)\}$. The inverse image of a subset $M \subset B$ is defined by $f^{-1}(M) = \{x|x \in A, \ f(x) \in M\}$. A map $f : A \longrightarrow B$ is called **injective**, or 1-1, if $f(x) = f(y)$ implies $x = y$. A map $f : A \longrightarrow B$ is called **surjective**, or onto, if $f(A) = B$, for example, for each $y \in B$ there exists an $x \in A$ such that $y = f(x)$.

Given maps $f : A \longrightarrow B$ and $g : B \longrightarrow C$, we can define a map $h : A \longrightarrow C$ by letting $h(x) = g(f(x))$. We call this map h the **composition** or **product** of the maps f and g. This will be denoted by $h = g \circ f$. Given three maps $A \xrightarrow{f} B \xrightarrow{g} C \xrightarrow{h} D$, we compute

$$h \circ (g \circ f)(x) = h(g(f(x)))$$

and

$$(h \circ g) \circ f(x) = h(g(f(x))).$$

So the product of maps is associative, that is,

$$h \circ (g \circ f) = (h \circ g) \circ f.$$

Due to the associative law of composition, we can write $h \circ g \circ f$ and, more generally, $f_n \circ \cdots \circ f_1$, unambiguously.

Given a map $f : A \longrightarrow B$, we define an equivalence relation R in A by letting

$$x_1 \simeq x_2 \Leftrightarrow f(x_1) = f(x_2).$$

Thus the equivalence class of a is given by $A_a = \{x|x \in A, \ f(x) = f(a)\}$. We will denote by A/R the set of equivalence classes and refer to this as the quotient set by the equivalence relation.

Next we define three transformations,

$$A \xrightarrow{f_1} A/R \xrightarrow{f_2} f(A) \xrightarrow{f_3} B,$$

with the f_i defined by

$$\begin{aligned} f_1(a) &= A_a \\ f_2(A_a) &= f(a) \\ f_3(b) &= b, \ b \in f(A). \end{aligned}$$

Clearly the map f_1 is surjective, f_2 is bijective, and f_3 is injective. Moreover, we have

$$f = f_3 \circ f_2 \circ f_1.$$

This factorization of f is referred to as the **canonical factorization**. The canonical factorization also can be described via the following commutative diagram:

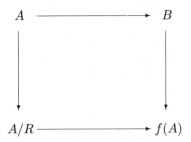

We note that $f_2 \circ f_1$ is surjective whereas $f_3 \circ f_2$ is injective.

1.2   Groups

Given a set M, a **binary operation** is a map from $M \times M$ into M. Thus an ordered pair (a, b) is mapped into an element of M denoted by ab.

A set M with an associative binary operation is called a **semigroup**. Thus, if $a, b \in M$, we have $ab \in M$ and the associative rule is $a(bc) = (ab)c$. Thus the product $a_1 \cdots a_n$ of elements of M is unambiguously defined.

We proceed to define the notion of a **group**, which is the cornerstone of most mathematical structures.

Definition 1.2.1 *A* **group** *is a set G with a binary operation, called multiplication, that satisfies*

1. *$a(bc) = (ab)c$, that is, the associative law.*

2. *There exists a left identity $e \in G$, that is, $ea = a$ for all $a \in G$.*

3. *For each $a \in G$ there exists a left inverse, denoted by a^{-1}, that satisfies $a^{-1}a = e$.*

4. *A group G is called **abelian** if the group operation is commutative, that is, if $ab = ba$ holds for all $a, b \in G$.*

Theorem 1.2.1

1. *Let G be a group and let a be an element of G. Then a left inverse a^{-1} of a is also a right inverse.*

2. *A left identity is also a right identity.*

3. *The identity element of a group is unique.*

Proof:

1. We compute

$$(a^{-1})^{-1}a^{-1}aa^{-1} = ((a^{-1})^{-1}a^{-1})(aa^{-1}) = e(aa^{-1})$$
$$= aa^{-1} = (a^{-1})^{-1}(a^{-1}a)a^{-1} = (a^{-1})^{-1}(ea^{-1}) = (a^{-1})^{-1}a^{-1} = e.$$

So, in particular, $aa^{-1} = e$.

2. Let $a \in G$ be arbitrary and let e be a left identity. Then

$$aa^{-1}a = a(a^{-1}a) = ae = (aa^{-1})a = ea = a.$$

Thus $ae = a$ for all a. So e is also a right identity.

3. Let e, e' be two identities in G. Then, using the fact that e is a left identity and e' a right identity, we get

$$e = ee' = e'. \qquad \square$$

In a group G, equations of the type $axb = c$ are easily solvable with the solution given by $x = a^{-1}cb^{-1}$. Also, it is easily checked that we have the following rule for inversion:

$$(a_1 \cdots a_n)^{-1} = a_n^{-1} \cdots a_1^{-1}.$$

Definition 1.2.2 *A subset H of a group G is called a **subgroup** of G if it is a group with the composition rule inherited from G. Thus H is a subgroup if, with $a, b \in H$, we have $ab \in H$ and $a^{-1} \in H$.*

This can be made a bit more concise.

Lemma 1.2.1 *A subset H of a group G is a subgroup if and only if, with $a, b \in H$, $ab^{-1} \in H$ also.*

Proof: If H is a subgroup, then with $a, b \in H$ it also contains b^{-1} and hence also ab^{-1}.

Conversely, if $a, b \in H$ implies $ab^{-1} \in H$, then $b^{-1} = eb^{-1} \in H$ and hence also $ab = a(b^{-1})^{-1} \in H$, $a, b \in H$. $\qquad \square$

Given a subgroup H of a group G, we say that two elements $a, b \in G$ are **equivalent**, and we write $a \simeq b$ if $b^{-1}a \in H$. It is easily checked that this is a bona fide equivalence relation in G, that is, it is a reflexive, symmetric, and transitive relation. We denote by M_a the equivalence class of a, that is,

$$M_a = \{x | x \in G, x \simeq a\}.$$

If we denote by aH the set $\{x | ah, h \in H\}$, then $M_a = aH$. We will refer to these as **right equivalence classes** or as **right cosets**. Left equivalence classes or **left cosets** Ha are defined in a completely analogous way.

Given a subgroup H of G, it is not usually the case that the sets of left and right cosets coincide. If this is the case, then we say that H is a **normal subgroup** of G.

Assuming that a left coset aH is also a right coset Hb, we clearly have $a = ae \in H$ and hence $a \in Hb$, so necessarily $Hb = Ha$. Thus for a normal subgroup, $aH = Ha$ for all $a \in G$. Equivalently, H is normal if and only if for all $a \in G$ we have $aHa^{-1} = H$.

Given a subgroup H of a group G, any two cosets can be mapped bijectively onto each other. In fact, the map $\phi(ah) = bh$ is such a bijection between aH and bH.

We will define the **index** of a subgroup H in G, and denote it by $i_G(H)$, as the number of left cosets. The index will also be denoted by $[G : H]$. Given the trivial subgroup of G given by $E = e$, the left and right cosets contain single elements. Thus $[G : E]$ is just the number of elements of the group G. $[G : E]$ also will be referred to as the **order** of the group G.

We now can proceed to the connection between index and order.

Theorem 1.2.2 (Lagrange) *Given a subgroup H of a group G, we have*

$$[G : H][H : E] = [G : E]$$

or

$$o(G) = i_G(H)o(H). \qquad \square$$

Homomorphisms are maps that preserve given structures. In the case of groups, given groups G and G_1, a map $\phi : G \longrightarrow G_1$ is called a **homomorphism** if, for all $g_1, g_2 \in G$, we have

$$\phi(g_1 g_2) = \phi(g_1)\phi(g_2).$$

Lemma 1.2.2 *Let G and G_1 be groups with unit elements e and e', respectively. Let $\phi : G \longrightarrow G_1$ be a homomorphism. Then*

1. $\phi(e) = e'$.

2. $\phi(x^{-1}) = (\phi(x))^{-1}$.

Proof:

1. Let $g \in G$. Then we compute

$$\phi(x)e' = \phi(x) = \phi(xe) = \phi(x)\phi(e).$$

   Multiplying by $\phi(x)^{-1}$, we get $\phi(e) = e'$.

2. We compute

$$e' = \phi(e) = \phi(xx^{-1}) = \phi(x)\phi(x^{-1}).$$

   This shows that $\phi(x^{-1}) = (\phi(x))^{-1}$. $\qquad \square$

A homomorphism $\phi : G \longrightarrow G_1$ that is both injective and surjective is called an **isomorphism**. In this case, G and G_1 will be called isomorphic.

The general canonical factorization of maps discussed in Section 1.1 now can be applied to the special case of group homomorphisms. To this end we define, for a homomorphism $\phi : G \longrightarrow G_1$, the kernel and image of ϕ by

$$Ker\,\phi = \phi^{-1}\{e'\} = \{g \in G | \phi(g) = e'\}$$

and

$$\text{Im}\phi = \phi(G) = \{g' \in G_1 | \text{there exists a } g \in G, \phi(g) = g'\}.$$

The kernel of a group homomorphism has a special property.

Lemma 1.2.3 *Let $\phi : G \longrightarrow G'$ be a group homomorphism, and let $N = \text{Ker}\,\phi$. Then N is a normal subgroup of G.*

Proof: Let $x \in G$ and $n \in N$. Then

$$\begin{aligned} \phi(xnx^{-1}) &= \phi(x)\phi(n)\phi(x^{-1}) = \phi(x)e'\phi(x^{-1}) \\ &= \phi(x)\phi(x^{-1}) = e'. \end{aligned}$$

So $xnx^{-1} \in N$. This implies that $xNx^{-1} \subset N$ for every $x \in G$, which implies that $N \subset x^{-1}Nx$. Since this inclusion holds for all $x \in G$, we get $N = xNx^{-1}$ or $Nx = xN$, that is, the left and right cosets are equal. So N is a normal subgroup. $\qquad\square$

Note now that given $x, y \in G$ and a normal subgroup N of G, we can define a product in the set of all cosets by

$$xN \cdot yN = xyN. \tag{1.1}$$

Theorem 1.2.3 *Let $N \subset G$ be a normal subgroup. Denote by G/N the set of all cosets and define the product of cosets by Eq. (1.1). Then G/N is a group called the **factor group** of G by N.*

Proof: Clearly Eq. (1.1) shows that G/N is closed under multiplication. To check associativity, we note that

$$\begin{aligned} (xN \cdot yN) \cdot zN &= (xyN) \cdot zN = (xy)zN \\ &= x(yz)N = xN \cdot (yzN) \\ &= xN \cdot (yN \cdot zN). \end{aligned}$$

For the unit element e of G, we have $eN = N$ and $eN \cdot xN = (ex)N = xN$. So $eN = N$ is the unit element in G/N. Finally, given $x \in G$, we check that $(xN)^{-1} = x^{-1}N$. $\qquad\square$

Theorem 1.2.4 *N is a normal subgroup of the group G if and only if N is the kernel of a group homomorphism.*

Proof: By Lemma 1.2.3 it suffices to show that, if N is a normal subgroup, it is the kernel of a group homomorphism. We do this by constructing such a homomorphism. Let G/N be the factor group of G by N. Define $\pi : G \longrightarrow G/N$ by

$$\pi(g) = gN. \tag{1.2}$$

Clearly Eq. (1.1) shows that π is a group homomorphism. Moreover, $\pi(g) = N$ if and only if $gN = N$, and this holds if and only if $g \in N$. Thus, $\mathrm{Ker}\,\pi = N$. $\qquad\square$

The map $\pi : G \longrightarrow G/N$ defined by Eq. (1.2) is called the **canonical projection**.

A homomorphism ϕ whose kernel contains a normal subgroup can be factored through the factor group.

Proposition 1.2.1 *Let $\phi : G \longrightarrow G'$ be a homomorphism with $\mathrm{Ker}\,\phi \supset N$, where N is a normal subgroup of G. Let π be the canonical projection of G onto G/N. Then there exists a unique homomorphism $\overline{\phi} : G/N \longrightarrow G'$ for which $\phi = \overline{\phi} \circ \pi$, or, equivalently, the following diagram is commutative:*

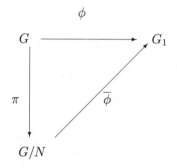

Proof: We define $\overline{\phi}(xN) = \phi(x)$. This map is well defined as $Ker\,\phi \supset N$. It is a homomorphism as is ϕ, and of course

$$\phi(x) = \overline{\phi}(xN) = \overline{\phi}(\pi(x)) = (\overline{\phi} \circ \pi)(x).$$

Finally, $\overline{\phi}$ is uniquely defined by $\overline{\phi}(xN) = \phi(x)$. $\qquad\square$

We call $\overline{\phi}$ the **induced map** by ϕ on G/N. We derive next an important result, the prototype of many others, that classifies images of group homomorphisms.

Theorem 1.2.5 *Let $\phi : G \longrightarrow G'$ be a surjective group homomorphism with $\mathrm{Ker}\,\phi = N$. Then G' is isomorphic to the factor group G/N.*

Proof: The induced map is clearly injective. In fact, if $\overline{\phi}(xN) = e'$, we get $\phi(x) = e'$ or $x \in Ker\,\phi = N$. It is also surjective by the assumption that ϕ is surjective. So we conclude that $\overline{\phi}$ is an isomorphism. $\qquad\square$

1.3 Rings and Fields

In this section we collect some basic algebraic material that is necessary for the understanding of most of what follows.

Definition 1.3.1 *A* **ring** *is a set with two laws of composition called* **addition** *and* **multiplication** *which satisfy for all elements of R the following:*

1. *Laws of addition*

    (a) *Associative law:* $a + (b + c) = (a + b) + c$.

    (b) *Commutative law:* $a + b = b + a$.

    (c) *Solvability of the equation* $a + x = b$.

2. *Laws of multiplication*

    (a) *Associative law:* $a(bc) = (ab)c$.

    (b) *Distributive laws:*
    $$a(b + c) = ab + ac$$
    $$(b + c)a = ba + ca.$$

    R is called **commutative** *if*

3. *Commutative law:* $ab = ba$ *holds for all* $a, b \in R$.

Law 1(c) implies the existence of a unique zero element. Also, R is a ring with **identity** if there exists an element $e \in R$ such that for all $a \in R$ $ea = ae = a$. An element a in a ring with identity has a **right inverse** b if $ab = e$ and a **left inverse** if $ba = e$. If a has both left and right inverses, they must be equal, and then we say that a is invertible and denote its inverse by a^{-1}.

A **field** is a commutative ring in which every nonzero element is invertable.

Definition 1.3.2 *Let R and R_1 be two rings. A* **ring homomorphism** *is a function $\phi : R \longrightarrow R_1$ that satifies*
$$\begin{cases} \phi(x + y) = \phi(x) + \phi(y) \\ \phi(xy) = \phi(x)\phi(y). \end{cases}$$

If R and R_1 are rings with identities, then we also require that $\phi(e) = e'$.

The kernel of a ring homomorphism has special properties. In fact, if $x, y \in Ker\,\phi$ and $r \in R$, then $x + y, rx \in Ker\,\phi$ also. This leads to the following definition.

Definition 1.3.3 *A subset J of a ring R is called an* **ideal** *if $x, y \in J$ and $r \in R$ implies that $x + y, rx \in J$.*

Thus, a subset J of R is a **left ideal** if it is an additive subgroup of R and $RJ \subset J$. If R contains an identity, then $RJ = J$. For right ideals, we replace the second condition by $JR = J$. A two-sided ideal, or just an **ideal**, is a subset of R that is both a left and a right ideal.

Proposition 1.3.1

1. *The sum of a finite number of left ideals is a left ideal.*

2. *The intersection of any set of left ideals in R is a left ideal.*

Proof:

1. Let $J_1, \ldots, J_k$ be left ideals in R. Then $J = J_1 + \cdots + J_k = \{a_1 + \cdots + a_k | a_i \in J_i\}$. Clearly, if $a_i, b_i \in J_i$ and $r \in R$, then

$$\sum_{i=1}^{k} a_i + \sum_{i=1}^{k} b_i = \sum_{i=1}^{k} (a_i + b_i) \in J$$

and

$$r \sum_{i=1}^{k} a_i = \sum_{i=1}^{k} r a_i \in J.$$

2. Let $J = \cap_\alpha J_\alpha$ with J_α left ideals. If $a, b \in J$, then $a, b \in J_\alpha$ for all α. Hence $a + b \in J_\alpha$ for all α, which implies that $a + b \in J$. A similar argument holds to show that $ra \in J$. $\qquad\square$

Given a two-sided ideal J in a ring R, we can construct the **quotient ring**, denoted by R/J, whose elements are the cosets $a + J$ by defining the operations of addition and multiplication by

$$\begin{cases} (a + J) + (b + J) &= (a + b) + J \\[2mm] (a + J)(b + J) &= ab + J. \end{cases}$$

It is easy to check that, with the arithmetic operations so defined, R/J is indeed a ring.

The following theorem gives a complete characterization of ideals. It is the counterpart, in the setting of rings, of Theorem 1.2.4.

Theorem 1.3.1 *Let R be a ring. A subset $J \subset R$ is a two-sided ideal if and only if it is the kernel of a ring homomorphism.*

Proof: We saw already that the kernel of a ring homomorphism is a two-sided ideal. Now let J be a two-sided ideal. We define the canonical projection $\pi : R \longrightarrow R/J$ by $\pi(a) = a + J$. It is easy to check that π is a surjective ring homomorphism, with $Ker\,\pi = J$. $\qquad\square$

An ideal in a ring R that is generated by a single element, that is, of the form $J = \{rd|r \in R\}$, is called a **principal ideal**. The element d is called the **generator** of the ideal. More generally, given $a_1, \ldots, a_n \in R$, the set $J = \{\sum_{i=1}^{n} r_i a_i | r_i \in R\}$ is obviously an ideal. We say that $a_1, \ldots, a_n$ are the generators of this ideal.

A commutative ring R with no zero divisors is called a **principal ideal domain** if every ideal in R is principal.

In a ring R, we have a division relation. If $c = ab$, we say that a **divides** c or that a is a **divisor** or **factor** of c. Given $a_1, \ldots, a_n \in R$, we say that a is a common divisor of the a_i if it is a divisor of all a_i. a is a **greatest common divisor**, or g.c.d., if it is a common divisor and is divisible by any other common divisor. Two greatest common divisors differ by a factor that is invertible in R. We say that $a_1, \ldots, a_n \in R$ are **coprime** if their greatest common divisor is 1.

Proposition 1.3.2 *Let R be a principal ideal domain. Then $a_1, \ldots, a_n \in R$ are coprime if and only if there exist $b_i \in R$ for which the **Bezout identity***

$$a_1 b_1 + \cdots + a_n b_n = 1$$

holds.

Proof: If there exist $b_i \in R$ for which the Bezout identity holds, then any common divisor of the a_i is a divisor of 1 and hence necessarily invertible.

Conversely, we consider the ideal J generated by the a_i. Since R is a principal ideal domain, J is generated by a single element d that necessarily is invertible. So $1 \in J$, and hence there exist b_i such that the Bezout identity holds. $\qquad\square$

We present now a few examples of rings. We pay special attention to the ring of polynomials, due to the central role it plays in this book. Much of the results concerning ideals, factorizations, the Chinese remainder theorem, and so on, hold also in the ring of integers. We do not give separate proofs for those nor, for the sake of concreteness, do we give proofs in the general context of Euclidean domains.

The integers. The set of integers $\mathbf{Z}$ is a commutative ring under the usual operations of addition and multiplication.

The polynomial ring. A **polynomial** is an expression of the form

$$p(z) = \sum_{i=0}^{n} a_i z^i.$$

We shall denote by $F[z]$ the set of all polynomials with coefficients in F, that is, $F[z] = \{\sum_{i=0}^n a_i z^i | a_i \in F, 0 \le n \in \mathbf{Z}\}$. Two polynomials are called **equal** if all of their coefficients coincide. If $p(z) = \sum_{i=0}^n a_i z^i$ and $a_n \ne 0$, then we say that n is the **degree** of the polynomial p, which we denote by $\deg p$. We define the degree of the zero polynomial to be $-\infty$.

Given two polynomials $p(z) = \sum_{i=0}^n a_i z^i$ and $q(z) = \sum_{i=0}^m b_i z^i$, we define their sum $p + q$ by

$$(p+q)(z) = \sum_{i=0}^{max(m,n)} (a_i + b_i)z^i.$$

The product, pq, is defined by

$$(pq)(z) = \sum_{i=0}^{m+n} c_i z^i,$$

where

$$c_i = \sum_{j=0}^{i} a_j b_{i-j}.$$

It is easily checked that, with these operations of addition and multiplication, $F[z]$ is a commutative ring with an identity.

The next theorem sums up the most elementary properties of polynomials.

Theorem 1.3.2 *Let p, q be polynomials in $F[z]$. Then*

1. $\deg(pq) = \deg p + \deg q$.

2. $\deg(p + q) \le \max\{\deg p, \deg q\}$.

Proof:

1. If p or q is the zero polynomial, then both sides of the equality are equal to $-\infty$. So we assume that both p and q are nonzero. Let $p(z) = \sum_{i=0}^n a_i z^i$ and $q(z) = \sum_{i=0}^m b_i z^i$, with $a_n, b_m \ne 0$. Then $c_{n+m} = a_n b_m \ne 0$ but $c_k = 0$ for $k > m + n$.

2. This is immediate.   □

Corollary 1.3.1 *If $p, q \in F[z]$ and $pq = 0$, then $p = 0$ or $q = 0$.*

Proof: If $pq = 0$, then

$$-\infty = \deg pq = \deg p + \deg q.$$

So either $\deg p = -\infty$ or $\deg q = -\infty$.   □

In $F[z]$, as in $\mathbf{Z}$, we have a process of division with remainder.

Lemma 1.3.1 *Given a nonzero polynomial $p(z) = \sum_{i=0}^{m} p_i z^i$ with $p_m \neq 0$, an arbitrary polynomial $q \in F[z]$ can be written uniquely in the form*

$$q(z) = a(z)p(z) + r(z) \tag{1.3}$$

with $\deg(r) < \deg(p)$.

Proof: If degree of q is less than the degree of p, then we write $q = 0 \cdot p + q$, and this is the required representation. So we may assume that $\deg(q) = n \geq m = \deg(p)$. The proof will proceed by induction on the degree of q. We assume that for all polynomials of degree less than n such a representation exists. Clearly,

$$q_1(z) = q(z) - q_n p_m^{-1} z^{n-m} p(z)$$

is a polynomial of degree $\leq n - 1$. Hence, by the induction hypothesis,

$$q_1(z) = a_1(z)p(z) + r_1(z)$$

with $\deg(r_1) < \deg(p)$. But this implies that

$$q(z) = (q_n p_m^{-1} z^{n-m} + a_1(z))p(z) + r(z) = a(z)p(z) + r(z)$$

with

$$a(z) = q_n p_m^{-1} z^{n-m} + a_1(z).$$

To show uniqueness let

$$q(z) = a_1(z)p(z) + r_1(z) = a_2(z)p(z) + r_2(z).$$

This implies that

$$(a_1 - a_2)p = r_2 - r_1.$$

A consideration of the degrees of both sides shows that necessarily they both are equal to zero. Hence the uniqueness of the representation in Eq. (1.3). □

The properties of the degree function in the ring $F[z]$ can be abstracted to general rings.

Definition 1.3.4 *A ring R is called a **Euclidean ring** if there exists a function δ from the set of nonzero elements in R into the set of nonnegative integers that satisfies:*

1. *For all $a, b \neq 0$ we have $\delta(ab) \geq \delta(a)$.*

2. *For all $f, g \in R$, with $g \neq 0$, there exist $a, r \in R$ such that $f = ag + r$ and $\delta(r) < \delta(g)$.*

We can define $\delta(0) = -\infty$.

Obviously, with this definition, the ring of polynomials $F[z]$ is a Euclidean ring. We note that in a Euclidean ring there are no zero divisors.

It is convenient to have a notation for the remainder of a polynomial f divided by q. If $f = aq + r$ with $\deg r < \deg q$, we shall write $r = \pi_q f$.

We give several properties of the operation of taking a remainder.

Lemma 1.3.2 *Let $q, a, b \in F[z]$ with q nonzero. Then*

$$\pi_q(a\pi_q b) = \pi_q(ab). \tag{1.4}$$

Proof: Let $b = b_1 q + \pi_q b$. Then $ab = ab_1 q + a\pi_q b$. Obviously, $\pi_q ab_1 q = 0$ and hence Eq. (1.4) follows. □

Corollary 1.3.2 *Given polynomials $a_i \in F[z], i = 1, \ldots, k$,*

$$\pi_q(a_1 \cdots a_k) = \pi_q(a_1 \cdots \pi_q a_k).$$

Proof: By induction. □

The following result simplifies in some important cases the computation of the remainder.

Lemma 1.3.3 *Let $f, p, q \in F[z]$, with p, q nonzero. Then*

$$\pi_{pq} pf = p\pi_q f. \tag{1.5}$$

Proof: Let $r = \pi_q f$, that is, for some polynomial a, we have $f = aq + r$ and $\deg r < \deg q$. The representation of f implies $pf = a(pq) + pr$. Since

$$\deg pr = \deg p + \deg r < \deg p + \deg q = \deg(pq),$$

it follows that $\pi_{pq} pf = pr = p\pi_q f$, and hence Eq. (1.5) holds. □

Definition 1.3.5 *Let $p, q \in F[z]$. We say that p **divides** q, or that p is a **factor** of q, and write $p|q$ if there exists a polynomial a such that $q = pa$.*

If $p \in F[z]$ and $p(z) = \sum_{i=0}^{n} a_i z^i$, then p defines a function on F given by

$$p(\alpha) = \sum_{i=0}^{n} a_i \alpha^i, \quad \alpha \in F.$$

$p(\alpha)$ is called the **value** of p at α. A scalar $\alpha \in F$ is called a **zero** of p if $p(\alpha) = 0$. We never identify the polynomial with the function defined by it.

Theorem 1.3.3 *Let $p \in F[z]$. Then α is a zero of p if and only if $(z-\alpha)|p$.*

Proof: If $(z-\alpha)|p$, then $p(z) = (z-\alpha)a(z)$, and hence $p(\alpha) = 0$. Conversely, by the division rule, we have

$$p(z) = a(z)(z - \alpha) + r(z),$$

with r necessarily a constant. Substituting α in this equality implies that $r = 0$. □

Theorem 1.3.4 *Let $p \in F[z]$ be a polynomial of degree n. Then p has at most n zeroes in F.*

Proof: The proof is by induction. The statement is certainly true for zero-degree polynomials. Assume that we have proved it for all polynomials of degree less than n. Suppose that p is a polynomial of degree n. Either it has no zeroes and the statement holds, or there exists a zero α. But then $p(z) = (z - \alpha)a(z)$, and a has, by the induction hypothesis, at most $n - 1$ zeroes. $\qquad\square$

Theorem 1.3.5 *Let F be a field and $F[z]$ the ring of polynomials over F. Then $F[z]$ is a principal ideal domain.*

Proof: Let J be any ideal in $F[z]$. If $J = \{0\}$, then J is generated by 0. So let us assume that $J \neq \{0\}$. Let d be any nonzero polynomial in J of minimal degree. We will show that $J = dF[z]$.

To this end, let $f \in J$ be an arbitrary element. By the division rule of polynomials we have $f = ad + r$ with $\deg r < \deg d$. Now $r = f - ad \in J$ as both f and D are in J. Since d was a nonzero element of smallest degree, we must have $r = 0$. So $f \in dF[z]$, and hence $J \subset dF[z]$. Conversely, since $d \in J$, we have $dF[z] \subset J$, and so equality follows. $\qquad\square$

The availability of the division process in $F[z]$ leads directly to the **Euclidean algorithm**. This gives an algorithm for the compuation of a g.c.d. of two polynomials.

Theorem 1.3.6 *Let $p, q \in F[z]$. Set $q_{-1} = q$ and $q_0 = p$. Define inductively, using the division rule for polynomials, a sequence of polynomials q_i by*

$$q_{i+1} = a_{i+1}q_i - q_{i-1} \tag{1.6}$$

with $\deg q_{i+1} < \deg q_i$. Let q_s be the last nonzero remainder, that is, $q_{s+1} = 0$. Then q_s is a g.c.d. of q and p.

Proof: First we show that q_s is a common divisor of q and p. Indeed, since $q_{s+1} = 0$, we have $q_{s-1} = a_{s+1}q_s$, that is, $q_s | q_{s-1}$. Assume that we have proved $q_s | q_{s-1}, \dots, q_i$. Since $q_{i-1} = a_{i+1}q_i - q_{i+1}$, it follows that $q_s | q_{i-1}$ and, hence, by induction q_s divides all q_i, in particular q_0, q_{-1}. So q_s is a common divisor of q and p.

Let $J(q_i, q_{i-1}) = \{aq_i + bq_{i-1} | a, b \in F[z]\}$ be the ideal generated by q_i, q_{i-1}. Clearly, Eq. (1.6) shows, again using an induction argument, that

$$q_{i+1} \in J(q_i, q_{i-1}) \subset J(q_{i-1}, q_{i-2}) \subset \cdots \subset J(q_0, q_{-1}).$$

In particular, $q_s \in J(q_0, q_{-1})$. So there exist polynomials a, b such that $q_s = ap + bq$. This shows that any common divisor r of p and q also divides q_s. Thus q_s is a g.c.d. of p and q. $\qquad\square$

We remark that the polynomials a, b in the representation $q_s = ap + bq$ can be calculated easily from the polynomials a_i. We will return to this in Chapter 8.

Corollary 1.3.3 *Let $p, q \in F[z]$. Then p and q are coprime if and only if the* **Bezout equation**

$$a(z)p(z) + b(z)q(z) = 1$$

is solvable in $F[z]$.   □

In view of Theorem 1.3.1, the easiest way to construct ideals is by taking sums and intersections of kernels of ring homomorphisms. The case of interest for us is for the ring of polynomials.

Definition 1.3.6 *We define for each $\alpha \in F$ a map $\phi_\alpha : F[z] \longrightarrow F$ by*

$$\phi_\alpha(p) = p(\alpha). \tag{1.7}$$

Theorem 1.3.7 *A map $\phi : F[z] \longrightarrow F$ is a ring homomorphism if and only if $\phi(p) = p(\alpha)$ for some $\alpha \in F$.*

Proof: Let $\phi : F[z] \longrightarrow F$ be a ring homomorphism. Set $\phi(z) = \alpha$. Then, given $p(z) = \sum_{i=0}^{k} p_i z^i$, we have

$$
\begin{aligned}
\phi(p) &= \phi \sum_{i=0}^{k} p_i z^i = \sum_{i=0}^{k} p_i \phi(z^i) \\
&= \sum_{i=0}^{k} p_i \phi(z)^i = \sum_{i=0}^{k} p_i \alpha^i = p(\alpha).
\end{aligned}
$$
□

Corollary 1.3.4 *Given $\alpha_1, \ldots, \alpha_n \in F$, the set*

$$J_{\alpha_1,\ldots,\alpha_n} = \{p \in F[z] | p(\alpha_1) = \cdots = p(\alpha_n) = 0\}$$

is an ideal in $F[z]$. Moreover, $J_{\alpha_1,\ldots,\alpha_n} = dF[z]$, where $d(z) = \Pi_{i=1}^{n}(z - \alpha_i)$.

Proof: For ϕ_α defined by Eq. (1.7), we have $Ker\, \phi_\alpha = \{p \in F[z] | p(\alpha) = 0\} = J_\alpha$, which is an ideal. Clearly $J_{\alpha_1,\ldots,\alpha_n} = \cap_{i=1}^{n} J_{\alpha_i}$, and the intersection of ideals is an ideal.

Obviously, for d defined as above and an arbitrary polynomial f, we have $(df)(\alpha_i) = d(\alpha_i)f(\alpha_i) = 0$, so $df \in J_{\alpha_1,\ldots,\alpha_n}$. Conversely, if $g \in J_{\alpha_1,\ldots,\alpha_n}$, we have $g(\alpha_i) = 0$, and hence g is divisible by $z - \alpha_i$. Since the α_i are distinct, g is divisible by d, or $g = df$.   □

Proposition 1.3.3 *Let $d \in F[z]$. Then*

$$dF[z] = \{dp | p \in F[z]\}$$

is an ideal.

Definition 1.3.7

1. *Given polynomials* $p_1, \ldots, p_n \in F[z]$, *a polynomial* $d \in F[z]$ *will be called a* **greatest common divisor** *of* $p_1, \ldots, p_n \in F[z]$ *if*

   (a) *We have the division relation* $d|p_i$ *for all* $i = 1, \ldots, n$.

   (b) *If* $d_1|p_i$ *for all* $i = 1, \ldots, n$, *then* $d_1|d$.

2. *Given polynomials* $p_1, \ldots, p_n \in F[z]$, *a polynomial* $d \in F[z]$ *will be called a* **least common multiple** *(l.c.m.) of* $p_1, \ldots, p_n \in F[z]$, *if*

   (a) *We have the division relation* $p_i|d$ *for all* $i = 1, \ldots, n$.

   (b) *If* $p_i|d'$ *for all* $i = 1, \ldots, n$, *then* $d|d'$.

It is easily shown that a greatest common divisor is unique up to a constant multiple.

We will say that polynomials $p_1, \ldots, p_n \in F[z]$ are **coprime** if their greatest common divisor is 1.

The next, important, result relates the generator of an ideal to division properties.

Theorem 1.3.8 *Let* $p_1, \ldots, p_n \in F[z]$. *Then the ideal* J *generated by* $p_1, \ldots, p_n$, *namely,*

$$J = \{\sum_{i=1}^{n} r_i p_i | r_i \in F[z]\} \tag{1.8}$$

has the representation $J = dF[z]$ *if and only if* d *is a greatest common divisor of* $p_1, \ldots, p_n \in F[z]$.

Proof: By Theorem 1.3.5 there exists a $d \in J$ such that $J = (d)$. Since $p_i \in J$, there exist polynomials q_i such that $p_i = dq_i$ for $i = 1, \ldots, n$. So d is a common divisor af the p_i. We will show that it is maximal. Assume that d' is another common divisor of the p_i, that is, $p_i = d's_i$. As $d \in J$, there exist polynomials a_i such that $d = \sum_{i=1}^{n} a_i p_i$. Therefore,

$$d = \sum_{i=1}^{n} a_i p_i = \sum_{i=1}^{n} a_i d' q_i = d' \sum_{i=1}^{n} a_i q_i.$$

But this means that $d'|d$ and so d is a g.c.d.   □

Corollary 1.3.5 *Let* $p_1, \ldots, p_n \in F[z]$, *and let* d *be their greatest common divisor. Then there exist polynomials* $a_1, \ldots, a_n \in F[z]$ *such that*

$$d(z) = \sum_{i-1}^{n} a_i(z) p_i(z). \tag{1.9}$$

Proof: Let d be the g.c.d. of the p_i. Obviously, $d \in J = \{\sum_{i=1}^{n} r_i p_i | r_i \in F[z]\}$. Therefore there exist $a_i \in F[z]$ for which Eq. (1.9) holds.   □

Corollary 1.3.6 *Polynomials $p_1, \ldots, p_n \in F[z]$ are coprime if and only if there exist polynomials $a_1, \ldots, a_n \in F[z]$ such that*

$$\sum_{i=1}^{n} a_i(z) p_i(z) = 1. \tag{1.10}$$

Equation (1.10) is one of the most important equations in mathematics. We will refer to it as the **Bezout equation**.

The importance of polynomials in linear algebra stems from the strong connection between the factorization of polynomials and the structure of linear transformations. The primary decomposition theorem is of particular applicability.

Definition 1.3.8 *A polynomial $p \in F[z]$ is **factorizable** or **reducible** if there exist polynomials $f, g \in F[z]$ of degree ≥ 1 such that $p = fg$. If p is not factorizable, it is called a **prime** or an **irreducible** polynomial.*

Note that the reducibility of a polynomial is dependent on the field F.

Theorem 1.3.9 *Let $p, f, g \in F[z]$, and assume that p is irreducible and $p|(fg)$. Then either $p|f$ or $p|g$.*

Proof: Assume that $p|(fg)$ but p does not divide f. Then the g.c.d. of p and f is 1. There exist therefore polynomials a, b such that the Bezout equation $1 = af + bp$ holds. From this it follows that

$$g = a(fg) + (bg)p.$$

This implies that $p|g$. □

Corollary 1.3.7 *Let p be an irreducible polynomial and assume that $p|(f_1 \cdots f_n)$. Then there exists an index i for which $p|f_i$.*

Proof: By induction. □

Lemma 1.3.4 *Let p and q be coprime. Then, if $p|qs$, it follows that $p|s$.*

Proof: By coprimeness, there exist polynomials a, b such that $ap + bq = 1$ and hence $s = aps + bqs$. This shows that $p|s$. □

Lemma 1.3.5 *Let p, q be coprime. Then pq is their l.c.m.*

Proof: Clearly, pq is a common multiple. Let s be an arbitrary common multiple. In particular, we can write $s = qt$. Since p and q are coprime and $p|s$, it follows that $p|t$ or that $t = pt'$. Thus $s = (pq)t'$, and therefore pq is a least common multiple. □

A polynomial $p \in F[z]$ is called **monic** if its highest nonzero coefficient is 1.

Theorem 1.3.10 *Let p be a monic polynomial in $F[z]$. Then p has a unique, up to ordering, factorization into a product of prime polynomials.*

Proof: We prove the theorem by induction. If $\deg p = 1$, the statement is trivially true.

Assume that we have proved the statement for all polynomials of degree $< n$. Let p be of degree n. Then either p is prime or $p = fg$ with $1 < \deg f, \deg g < n$. By the induction hypothesis both f and g are decomposable into the product of prime polynomials. This implies the existence of a decomposition of p into the product of primes.

It remains to prove uniqueness. Suppose that $p_1, \ldots, p_m$ and $q_1, \ldots, q_n$ are all prime and that

$$p_1 \cdots p_m = q_1 \cdots q_n.$$

Clearly, $p_m | q_1 \cdots q_n$; so, by Corollary 1.3.7, there exists an i such that $p_m | q_i$. Since both are monic and irreducible, it follows that $p_m = q_i$. Without loss of generality we may assume that $p_m = q_n$. Since there are no zero divisors in $F[z]$, we get

$$p_1 \cdots p_{m-1} = q_1 \cdots q_{n-1}.$$

We finish by using the induction hypothesis. □

Corollary 1.3.8 *Given a nonzero, monic polynomial $p \in F[z]$, there exist monic primes p_i and positive integers n_i, $i = 1, \ldots, s$ such that*

$$p = p_1^{n_1} \cdots p_s^{n_s}. \tag{1.11}$$

The primes p_i and the integers n_i are uniquely determined.

Proof: Follows from the previous theorem. □

The factorization in Eq. (1.11) is called the **primary decomposition** of p. The monicity assumption is only necessary to get uniqueness. Without it, the theorem still holds, but the primes are only determined up to constant factors.

The next result relates division in the ring of polynomials to the geometry of submodules.

Proposition 1.3.4

1. Let $p, q \in F[z]$. Then $qF[z] \subset pF[z]$ if and only if $p|q$.

2. Let $p_i \in F[z]$ for $i = 1, \ldots, n$. Then

$$\bigcap_{i=1}^{n} p_i F[z] = pF[z],$$

where p is the l.c.m. of the p_i.

3. *Let $p_i \in F[z]$ for $i = 1, \ldots, n$. Then*

$$\sum_{i=1}^{n} p_i F[z] = pF[z],$$

where p is the g.c.d. of the p_i.

Proof:

1. Assume that $qF[z] \subset pF[z]$. Thus there exists a polynomial f for which $q = pf$, that is, $p|q$.

   Conversely, assume that $p|q$, that is, $q = pf$ for some polynomial f. Then

   $$qF[z] = \{q \cdot g | g \in F[z]\} = \{pfq | g \in F[z]\} \subset \{ph | h \in F[z]\} = pF[z].$$

2. Assume that $\cap_{i=1}^{m} p_i F[z] = pF[z]$. Clearly, $pF[z] \subset p_i F[z]$ for all i. By part (1) this implies that $p_i|p$. So p is a common multiple of the p_i. Then $q = p_i q_i$, and so $qF[z] \subset p_i F[z]$ for all i, which implies that $qF[z] \subset \cap_{i=1}^{m} p_i F[z] = pF[z]$. But this inclusion shows that $p|q$ and hence p is the l.c.m of the p_i.

   Conversely, let p be the l.c.m of the p_i, $i = 1, \ldots, m$. Thus $p = p_i g_i$ for some $g_i \in F[z]$. This implies that $pF[z] \subset p_i F[z]$ and hence $pF[z] \subset \cap_{i=1}^{m} p_i F[z]$. Since the intersection of ideals is an ideal, there exists a polynomial q for which $qF[z] = \cap_{i=1}^{m} p_i F[z]$. This implies that $q|p$. But p is the l.c.m of the p_i so $p|q$. Hence p and q differ by a constant, nonzero factor.

3. Assume now that $\sum_{i=1}^{m} p_i F[z] = pF[z]$. Obviously, $p_i F[z] \subset pF[z]$ for all i and hence $p|p_i$. Thus q is a common divisor for the p_i. Let q be any other common divisor. Then $p_i F[z] \subset qF[z]$, and hence $pF[z] = \sum_{i=1}^{m} p_i F[z] \subset qF[z]$, which shows that $q|p$, that is p is the g.c.d.

   Conversely, let $\sum_{i=1}^{m} p_i F[z] = pF[z]$. Then $p_i F[z] \subset pF[z]$ for all i. This shows that $p|p_i$, that is, p is a common divisor. Let q be any other common divisor. The relation $q|p_i$ implies that $p_i F[z] \subset qF[z]$, and hence

   $$pF[z] = \sum_{i=1}^{m} p_i F[z] \subset qF[z],$$

   which shows that $q|p$. We conclude that p is the g.c.d of the p_i. $\square$

For the case of two polynomials we have the following.

Proposition 1.3.5 *Let p, q be nonzero polynomials, and let r and s be their g.c.d. and l.c.m., respectively. Then we have $pq = rs$.*

Proof: Write $p = rp_1, q = rq_1$ with p_1, q_1 coprime. Clearly $s = rp_1q_1$ is a common multiple of p, q. Let s' be any other multiple. As $p|s'$, we have $s' = rp_1t$ for some polynomial t. Since $q|s'$, we have $q_1|p_1t$. Since p_1, q_1 are coprime, it follows from Lemma 1.3.4 that $q_1|t$. This shows that $s = rp_1q_1$ is the l.c.m. of p and q. The equality $pq = rs$ is now obvious.   □

Formal power series. For a given field we denote by $F[[z]]$ the set of all formal power series, that is, the set of formal sums of the form $f(z) = \sum_{j=0}^{\infty} f_j z^j$. Addition and multiplication are defined by

$$\sum_{j=0}^{\infty} f_j z^j + \sum_{j=0}^{\infty} g_j z^j = \sum_{j=0}^{\infty} (f_j + g_j) z^j$$

and

$$\sum_{j=0}^{\infty} f_j z^j \cdot \sum_{j=0}^{\infty} g_j z^j = \sum_{k=0}^{\infty} h_k z^k$$

with

$$h_k = \sum_{j=0}^{k} f_j g_{k-j}.$$

$F[[z]]$ is a ring.

An element $f(z) = \sum_{j=0}^{\infty} f_j z^j \in F[[z]]$ is invertible if and only if $f_0 \neq 0$. To see this, let $g(z) = \sum_{j=0}^{\infty} g_j z^j$. Then g is an inverse of f if and only if

$$1 = (fg)(z) = \sum_{k=0}^{\infty} \left\{ \sum_{j=0}^{k} f_j g_{k-j} \right\} z^k.$$

This is equivalent to the solvability of the infinite system of equations

$$\sum_{j=0}^{k} f_j g_{k-j} = \begin{cases} 1 & k = 0 \\ 0 & k > 0. \end{cases}$$

The first equation is $f_0 g_0 = 1$, which shows the necessity of the condition $f_0 \neq 0$. This is also sufficient as the system of equations can be solved recursively.

The following result analyzes the ideal structure in $F[[z]]$.

Proposition 1.3.6 $J \subset F[[z]]$ *is a nonzero ideal if and only if, for some nonnegative integer* n, *we have* $J = z^n F[[z]]$. *Thus* $F[[z]]$ *is a principal ideal domain.*

Proof: Clearly, any set of the form $z^n F[[z]]$ is an ideal. To prove the converse, we set, for $f(z) = \sum_{j=0}^{\infty} f_j z^j \in F[[z]]$,

$$\delta(f) = \begin{cases} \min\{n | f_n \neq 0\} & f \neq 0 \\ -\infty & f = 0. \end{cases}$$

Now let $f \in J$ be any nonzero element that minimizes $\delta(f)$. Then $f(z) = z^n h(z)$ with h invertible. Thus $z^n \in J$ and generates it.   □

In the sequel we find it convenient to work with the ring $F[[z^{-1}]]$ of formal power series in z^{-1}.

We now study an important construction that allows us to construct some ring from larger rings. The prototype of this situation is the construction of the field of rational numbers out of the ring of integers.

Given rings R and $\bar{R}$, we say that R is **embedded** in $\bar{R}$ if there exists an injective homomorphism of R into $\bar{R}$. In a ring R, a nonzero element a is called a **zero divisor** if there exists another nonzero element $b \in R$ such that $ab = 0$. A commutative ring without a zero divisor is called an **entire ring** or an **integral domain**.

A set S in a ring R with identity is called a **multiplicative set** if $0 \notin S$, $1 \in S$, and $a, b \in S$ also imply that $ab \in S$.

Given a commutative ring with identity R and a multiplicative set S in R, we proceed to construct a new ring. Let $\mathcal{M}$ be the set of ordered pairs (r, s) with $r \in R$ and $s \in S$. We introduce a relation in $\mathcal{M}$ by saying that $(r, s) \simeq (r', s')$ if there exists a $\sigma \in S$ for which $\sigma(s'r - sr') = 0$. We claim that this is indeed an equivalence relation. Reflexivity and symmetry are trivial to check. To check transitivity, assume that $\sigma(s'r - sr') = 0$ and $\tau(s"r' - s'r") = 0$ with $\sigma, \tau \in S$. We compute

$$
\begin{aligned}
\tau \sigma s'(s"r) &= \tau s" \sigma(s'r) = \tau s" \sigma(sr') \\
&= \sigma s \tau(s"r') = \sigma s \tau(s'r") \\
&= \sigma s'\tau(sr")
\end{aligned}
$$

or $\tau \sigma s'(s"r - sr") = 0$.

We denote by r/s the equivalence class of the pair (r, s). We denote by R/S the set of all equivalence classes in $\mathcal{M}$. In R/S we define operations of addition and multiplication by

$$
\begin{cases}
\dfrac{r}{s} + \dfrac{r'}{s'} &= \dfrac{rs' + sr'}{ss'} \\[3mm]
\dfrac{r}{s} \cdot \dfrac{r'}{s'} &= \dfrac{rr'}{ss'}.
\end{cases}
\tag{1.12}
$$

It can be checked that these operations are well defined, that is, they are independent of the equivalence class representatives. Also, it is straightforward to verify that, with these operations, R/S is a commutative ring. This is called a **ring of quotients** of R. Of course, there may be many such rings, depending on the multiplicative sets we are taking.

In case R is an entire ring, then the map $\phi : R \longrightarrow R/S$ defined by $\phi(r) = r/1$ is an injective ring homomorphism. More can be said about this case.

Theorem 1.3.11 *Let R be an entire ring. Then R can be embedded in a field.*

Proof: Let $S = R - \{0\}$, that is, S is the set of all nonzero elements in R. Obviously, S is a multiplicative set. We let $F = R/S$. F is a commutative ring with identity. To show that F is a field, it suffices to show that any nonzero element is invertible. If $a/b \neq 0$, this implies that $a \neq 0$ and hence $(a/b)^{-1} = b/a$. Moreover, the map ϕ given before provides an embedding of R in F.   □

The field F constructed by the previous theorem is called the **field of quotients** of R.

For our purposes, the most important example of a field of quotients is that of the field of rational functions, denoted by $F(z)$, which is obtained as the field of quotients of the ring of polynomials $F[z]$.

Let us consider now an entire ring that is also a principal ideal domain. Let F be the field of fractions of R. Given any $f \in F$, we can consider $J = \{r \in R | rf \in R\}$. Obviously, J is an ideal and hence generated by an element $q \in R$ which is uniquely defined up to an invertible factor. Thus $qf = p$ for $p \in R$ and $f = p/q$. Obviously, p, q are coprime, and $f = p/q$ is called a **coprime factorization** of f.

We give now a few examples of rings and fields that are products of the process of taking a ring of quotients.

Rational functions. For a field F we saw that the ring of polynomials $F[z]$ is a principal ideal domain. Its field of quotients is called the **field of rational functions** and is denoted by $F(z)$. Its elements are called **rational functions**.

Every rational function has a representation of the form $p(z)/q(z)$ with p, q coprime. We can make the coprime factorization unique if we require the polynomial q to be monic. By Lemma 1.3.1, every polynomial p has a unique representation in the form $p = aq + r$ with $\deg r < \deg q$. This implies that

$$\frac{p}{q} = a + \frac{r}{q}. \tag{1.13}$$

A rational function r/q is called **proper** if $\deg r \leq \deg q$ and **strictly proper** if $\deg r < \deg q$. The set of all strictly proper rational functions is denoted by $F_-(z)$. Thus we have

$$F(z) = F[z] \oplus F_-(z).$$

Here the direct sum representation refers to the uniqueness of the representation in Eq. (1.13).

Proper rational functions. We denote by $F_{pr}(z)$ the subring of $F(z)$ defined by

$$F_{pr}(z) = \{f \in F(z) | f = \frac{p}{q}, \deg p < \deg q\}. \tag{1.14}$$

It is easily checked that $F_{pr}(z)$ is a commutative ring with identity. An element f in $F_{pr}(z)$ is a unit if and only if in any representation $f = p/q$ we have $\deg p < \deg q$. We define the **relative degree** ρ by $\rho(p/q) = \deg q - \deg p$ and $\rho(0) = -\infty$.

Theorem 1.3.12 $F_{pr}(z)$ *is a Euclidean domain with respect to the relative degree.*

Proof: We verify the axioms of a Euclidean ring as given in Definition 1.3.4. Let ρ be the relative degree. Let $f = p_1/q_1$ and $g p_2/q_2$. Then

$$
\begin{aligned}
\rho(fg) &= \rho\left(\frac{p_1 p_2}{q_1 q_2}\right) = \deg(q_1 q_2) - \deg(p_1 p_2) \\
&= \deg q_1 + \deg q_2 - \deg p_1 - \deg p_2 \\
&= (\deg q_1 - \deg p_1) + (\deg q_2 - \deg p_2) \\
&= \rho(f) + \rho(g) \geq \rho(f).
\end{aligned}
$$

Next we turn to division. If $\rho(f) < \rho(g)$, we write $f = 0 \cdot g + f$. If $g \neq 0$ and $\rho(f) \geq \rho(g)$, we show that g divides f. Let σ and τ be the relative degrees of f and g. So we can write

$$f = \frac{1}{z^\sigma} f_1, \qquad g = \frac{1}{z^\tau} g_1,$$

with f_1, g_1 units in $F_{pr}(z)$, that is, satisfying $\rho(f_1) = \rho(g_1) = 0$. Then we can write

$$f = (\frac{1}{z^\tau} g_1)(\frac{1}{z^{\sigma-\tau}} g_1^{-1} f_1).$$

Of course, $1/z^{\sigma-\tau} g_1^{-1} f_1 \in F_{pr}(z)$ and has relative degree $\sigma - \tau$. $\square$

Since $F_{pr}(z)$ is Euclidean, it is a principal ideal domain. We proceed to characterize all ideals in $F_{pr}(z)$.

Theorem 1.3.13 *A subset $J \subset F_{pr}(z)$ is a nonzero ideal if and only if it is of the form $J = 1/z^\sigma F_{pr}(z)$ for some nonnegative integer σ.*

Proof: Clearly, $J = 1/z^\sigma F_{pr}(z)$ is an ideal.

Conversely, let J be a nonzero ideal. Let f be a nonzero element in J of least relative degree, say σ. We may assume without loss of generality that $f = 1/z^\sigma$. By the proof of Theorem 1.3.12, f divides any element with relative degree greater than or equal to σ. So f is a generator of J. $\square$

Heuristically speaking, and adopting the language of complex analysis, we see that ideals are determined by the zeroes at infinity, with multiplicities counted. In terms of singularities, $F_{pr}(z)$ is the set of all rational functions that have no singularity (pole) at infinity. In comparison, $F[z]$ is the set of all rational functions whose only singularity is at infinity. Of course, there are many intermediate situations, and we turn to these next.

Stable rational functions. Fix the field to be the real field $\mathbf{R}$. We can identify many subrings of $\mathbf{R}(z)$ simply by taking the ring of quotients of $\mathbf{R}[z]$ with respect to multiplicative sets that are smaller than the set of all nonzero polynomials. We shall say that a polynomial is stable if all of its (complex) zeros are in a given subset Σ of the complex plane. It is antistable if all of its zeros lie in the complement of Σ. We will assume now that the domain of stability is the open left half-plane. Let S be the set of all stable polynomials. S is obviously a multiplicative subset of $\mathbf{R}[z]$. The ring of quotients $\mathcal{S} = \mathbf{R}[z]/S \subset \mathbf{R}(z)$ is called the **ring of stable rational functions**. As a ring of fractions, it is a commutative ring with identity. An element $f \in \mathcal{S}$ is a unit if in an irreducible representation $f = p/q$ the numerator p is stable. Thus we may say that $f \in \mathcal{S}$ is a unit if it has no zeros in the closed left half-plane. Such functions sometimes are called **minimum phase** functions.

From the point of view of complex analysis, the ring of stable rational functions is the set of rational functions that have all of their singularities in the open left half-plane or at the point at infinity.

In $\mathcal{S}$ we define a degree function δ as follows. Let $f = p/q$ with p, q coprime. We let $\delta(f)$ be the number of antistable zeros of p and hence of f. Note that δ does not take into account zeros at infinity.

Theorem 1.3.14 $\mathcal{S}$ *is a Euclidean ring with respect to the degree function* δ.

Proof: That $\delta(fg) = \delta(f) + \delta(g) \geq \delta(f)$ is obvious. Now let $f, g \in \mathcal{S}$ with $g \neq 0$. Assume without loss of generality that $\delta(f) \geq \delta(g)$. Let $g = \alpha_g/\beta_g$ with $\alpha_g, beta_g$ coprime polynomials. Similarly, let $f = \alpha_f/\beta_f$ with $\alpha_f, beta_f$ coprime. Factor $\alpha_g = \alpha_+\alpha_-$, with α_- stable and α_+ antistable. Then

$$g = \frac{\alpha_+\alpha_-}{\beta_g} = \frac{\alpha_-(z+1)^\nu}{\beta_g} \cdot \frac{\alpha_+}{(z+1)^\nu} = e(z) \cdot \frac{\alpha_+}{(z+1)^\nu},$$

with e a unit and $\nu = \delta(g)$. As β_f is stable, β_f, α_+ are coprime in $\mathbf{R}[z]$ and there exist polynomials ϕ, ψ for which $\phi\alpha_+ + \psi\beta_f = \alpha_f(z+1)^{\nu-1}$, and we may assume without loss of generality that $\deg \psi < \deg \alpha_+$. Dividing both sides by $\beta_f(z+1)^{\nu-1}$, we get

$$\frac{\alpha_f}{\beta_f} = \left(\frac{\phi(z+1)}{\beta_f}\right)\left(\frac{\alpha_+}{(z+1)^{\nu-1}}\right) + \left(\frac{\psi}{(z+1)^{\nu-1}}\right),$$

and we check that

$$\delta\left(\frac{\alpha_+}{(z+1)^{\nu-1}}\right) \le \deg\psi < \deg\alpha_+ = \delta(g). \qquad \square$$

Ideals in $\mathcal{S}$ can be represented by generators that are, up to unit factors, antistable polynomials. For two antistable polynomials p_1, p_2, we have $p_2\mathcal{S} \subset p_1\mathcal{S}$ if and only if $p_1|p_2$. More generally, for $f_1, f_2 \in \mathcal{S}$, we have $f_2\mathcal{S} \subset f_1\mathcal{S}$ if and only if $f_1|f_2$. However, the division relation $f_1|f_2$ means that every zero of f_1 in the closed right half-plane is also a zero of f_2 with at least the same multiplicity. Here zeros should be interpreted as complex zeros.

Since the intersection of a subring is itself a ring, we can consider the ring $\mathbf{R}_{pr}(z) \cap \mathcal{S}$. This is equivalent to excluding from $\mathcal{S}$ elements with a singularity at infinity. We denote this ring by $\mathbf{RH}_+^\infty$ and proceed to discuss it.

Bounded, stable, real rational functions. We define $\mathbf{RH}_+^\infty$, with notation and terminology following the customary ones in functional analysis and control theory, to be the subset of the ring of stable rational functions that are also proper. So

$$\mathbf{RH}_+^\infty = \left\{\frac{p}{q} \mid q \in \mathcal{S}, \deg p \le \deg q\right\}.$$

Thus $\mathbf{RH}_+^\infty$ is the set of all rational functions that are uniformly bounded in the closed right half-plane. This is clearly a commutative ring with an identity.

Given $f = p/q \in \mathbf{RH}_+^\infty$ and a factorization $p = p_+p_-$ into the stable factor p_- and antistable factor p_+, we use the notation $\nabla f = p_+$ and set $\pi_+ = \deg\nabla f$. We define a degree function $\delta : \mathbf{RH}_+^\infty \longrightarrow \mathbf{Z}_+$ by

$$\delta(f) = \begin{cases} \deg\nabla f + \rho(f) & f \ne 0 \\ -\infty & f = 0. \end{cases}$$

Obviously, $f \in \mathbf{RH}_+^\infty$ is invertible in $\mathbf{RH}_+^\infty$ if and only if it can be represented as the quotient of two stable polynomials of equal degree. This happens if and only if $\delta(f) = 0$.

Let us now fix a monic, stable polynomial σ. To be specific, we can choose $\sigma(z) = z + 1$. Given $f = p/q \in \mathbf{RH}_+^\infty$, f can be brought, by multiplication by an invertible element, to the form $p_+/\sigma^{\nu+\pi}$, where $\nu = \deg q - \deg p$.

Proposition 1.3.7 *Given $f_i = p_i/q_i \in \mathbf{RH}_+^\infty, i = 1, 2$ with $\nabla(p_i/q_i) = p_i^+$ and $\pi_{i,+} = \deg p_i^+$, we set $\nu_i = \deg q_i = \deg p_i$. Let p_+ be the greatest common divisor of p_1^+, p_2^+ and set $\pi_+ = \deg p_+$ and $\nu = \min\{\nu_1, \nu_2\}$. Then $p_+/\sigma^{\nu+\pi_+}$ is a greatest common divisor of $p_i/q_i, i = 1, 2$.*

Proof: Without loss of generality we can assume that

$$f_1 = \frac{p_1^+}{\sigma^{\nu_1+\pi_{1,+}}}, f_2 = \frac{p_2^+}{\sigma^{\nu_2+\pi_{2,+}}}.$$

Writing $p_i^+ = p_+ \hat{p}_i^+$, we have $f_i = \frac{p_+}{\sigma^{\nu+\pi_+}} \cdot \frac{\hat{p}_i^+}{\sigma^{\nu_i-\nu+\pi_{i,+}-\pi_+}}$. Now

$$\rho\left(\frac{\hat{p}_i^+}{\sigma^{\nu_i-\nu+\pi_{i,+}-\pi_+}}\right) = \nu_i - \nu + \pi_{i,+} - \pi_+ - \pi_{i,+} + \deg p_+ = \nu_i - \nu \geq 0.$$

This shows that $\hat{p}_i^+/\sigma^{\nu_i-\nu+\pi_{i,+}-\pi_+}$ is proper and hence $p_+/\sigma^{\nu+\pi_+}$ is a common factor.

Now let r/s be any other common factor of $p_i/q_i, i = 1, 2$. Without loss of generality we can assume that r is antistable. Let

$$\frac{p_i}{q_i} = \frac{r}{s} \cdot \frac{u_i}{t_i}.$$

From the equality $st_i p_i^+ = ru_i \sigma^{\nu_i+\pi_{i,+}}$, it follows that r_+, the antistable factor of r, divides p_i^+ and hence also divides p_+, the greatest common divisor of the p_i^+. Let us write $p_+ = r_+ \hat{r}_+$. Since u/t is proper, we have $\deg s - \deg r \leq \deg q_i - \deg p_i = \nu_i$ and hence $\deg s - \deg r \leq \nu$. Now we write

$$\frac{p_+}{\sigma^{\nu+\pi_+}} = \frac{r_+\hat{r}_+}{\sigma^{\nu+\pi_+}} = \frac{r_+\hat{r}_+r_-}{s} \cdot \frac{s}{\sigma^{\nu+\pi_++r_-}}$$
$$= \frac{r}{s} \cdot \frac{s}{\sigma^{\nu+\pi_++r_-}}.$$

We compute now the relative degree of the right factor:

$$\nu + \pi_+ + \deg r_- - \deg s = \nu + \deg r_- + \deg p_+ - \deg s + (\pi_+ - \deg p_+)$$
$$= \nu + \deg r - \deg s \geq 0.$$

This shows that $s/\sigma^{\nu+\pi_++r_-}$ is proper and hence $p_+/\sigma^{\nu+\pi_+}$ is indeed a greatest common divisor of p_1/q_1 and p_2/q_2. □

Theorem 1.3.15 *Let $f_1, f_2 \in \mathbf{RH}_+^\infty$ be coprime. Then there exist $g_i \in \mathbf{RH}_+^\infty$ such that*

$$g_1 f_1 + g_2 f_2 = 1. \tag{1.15}$$

Proof: We may assume without loss of generality that

$$f_1 = \frac{p_1}{\sigma^{\pi_1}}, f_2 = \frac{p_2}{\sigma^{\nu+\pi_2}}$$

with p_i coprime, antistable polynomials, and $\pi_i = \deg p_i$.

Clearly, there exist polynomials α_1, α_2 for which $\alpha_1 p_1 + \alpha_2 p_2 = \sigma^{\pi_1 + \pi_2 + \nu}$, and we may assume without loss of generality that $\deg \alpha_2 < \deg p_1$. Dividing by the right-hand side, we obtain

$$\frac{\alpha_1}{\sigma^{\pi_2 + \nu}} \cdot \frac{p_1}{\sigma^{\pi_1}} + \frac{\alpha_2}{\sigma^{\pi_1}} \cdot \frac{p_2}{\sigma^{\pi_2 + \nu}} = 1.$$

Now $p_2 / \sigma^{\pi_2 + \nu}$ is proper, whereas $\alpha_2 / \sigma^{\pi_1}$ is strictly proper by the condition $\deg \alpha_2 < \deg p_1$; so the product of these functions is strictly proper. Next, we note that the relative degree of p_1 / σ^{π_1} is zero, which forces $\alpha_1 / \sigma^{\pi_2 + \nu}$ to be proper. Defining

$$g_1 = \frac{\alpha_1}{\sigma^{\pi_2 + \nu}}, \qquad g_2 = \frac{\alpha_2}{\sigma^{\pi_1}},$$

we have, by the stability of σ, that $g_i \in \mathbf{RH}_+^\infty$ and that the Bezout identity in Eq. (1.15) is satisfied. $\qquad\square$

Corollary 1.3.9 *Let $f_1, f_2 \in \mathbf{RH}_+^\infty$ and let f be a greatest common divisor of f_1, f_2. Then there exist $g_i \in \mathbf{RH}_+^\infty$ such that*

$$g_1 f_1 + g_2 f_2 = f. \tag{1.16}$$

Theorem 1.3.16 *The ring $\mathbf{RH}_+^\infty$ is a principal ideal domain.*

Proof: It suffices to show that any ideal $J \subset \mathbf{RH}_+^\infty$ is principal. This is clearly the case for the zero ideal. So we may as well assume that J is a nonzero ideal.

Now any nonzero element $f \in J$ can be wriiten as $f = p/q = p_+ p_- / q$ with p, q coprime, q stable, and p factored into its stable factor p_- and antistable factor p_+.

Of all nonzero elements in J, we choose an f for which $\pi_+ = \deg \nabla f$ is minimal. Let $\mathcal{C} = \{\{ \in \mathcal{J} \mid \deg \{ = \pi_+ \}$. In $\mathcal{C}$ we choose an arbitrary element g of minimal relative degree. We will show that g is a generator of J.

To this end, let f be an arbitrary element in J. Let h be a greatest common divisor of f and g. By Corollary 1.3.9, there exist $k, l \in \mathbf{RH}_+^\infty$ for which $kf + lg = h$.

Since $\deg \nabla g$ is minimal, it follows that $\deg \nabla g \le \deg \nabla h$. On the other hand, as h divides g, we have $\nabla h | \nabla g$ and hence $\deg \nabla g \ge \deg \nabla h$; so the equality $\deg \nabla g = \deg \nabla h$ follows. This shows that $h \in \mathcal{C}$ and hence $\rho(g) \le \rho(h)$, as g is the element of $\mathcal{C}$ of minimal relative degree. Again the division relation $h | g$ implies that $\rho(g) \ge \rho(h)$. Hence we have the equality $\rho(g) = \rho(h)$. It follows that g and h differ at most by a factor that is invertible in $\mathbf{RH}_+^\infty$, so g divides f. $\qquad\square$

Actually, it can be shown that $\mathbf{RH}_+^\infty$ is a Euclidean domain if we define $\delta(f) = \rho(f) + \deg \nabla f$. Thus δ counts the number of zeros of f in the closed right half-plane together with the zeros at infinity. We omit the details.

Truncated Laurent series. Let F be a field. We denote by $F((z^{-1}))$ the set of all formal sums of the form $f(z) = \sum_{j=-\infty}^{n_f} f_j z^j$ with $n_f \in \mathbf{Z}$. The operations of addition and multiplication are defined by

$$(f + g)(z) = \sum_{j=-\infty}^{max\{n_f, n_g\}} (f_j + g_j) z^j$$

and

$$(fg)(z) = \sum_{k=-\infty}^{\{n_f + n_g\}} h_k z^k,$$

with

$$h_k = \sum_{j=-\infty}^{\infty} f_j g_{k-j}.$$

Notice that the last sum is well defined as it contains only a finite number of nonzero terms. We can check that all of the field axioms are satisfied, in particular, that all nonzero elements are invertible. We call this the **field of truncated Laurent series.** Note that $F((z^{-1}))$ is the field of fractions of $F[[z^{-1}]]$.

We introduce for later use the maps

$$\pi_+ \sum_{j=-\infty}^{n_f} f_j z^j = \sum_{j=0}^{n_f} f_j z^j \qquad (1.17)$$

and

$$\pi_- \sum_{j=-\infty}^{n_f} f_j z^j = \sum_{j=-\infty}^{-1} f_j z^j. \qquad (1.18)$$

Clearly, $F(z)$ can be considered a subfield of $F((z^{-1}))$.

1.4   Modules

The module structure is one of the most fundamental algebraic concepts. Most of the rest of this book is to a certain extent an elaboration on the module theme. This is particularly true for the case of linear transformations and linear systems.

A **left module** M over the ring R is a commutative group together with an operation of R on M that satisfies

$$r(x + y) = rx + ry$$
$$(r + s)x = rx + sx$$
$$r(sx) = (rs)x$$
$$1x = x.$$

Right modules are defined similarly. Let M be a left R-module. A subset N of M is a **submodule** of M if it is an additive subgroup of M that further satisfies $RN \subset M$.

Given two left R-modules M and M_1, a map $\phi : M \longrightarrow M_1$ is an R-module homomorphism if for all $x, y \in M$ and $r \in R$

$$\phi(x + y) = \phi x + \phi y$$
$$\phi(rx) = r\phi(x).$$

Given an R-module homomorphism $\phi : M \longrightarrow M_1$, $Ker\,\phi$ and $Im\phi$ are submodules of M and M_1, respectively. Given R-modules $M_0, \ldots, M_n$, a sequence of R-module homomorphisms

$$\longrightarrow M_{i-1} \xrightarrow{\phi_{i-1}} M_i \xrightarrow{\phi_i} M_{i+1} \longrightarrow$$

is called an **exact sequence** if $Im\phi_{i-1} = Ker\,\phi_i$. An exact sequence of the form

$$0 \longrightarrow M_1 \xrightarrow{\phi} M_2 \xrightarrow{\psi} M_3 \longrightarrow 0$$

is called a **short exact sequence**. This means that ϕ is injective and ψ is surjective.

Given modules M_i over the ring R, we can make the cartesian product $M_1 \times \cdots \times M_k$ into an R-module by defining, for $m_i, n_i \in M_i$ and $r \in R$,

$$\left\{ \begin{array}{rcl} (m_1, \ldots, m_k) + (n_1, \ldots, n_k) & = & (m_1 + n_1, \ldots, m_k + n_k) \\ r(m_1, \ldots, m_k) & = & (rm_1, \ldots, rm_k). \end{array} \right. \tag{1.19}$$

Given a module M and submodules M_i, we say that M is the **direct sum** of the M_i and write $M = M_1 \oplus \cdots \oplus M_k$ if every $m \in M$ has a unique representation of the form $m = m_1 + \cdots + m_k$ with $m_i \in M_i$.

Given a submodule N of a left R-module M we can construct a module structure in the same manner in which we constructed quotient groups.

We say that two elements $x, y \in M$ are equivalent if $x - y \in N$. The equivalence class of x is denoted by $[x] = x + N$. The set of equivalence classes is denoted by M/N. We make M/N into an R-module by defining

$$\left\{ \begin{array}{rcl} [x] + [y] & = & [x + y] \\ r[x] & = & [rx]. \end{array} \right. \tag{1.20}$$

It is easy to check that these operations are well defined, that is, independent of the equivalence class representative. We state without proof the following.

Proposition 1.4.1 *With the operations defined in Eq. (1.20), the set of equivalence classes M/N is a module over R. This is called the* **quotient module** *of M by N.*

We shall make use of the following result. This is the counterpart of Theorem 1.2.4.

Proposition 1.4.2 *Let M be a module over the ring R. Then $N \subset M$ is a submodule if and only if it is the kernel of an R-module homomorphism.*

Proof: Clearly, the kernel of a module homomorphism is a submodule.

Conversely, let $\pi : M \longrightarrow M/N$ be the canonical projection. Then π is a module homomorphism and $Ker\,\pi = N$. $\qquad\qquad\qquad\qquad\qquad\square$

Note that with $i : N \longrightarrow M$, the natural embedding,

$$0 \longrightarrow N \overset{i}{\longrightarrow} M \overset{\pi}{\longrightarrow} M/N \longrightarrow 0$$

is a short exact sequence.

Proposition 1.4.3 *Let M, M_1 be R-modules and let $\phi : M \longrightarrow M_1$ be a surjective R-module homomorphism. Then we have the module isomorphism*

$$M_1 \simeq M/Ker\,\phi.$$

Proof: Let π be the canonical projection of M_1 onto $M_1/Ker\,\phi$, that is, $\pi(m) = m + Ker\,\phi$. We now define a map $\tau : M_1/Ker\,\phi \longrightarrow M$ by $\tau(m + Ker\,\phi) = \phi(m)$. The map τ is well defined. For, if m_1, m_2 are representatives of the same coset, then $m_1 - m_2 \in Ker\,\phi$. This implies that $\phi(m_1) = \phi(m_2)$. That $\tau\phi$ is a homomorphism follows from the fact that ϕ is one. Clearly, the surjectivity of ϕ implies that of τ. Finally, $\tau(m + Ker\,\phi) = 0$ if and only if $\phi(m) = 0$, that is, $m \in Ker\,\phi$. Thus, τ is also injective and hence an isomorphism. $\qquad\qquad\qquad\qquad\square$

We end by giving a few examples of important module structures. Every abelian group G is a module over the ring of integers $\mathbf{Z}$. Every ring is a module over itself, as well as over any subring. Thus $F(z)$ is a module over $F[z]$ and $F((z^{-1}))$ is a module over each of the subrings $F[z]$ and $F[[z^{-1}]]$. $z^{-1}F[[z^{-1}]]$ has an induced $F[z]$-module structure, being homomorphic to $F((z^{-1}))/F[z]$. This module structure is defined by

$$z \cdot \sum_{j=1}^{\infty} \frac{h_j}{z^j} = \sum_{j=1}^{\infty} \frac{h_{j+1}}{z^j}.$$

1.5   Exercises

1. Let q, p_1, p_2 be nonzero polynomials. Show that $p_1 - p_2 | q(p_1) - q(p_2)$.

2. Given a polynomial $p(z) = \sum_{k=0}^{n} p_k z^k$, we define its **formal deriva-tive** by $p'(z) = \sum_{k=0}^{n} k p_k z^{k-1}$. Show that

$$
\begin{aligned}
(p+q)' &= p' + q' \\
(pq)' &= pq' + qp' \\
(p^m)' &= mp'p^{m-1}.
\end{aligned}
$$

Show that, over a field of characteristic 0, a polynomial p factors into the product of distinct, irreducible factors if and only if the greatest common divisor of p and p' is 1.

3. Let M, M_1 be R-modules and $N \subset M$ a submodule. Let $\pi : M \longrightarrow M/N$ be the canonical projection and $\phi : M \longrightarrow M_1$ be an R-homomorphism. Show that if $Ker\,\phi \supset N$, there exists a unique R-homomorphism, called the induced homomorphism, $\phi|_{M/N} : M/N \longrightarrow M_1$ for which $\phi|_{M/N} = \phi \circ \pi$. Show that

$$
\begin{cases}
Ker\,\phi|_{M/N} &= (Ker\,\phi)/N \\[2mm]
Im\,\phi|_{M/N} &= Im\,\phi.
\end{cases}
$$

4. Let M be a module and M_i submodules. Show that if

$$
\begin{cases}
M &= M_1 + \cdots + M_k \\
M_i \cap \sum_{j \neq i} M_j &= 0,
\end{cases}
$$

then the map $\phi : M_1 \times \cdots \times M_k \longrightarrow M_1 + \cdots + M_k$ defined by

$$
\phi(m_1, \ldots, m_k) = m_1 + \cdots + m_k
$$

is a module isomorphism.

5. Let M be a module and M_i submodules. Show that if $M = M_1 + \cdots + M_k$ and

$$
\begin{aligned}
M_1 \cap M_2 &= 0 \\
(M_1 + M_2) \cap M_3 &= 0 \\
&\vdots \\
(M_1 + \cdots + M_{k-1}) \cap M_k &= 0,
\end{aligned}
$$

then $M = M_1 \oplus \cdots \oplus M_k$.

6. Let M be a module and K, L submodules.

   (a) Show that
   $$
   (K + L)/K \simeq L/(K \cap L).
   $$

(b) If $K \subset L \subset M$, then

$$M/L \simeq (M/K)/(L/K).$$

7. A module M over a ring R is called **free** if it is the zero-module or has a basis. Show that if M is a free module over a principal ideal domain R having n basis elements and N a submodule, then N is free and has at most n basis elements.

8. Show that $F^n[z]$, $z^{-1}F^n[[z^{-1}]]$, $F^n((z^{-1}))$ are $F[z]$-modules.

9. Show that every submodule M of $F^n[z]$ has a representation $M = DF^n[z]$ for some polynomial matrix D. Show that $DF^n[z] \subset D_1^n[z]$ if and only if $D = D_1 D_2$ for some polynomial matrix D_2.

1.6   Notes and Remarks

Modern expositions of algebra all stem from the classical book by Van der Waerden [1931], which in turn was based on lectures by Noether and Artin. For a modern, general book on algebra, Lang [1965] is recommended.

The emphasis on the ring of polynomials is not surprising, considering their well-entrenched role in the study of linear transformations in finite-dimensional vector spaces. The similar exposure given to the field of rational functions and in particular to the subrings of stable rational functions and bounded stable rational functions is motivated by the role they play in system theory. This will be treated in Chapters 10 and 11. For more information in this direction, the reader is advised to consult Vidyasagar [1985].

2

Linear Spaces

2.1 Linear Spaces

Definition 2.1.1 *Let F be a field. A **linear space** V over F is a set, whose elements are called **vectors**, which satisfies the following set of axioms:*

1. *For each pair of vectors $x, y \in V$ there exists a vector $x + y \in V$ called the **sum** of x and y, and the following hold:*

   *(a) The **commutative law**:*
   $$x + y = y + x.$$

   *(b) The **associative law**:*
   $$x + (y + z) = (x + y) + z.$$

   *(c) There exists a unique **zero vector** 0 that satisfies*
   $$0 + x = x + 0$$
   for all $x \in V$.

   (d) For each $x \in V$ there exists a unique element $-x$ such that
   $$x + (-x) = 0.$$

   That is, V is a commutative group under addition.

2. *For all $x \in V$ and $\alpha \in F$ there exists a vector $\alpha x \in V$ called the* **product** *of α and x, and the following are satisfied:*

(a) *The* **associative law:**

$$\alpha(\beta x) = (\alpha \beta) x.$$

(b) *For the unit $1 \in F$ and all $x \in V$, we have*

$$1 \cdot x = x.$$

3. *The* **distributive laws:**

(a) $(\alpha + \beta) x = \alpha x + \beta x$

(b) $\alpha(x + y) = \alpha x + \alpha y.$

Examples:

1. Let

$$F^n = \left\{ \begin{pmatrix} a_1 \\ \cdot \\ \cdot \\ \cdot \\ a_n \end{pmatrix} \Big| a_i \in F \right\}.$$

We define addition and multiplication by a scalar $\alpha \in F$ by

$$\begin{pmatrix} a_1 \\ \cdot \\ \cdot \\ a_n \end{pmatrix} + \begin{pmatrix} b_1 \\ \cdot \\ \cdot \\ b_n \end{pmatrix} = \begin{pmatrix} a_1 + b_1 \\ \cdot \\ \cdot \\ a_n + b_n \end{pmatrix}, \qquad \alpha \begin{pmatrix} a_1 \\ \cdot \\ \cdot \\ a_n \end{pmatrix} = \begin{pmatrix} \alpha a_1 \\ \cdot \\ \cdot \\ \alpha a_n \end{pmatrix}.$$

With these definitions, F^n is a linear space.

2. An $m \times n$ matrix over the field F is a set of mn elements a_{ij} arranged in rows and columns, that is,

$$A = \begin{pmatrix} a_{11} & \cdot & \cdot & \cdot & a_{1n} \\ \cdot & \cdot & \cdot & \cdot & \cdot \\ \cdot & \cdot & \cdot & \cdot & \cdot \\ \cdot & \cdot & \cdot & \cdot & \cdot \\ a_{m1} & \cdot & \cdot & \cdot & a_{mn} \end{pmatrix}.$$

We denote by $F^{m \times n}$ the set of all such matrices. We define in $F^{m \times n}$ addition and multiplication by scalars by

$$(a_{ij}) + (b_{ij}) = (a_{ij} + b_{ij}), \qquad \alpha(a_{ij}) = (\alpha a_{ij}).$$

These definitions make $F^{m \times n}$ into a linear space.

3. The rings $F[z], F((z^{-1})), F[[z]], F[[z^{-1}]]$ are all linear spaces over F. So is $F(z)$ the field of rational functions.

4. The space $\mathbf{RH}_+^\infty$ is a linear space over the field $\mathbf{R}$ of real numbers.

5. Many interesting examples are obtained by considering spaces of functions. A typical example is $C_{\mathbf{R}}(X)$, the space of all real-valued, continuous functions on a topological space X is a linear space. Addition and multiplication by scalars are defined by

$$(f + g)(x) = f(x) + g(x), \qquad (\alpha f)(x) = \alpha f(x).$$

Proposition 2.1.1 *Let V be a linear space over the field F. Let $\alpha \in F$ and $x \in V$. Then*

1. $0x = 0$.

2. $\alpha x = 0$ *implies* $\alpha = 0$ *or* $x = 0$.

2.2   Linear Combinations

Let V be a linear space over the field F. Let $x_1, \ldots, x_n \in V$ and $\alpha_1, \ldots, \alpha_n \in F$. The vector $\alpha_1 x_1 + \cdots + \alpha_n x_n$ is an element of V and is called a **linear combination** of the vectors $x_1, \ldots, x_n$. The scalars $\alpha_1, \ldots, \alpha_n$ are called the **coefficients** of the linear combination. The vector $x \in V$ is called a **linear combination** of the vectors $x_1, \ldots, x_n$ if there exist scalars $\alpha_1, \ldots, \alpha_n$ for which

$$x = \sum_{i=1}^{n} \alpha_i x_i.$$

The linear combination in which all coefficients are zero is called the **trivial linear combination**.

2.3   Subspaces

Definition 2.3.1 *Let V be a linear space over the field F. A nonempty subset M of V is called a **subspace** of V if for any pair of vectors $x, y \in V$ and any pair of scalars $\alpha, \beta \in F$ we have $\alpha x + \beta y \in V$.*

Thus a subset M of V is a subspace if and only if it is closed under linear combinations. An equivalent description of subspaces is subsets closed under addition and multiplication by scalars.

Examples:

1. For an arbitrary linear space V, V itself and $\{0\}$ are subspaces. These are called the **trivial subspaces**.

2. Let

$$M = \left\{ \begin{pmatrix} a_1 \\ \cdot \\ \cdot \\ \cdot \\ a_n \end{pmatrix} \Big| a_n = 0 \right\}.$$

Then M is a subspace of F^n.

3. Let A be an $m \times n$ matrix. Then

$$M = \{x \in F^n | Ax = 0\}$$

is a subspace of F^n. This is the space of solutions of a system of linear homogeneous equations.

Theorem 2.3.1 *Let $\{M_\alpha\}_{\alpha \in A}$ be a collection of subspaces of V. Then $M = \cap_{\alpha \in A} M_\alpha$ is a subspace of V.*

Proof: Let $x, y \in M$ and $\alpha, \beta \in F$. Clearly, $M \subset M_\alpha$ for all α; therefore, $x, y \in M_\alpha$ and as M_α is a subspace we have $\alpha x + \beta y \in M_\alpha$ for all α and hence to the intersection. So $\alpha x + \beta y \in M$. $\qquad\square$

Definition 2.3.2 *Let S be a subset of a linear space V. $L(S)$, or span (S), the **subspace spanned by** S, is defined as the intersection of the nonempty set of all subspaces containing S. This is therefore the smallest subspace of V containing S.*

Theorem 2.3.2 *Let S be a subset of a linear space V. $L(S)$, the subspace spanned by S, is the set of all finite linear combinations of elements of S.*

Proof: Let $M = \{\sum_{i=1}^n \alpha_i x_i | \alpha_i \in F, x_i \in S, n \in \mathbf{N}\}$. Clearly, $S \subset M$. M is a subspace of V as linear combinations of linear combinations of elements of S are also linear combinations of elements of S. Thus $L(S) \subset M$.

Conversely, we have $S \subset L(S)$. So necessarily $L(S)$ contains all finite linear combinations of elements of S. Hence $M \subset L(S)$, and equality follows. $\qquad\square$

We saw that if $M_1, \ldots, M_k$ are subspaces of V, then so is $M = \cup_{i=1}^k M_i$. This in general is not true for unions of subspaces that in general are not subspaces. The natural concept is that of sum of subspaces.

Definition 2.3.3 *Let $M_1, \ldots, M_k$ be subspaces of a linear space V. The sum of these subspaces, $M_1 + \cdots + M_k$, is defined by*

$$M_1 + \cdots + M_k = \{x_1 + \cdots + x_k | x_i \in M_i\} = L\left(\sum_{i=1}^n M_i\right).$$

2.4 Linear Dependence and Independence

Definition 2.4.1 *Vectors* $x_1, \ldots, x_k$ *in a linear space* V *are called* **linearly dependent** *if there exist* $\alpha_1, \ldots, \alpha_k \in F$, *not all zero, such that*

$$\alpha_1 x_1 + \cdots + \alpha_k x_k = 0.$$

$x_1, \ldots, x_k$ *in a linear space* V *are called* **linearly independent** *if they are not linearly dependent.*

Thus, $x_1, \ldots, x_k$ are linearly dependent if there exists a nontrivial vanishing linear combination. On the other hand, $x_1, \ldots, x_k$ are linearly independent if and only if $\alpha_1 x_1 + \cdots + \alpha_k x_k = 0$ implies $\alpha_1 = \cdots = \alpha_k = 0$. That is, the only vanishing linear combination of linearly independent vectors is the trivial one. We note that, as a consequence of Proposition 2.1.1, a set containing a single vector $x \in V$ is linearly independent if and only if $x \neq 0$.

Definition 2.4.2 *Vectors* $x_1, \ldots, x_k$ *in a linear space* V *are called a* **spanning set** *if* $V = L(x_1, \ldots, x_k) = L(x_1, \ldots, x_k)$.

Theorem 2.4.1 *Let* $S \subset V$.

1. *If* S *is a spanning set and* $S \subset S_1 \subset V$, *then* S_1 *is also a spanning set.*

2. *If* S *is a linearly independent set and* $S_0 \subset S$, *then* S_0 *is also a linearly independent set.*

3. *If* S *is linearly dependent and* $S \subset S_1 \subset V$, *then* S_1 *is also linearly dependent.*

4. *Every subset of* V *that includes the zero vector is linearly dependent.*

As a consequence, if we want a spanning set, it must be sufficiently large, whereas for linear independence the set must be sufficiently small. The case when these two properties are in balance is of special importance. This leads to the following.

Definition 2.4.3 *A subset* $\mathcal{B}$ *of vectors in* V *is called a* **basis** *if:*

1. *B is a spanning set.*

2. *B is linearly independent.*

V *is called a* **finite-dimensional space** *if there exists a basis in* V *having a finite number of elements.*

Note that we defined the concept of finite dimensionality before we defined dimension.

Example: Let $V = F^n$. Let

$$
e_1 = \begin{pmatrix} 1 \\ 0 \\ \cdot \\ \cdot \\ \cdot \\ 0 \end{pmatrix}, e_2 = \begin{pmatrix} 0 \\ 1 \\ 0 \\ \cdot \\ \cdot \\ 0 \end{pmatrix}, \ldots, e_n = \begin{pmatrix} 0 \\ \cdot \\ \cdot \\ \cdot \\ 0 \\ 1 \end{pmatrix}.
$$

Then $\mathcal{B} = \{e_1, \ldots, e_n\}$ is a basis for F^n.

The following result is the main technical instrument in the study of bases.

Theorem 2.4.2 *Let $x_1, \ldots, x_m \in V$ and let $e_1, \ldots, e_p$ be linearly independent vectors that satisfy $e_i \in L(x_1, \ldots, x_m)$ for all i. Then there exist p vectors in $\{x_i\}$. Without loss of generality, we may assume that they are the first p vectors, so that*

$$
L(e_1, \ldots, e_p, x_{p+1}, \ldots, x_m) = L(x_1, \ldots, x_m).
$$

Proof: We prove this by induction on p. For $p = 1$, we must have $e_1 \neq 0$ by linear independence. Therefore, there exist α_i such that $e_1 = \sum_{i=1}^m \alpha_i x_i$. Necessarily, $\alpha_i \neq 0$ for some i. Without loss of generality we assume that $\alpha_1 \neq 0$. Therefore, we can write

$$
x_1 = \alpha_1^{-1} e_1 - \alpha_1^{-1} \sum_{i=2}^m \alpha_i x_i.
$$

This means that $x_1 \in L(e_1, x_2, \ldots, x_m)$. Of course we also have $x_i \in L(e_1, x_2, \ldots, x_m)$ for $i = 2, \ldots, m$. Therefore,

$$
L(x_1, x_2, \ldots, x_m) \subset L(e_1, x_2, \ldots, x_m).
$$

On the other hand, by our assumption $e_1 \in L(x_1, x_2, \ldots, x_m)$, and hence

$$
L(e_1, x_2, \ldots, x_m) \subset L(x_1, x_2, \ldots, x_m).
$$

From these two inclusion relations the following equality follows:

$$
L(e_1, x_2, \ldots, x_m) = L(x_1, x_2, \ldots, x_m).
$$

Assume that we have proved the assertion for up to $p - 1$ elements and assume $e_1, \ldots, e_p$ to be linearly independent vectors that satisfy $e_i \in L(x_1, \ldots, x_m)$ for all i. By the induction hypothesis we have

$$
L(e_1, \ldots, e_{p-1}, x_p, \ldots, x_m) = L(x_1, x_2, \ldots, x_m).
$$

Therefore, $e_p \in L(e_1, \ldots, e_{p-1}, x_p, \ldots, x_m)$, hence there exist α_i such that

$$e_p = \alpha_1 x_1 + \cdots \alpha_{p-1} + \alpha_p x_p + \cdots + \alpha_m x_m.$$

It is impossible that $\alpha_p, \ldots, \alpha_m$ are all 0, for that implies that e_p is a linear combination of $e_1, \ldots, e_{p-1}$, contradicting the assumption of linear independence. So at least one of these numbers is nonzero, and without loss of generality, reordering the elements if necessary, we assume that $\alpha_p \neq 0$. Now

$$x_p = \alpha_p^{-1}(\alpha_1 e_1 + \cdots + \alpha_{p-1} e_{p-1} - e_p + \alpha_{p+1} x_{p+1} + \cdots + \alpha_n x_n).$$

That is, $x_p \in L(e_1, \ldots, e_p x_{p+1}, \ldots, x_m)$. Therefore,

$$L(x_1, x_2, \ldots, x_m) \subset L(e_1, \ldots, e_p x_{p+1}, \ldots, x_m) \subset L(x_1, x_2, \ldots, x_m),$$

and hence the equality

$$L(e_1, \ldots, e_p x_{p+1}, \ldots, x_m) = L(x_1, x_2, \ldots, x_m). \qquad \square$$

Corollary 2.4.1 *The following assertions hold:*

1. *Let $\{e_1, \ldots, e_n\}$ be a basis for the linear space V, and let $\{f_1, \ldots, f_m\}$ be linearly independent vectors in V. Then $p \leq n$.*

2. *Let $\{e_1, \ldots, e_n\}$ and $\{f_1, \ldots, f_m\}$ be two bases for V; then $n = m$.*

Proof:

1. Apply Theorem 2.4.2.

2. By the first part we have both $m \leq n$ and $n \leq m$, so equality follows.
$$\square$$

   Thus two different bases in a finite-dimensional linear space have the same number of elements. This leads us to the following definition.

Definition 2.4.4 *Let V be a linear space over the field F. The **dimension** of V is defined as the number of elements in an arbitrary basis. We denote the dimension of V by $\dim V$.*

Theorem 2.4.3 *Let V be a linear space of dimension n. Then,*

1. *Every subset of V containing more than n vectors is linearly dependent.*

2. *A set of $p < n$ vectors in V cannot be a spanning set.*

2.5 Subspaces and Bases

Theorem 2.5.1

1. Let V be a linear space of dimension n and let M be a subspace. Then $\dim M \leq \dim V$.

2. Let $\{e_1, \ldots, e_p\}$ be a basis for M. Then there exist vectors $\{e_{p+1}, \ldots, e_n\}$ in V so that $\{e_1, \ldots, e_n\}$ is a basis for V.

Proof: It suffices to prove the second assertion. Let $\{e_1, \ldots, e_p\}$ be a basis for M and $\{f_1, \ldots, f_n\}$ a basis for V. By Theorem 2.4.2 we can replace p of the f_i by the $e_j, j = 1, \ldots, p$, and get a spanning set for V. But a spanning set with n elements is necessarily a basis for V.   □

From two subspaces M_1, M_2 of a linear space V we can construct the subspaces $M_1 \cap M_2$ and $M_1 + M_2$. The next theorem studies the dimensions of these subspaces.

Definition 2.5.1 *Given subspaces $M_i, i = 1, \ldots, p$ of a linear space V, we define*

$$\sum_{i=1}^p M_i = \{\sum_{i=1}^p \alpha_i x_i | \alpha_i \in F, x_i \in M_i\}.$$

For the sum of two subspaces of a linear space V, we have the following.

Theorem 2.5.2 *Let M_1, M_2 be subspaces of a linear space V. Then*

$$\dim(M_1 + M_2) + \dim(M_1 \cap M_2) = \dim M_1 + \dim M_2.$$

Proof: Let $\{e_1, \ldots, e_r\}$ be a basis for $M_1 \cap M_2$. Then there exist vectors $\{f_{r+1}, \ldots, f_p\}$ and $\{g_{r+1}, \ldots, g_q\}$ such that the set $\{e_1, \ldots, e_r, f_{r+1}, \ldots, f_p\}$ is a basis for M_1 and the set $\{e_1, \ldots, e_r, g_{r+1}, \ldots, g_q\}$ is a basis for M_2. We will show that $\{e_1, \ldots, e_r, f_{r+1}, \ldots, f_p, g_{r+1}, \ldots, g_q\}$ is a basis for $M_1 + M_2$.

Clearly, $\{e_1, \ldots, e_r, f_{r+1}, \ldots, f_p, g_{r+1}, \ldots, g_q\}$ is a spanning set for $M_1 + M_2$, so it remains to show that it is linearly independent. Assume that they are linearly dependent, that is, there exist $\alpha_i, \beta_i, \gamma_i$ such that

$$\sum_{i=1}^r \alpha_i e_i + \sum_{i=r+1}^p \beta_i f_i + \sum_{i=r+1}^q \gamma_i g_i = 0 \qquad (2.1)$$

or

$$\sum_{i=r+1}^q \gamma_i g_i = - \left[\sum_{i=1}^r \alpha_i e_i + \sum_{i=r+1}^p \beta_i f_i \right].$$

So it follows that $\sum_{i=r+1}^q \gamma_i g_i \in M_1$. On the other hand, $\sum_{i=r+1}^q \gamma_i g_i \in M_2$ as a linear combination of some of the basis elements of M_2. Therefore,

$\sum_{i=r+1}^{q} \gamma_i g_i \in M_1 \cap M_2$ and hence can be expressed as a linear combination of the e_i. So there exist numbers $\epsilon_1, \ldots, \epsilon_r$ such that

$$\sum_{i=1}^{r} \epsilon_i e_i + \sum_{i=r+1}^{q} \gamma_i g_i = 0.$$

However, this is a linear combination of the basis elements of M_2; hence $\epsilon_i = \gamma_j = 0$. Now Eq. (2.1) reduces to

$$\sum_{i=1}^{r} \alpha_i e_i + \sum_{i=r+1}^{p} \beta_i f_i = 0.$$

From this we conclude by the same reasoning that $\alpha_i = \beta_j = 0$. This proves the linear independence of the vectors $\{e_1, \ldots, e_r, f_{r+1}, \ldots, f_p, g_{r+1}, \ldots, g_q\}$ and so they are a basis for $M_1 + M_2$. Now

$$\dim(M_1 + M_2) = p + q - r = \dim M_1 + \dim M_2 - \dim(M_1 \cap M_2). \quad \square$$

2.6   Direct Sums

Definition 2.6.1 *Let $M_i, i = 1, \ldots, p$ be subspaces of a linear space V. We say that $\sum_{i=1}^{p} M_i$ is a* **direct sum** *of the subspaces M_i and write $M = M_1 \oplus \cdots \oplus M_p$ if for every $x \in \sum_{i=1}^{p} M_i$ there exists a unique representation $x = \sum_{i=1}^{p} x_i$ with $x_i \in M_i$.*

Proposition 2.6.1 *Let M_1, M_2 be subspaces of a linear space V. Then $M = M_1 \oplus M_2$ if and only if $M = M_1 + M_2$ and $M_1 \cap M_2 = \{0\}$.*

Proof: Assume that $M = M_1 \oplus M_2$. Then for $x \in M$ we have $x = x_1 + x_2$, with $x_i \in M_i$. Suppose that there exists another representation of x in the form $x = y_1 + y_2$, with $y_i \in M_i$. From $x_1 + x_2 = y_1 + y_2$ we get $z = x_1 - y_1 = y_2 - x_2$. Now $x_1 - y_1 \in M_1$ and $y_2 - x_2 \in M_2$. So, since $M_1 \cap M_2 = \{0\}$, we have $z = 0$, that is, $x_1 = y_1$ and $x_2 = y_2$.

Conversely, suppose that every $x \in M$ has a unique representation $x = x_1 + x_2$, with $x_i \in M_i$. Thus $M = M_1 + M_2$. Let $x \in M_1 \cap M_2$; then $x = x + 0 = 0 + x$, which implies, by the uniqueness of the representation, that $x = 0$. $\square$

We consider next some examples.

1. The space $F((z^{-1}))$ of a truncated Laurent series has the spaces $F[z]$ and $F[[z^{-1}]]$ as subspaces. We clearly have the direct sum decomposition

$$F((z^{-1})) = F[z] \oplus z^{-1} F[[z^{-1}]]. \tag{2.2}$$

The factor of z^{-1} guarantees that the constant elements appear in one of the subspaces only.

2. We denote by $F_n[z]$ the space of all polynomials of degree $< n$, that is, $F_n[z] = \{p \in F[z] | \deg p < n\}$. The following result is based on Proposition 1.3.4.

Proposition 2.6.2 *Let* $p, q \in F[z]$ *with* $\deg p = m$ $\deg q = n$. *Let* r *be the g.c.d. of* p *and* q, *and* s *be their l.c.m. Let* $\deg r = \rho$. *Then*

(a) $pF_n[z] + qF_m[z] = rF_{m+n-\rho}[z]$.

(b) $pF_n[z] \cap qF_m[z] = sF_\rho[z]$.

Proof:

(a) We know that, with r the g.c.d. of p and q, $pF[z] + qF[z] = rF[z]$. So, given $f, g, \in F[z]$, there exists an $h \in F[z]$ such that $pf + qg = rh$. Now we take remainders after division by pq, that is, we apply the map π_{pq}. Now $\pi_{pq}pf = p\pi_q f$ and $\pi_{pq}qg = q\pi_p g$. Finally, since by Proposition 1.3.5 we have $pq = rs$, it follows that $\pi_{pq}rh = \pi_{rs}rh = r\pi_s h$. Now $\deg s = \deg p + \deg q - \deg r = n + m - \rho$, so we get the equality $pF_m[z] + qF_m[z] = rF_{m+n-\rho}[z]$.

(b) We have $pF[z] \cap qF[z] = sF[z]$. Again we apply π_{pq}. If $f \in pF[z]$, then $f = pf'$ and hence $\pi_{pq}pf' = p\pi_q f' \in pF_n[z]$. Similary, if $f \in qF[z]$, then $f = qf''$ and $\pi_{pq}f = \pi_{pq} \cdot qf'' = q\pi_p f'' \in qF_m[z]$. On the other hand, $f = sh$ implies $\pi_{pq}sh = \pi_{rs}sh = s\pi_r h \in sF_\rho[z]$.
□

Corollary 2.6.1 *Let* $p, q \in F[z]$ *with* $\deg p = m$ *and* $\deg q = n$. *Then* $p \wedge q = 1$ *if and only if*

$$F_{m+n}[z] = pF_n[z] \oplus qF_m[z].$$

We say that polynomials $p_1, \ldots, p_k$ are **mutually coprime** if, for $i \neq j$, p_i and p_j are coprime.

Theorem 2.6.1 *Let* $p_1, \ldots, p_k \in F[z]$ *with* $\deg p_i = n_i$ *with* $\sum_{i=1}^k n_i = n$. *Then the* p_i *are mutually coprime if and only if*

$$F_n[z] = \pi_1(z)F_{n_1}[z] \oplus \cdots \oplus \pi_k(z)F_{n_k}[z],$$

where the $\pi_j(z)$ *are defined by*

$$\pi_i(z) = \prod_{j \neq i} p_j(z).$$

Proof: The proof is by induction. Assume that the p_i are pairwise coprime. For $k = 2$, it was proved in Proposition 2.6.2. Assume that it was proved for positive integers $\leq k - 1$. Let $\tau_i(z) = \Pi_{\substack{j=1 \\ j \neq i}}^{k-1} p_j(z)$. Then

$$F_{n_1+\cdots+n_{k-1}}[z] = \tau_1(z)F_{n_1}[z] \oplus \cdots \oplus \tau_{k-1}(z)F_{n_{k-1}}[z].$$

Now $\pi_k(z) \wedge p_k(z) = 1$, so

$$
\begin{aligned}
F_n[z] &= \pi_k(z)F_{n_k}[z] \oplus p_k(z)F_{n_1+\cdots+n_{k-1}}[z] \\
&= p_k(z)\left\{ \tau_1(z)F_{n_1}[z] \oplus \cdots \oplus \tau_{k-1}(z)F_{n_{k-1}}[z] \right\} \\
&\quad \oplus \pi_k(z)F_{n_k}(z) \\
&= \pi_1(z)F_{n_1}[z] \oplus \cdots \oplus \pi_k(z)F_{n_k}[z].
\end{aligned}
$$

Conversely, if $F_n[z] = \pi_1(z)F_{n_1}[z] \oplus \cdots \oplus \pi_k(z)F_{n_k}[z]$, then there exist polynomials f_i such that $1 = \sum \pi_i f_i$. The coprimeness of the π_i implies the pairwise coprimeness of the p_i. $\qquad \square$

Corollary 2.6.2 *Let $p(z) = p_1(z)^{n_1} \cdots p_k(z)^{n_k}$ be the primary decomposition of p, with $\deg p_i = r_i$ and $n = \sum_{i=1}^{k} n_i$. Then*

$$
\begin{aligned}
F_n[z] &= \\
&p_2(z)^{n_2} \cdots p_k(z)^{n_k} F_{r_1 n_1}[z] \oplus \cdots \oplus p_1(z)^{n_1} \cdots p_{k-1}(z)^{n_{k-1}} F_{r_k n_k}[z].
\end{aligned}
$$

Proof: Follows from Theorem 2.6.1, replacing p_i by $p_i^{n_i}$. $\qquad \square$

2.7   Quotient Spaces

We begin by introducing the concept of codimension.

Definition 2.7.1 *We say that a subspace $M \subset U$ has **codimension** k, denoted by $\operatorname{codim} M = k$, if:*

1. *There exist k vectors $\{x_1, \ldots, x_k\}$, linearly independent over M, that is, for which $\sum_{i=1}^{k} \alpha_i x_i \in M$, if and only if $\alpha_i = 0$ for all $i = 1, \ldots, k$.*

2. *$U = L(M, x_1, \ldots, x_k)$.*

Now let X be a linear space over the field F and let M be a subspace. In U we define a relation

$$
x \simeq y \text{ if } x - y \in M. \tag{2.3}
$$

It is easy to check that this is indeed an equivalence relation, that is, it is reflexive, symmetric, and transitive. We denote by $[x] = x + M = \{x + m | m \in M\}$ the equivalence class of $x \in U$. We denote by X/M the set of equivalence classes with respect to the equivalence relation induced by M as in Eq. (2.3).

So far, X/M is just a set. We introduce in X/M two operations, addition and multiplication by a scalar, as follows:

$$
\begin{cases}
[x] + [y] = [x + y], & x, y \in U \\
\alpha[x] = [\alpha x].
\end{cases}
$$

Proposition 2.7.1

1. *The operation of addition and multiplication by scalar are well-defined, that is, independent of the representatives x, y.*

2. *With these operations, X/M as a vector space over F.*

3. *If M has codimension k in X, then* dim $X/M = k$.

Proof:

1. Let $x' \sim x$ and $y' \sim y$. Thus $[x] = [x']$. This means that $x' = x + m_1, y' = y + m_2$ with $m_i \in M$. Hence $x' + y' = x + y + (m_1 + m_2)$ which shows that $[x' + y'] = [x + y]$.

   Similarly $\alpha x' = \alpha x + \alpha m$. So $\alpha x' - \alpha x = \alpha m \in M$ hence $[\alpha x'] = [\alpha x]$.

2. The axioms of a vector space for X/M are easily shown to result from those in X.

3. Let $x_1, \ldots, x_k$ be linearly independent over M and such that $L(x_1, \ldots, x_k, M) = X$. We claim that $\{[x_1], \ldots, [x_k]\}$ are linearly independent, for if $\sum_{i=1}^{n} \alpha_i [x_i] = 0$, it follows that $[\sum \alpha_i x_i] = 0$, that is, $\sum_{i=1}^{n} \alpha_i x_i \in M$. Since $x_1, \ldots, x_n$ are linearly independent over M, necessarily, $\alpha_i = 0, i = 1, \ldots, k$. Now let $[x]$ be an arbitrary equivalence class in X/M. Assume that $x \in [x]$; then there exist $\alpha_i \in F$ s.t. $x = \alpha_1 x_1 + \ldots + \alpha_n x_n + m$ for some $m \in M$. This implies that

$$[x] = \alpha_1 [x_1] + \cdots + \alpha_k [x_k].$$

So indeed $\{[x_1], \ldots, [x_k]\}$ is a basis for X/M and hence

$$\dim\ X/M = k.$$

Corollary 2.7.1 *Let $q \in F[z]$ be a polynomial of degree n. Then,*

1. *$qF[z]$ is a subspace of $F[z]$ of codimension n.*

2. *dim $F[z]/qF[z] = n$.*

Proof: The polynomials $1, z, \ldots, z^{n-1}$ are obviously linearly independent over $qF[z]$. Moreover, applying the division rule of polynomials, it is clear that $1, z, \ldots, z^{n-1}$ together with $qF[z]$ span all of $F[z]$. □

Proposition 2.7.2 *Let $F((z^{-1}))$ be the space of a truncated Laurent series, and $F[z]$ and $z^{-1}F[[z^{-1}]]$ the corresponding subspaces. Then we have the isomorphisms*

$$F[z] \simeq F((z^{-1}))/z^{-1}F[[z^{-1}]],$$

and

$$z^{-1}F[[z^{-1}]] \simeq F((z^{-1}))/F[z].$$

Proof: Follows from the direct sum representation in Eq. (2.2). □

2.8   Coordinates

Lemma 2.8.1 *Let V be a finite-dimensional linear space of dimension n and let $B = \{e_1, \ldots, e_n\}$ be a basis for V. Then every vector $x \in V$ has a unique representation as a linear combination of the e_i. That is,*

$$x = \sum_{i=1}^{n} \alpha_i e_i. \tag{2.4}$$

Proof: Since B is a spanning set, such a representation exists. Since B is linearly independent, the representation in Eq. (2.4) is unique.   □

Definition 2.8.1 *The scalars $\alpha_1, \ldots, \alpha_n$ will be called the **coordinates** of x with respect to the basis B, and we will use the notation*

$$[x]^B = \begin{pmatrix} \alpha_1 \\ \cdot \\ \cdot \\ \cdot \\ \alpha_n \end{pmatrix}.$$

*The vector $[x]^B$ will be called the **coordinate vector** of x with respect to B. We will always write it in column form.*

The map $x \mapsto [x]^B$ is a map from V to F^n.

Proposition 2.8.1 *Let V be a finite-dimensional linear space of dimension n and let $B = \{e_1, \ldots, e_n\}$ be a basis for V. The map $x \mapsto [x]^B$ has the following properties:*

1. *$[x + y]^B = [x]^B + [y]^B$.*

2. *$[\alpha x]^B = \alpha [x]^B$.*

3. *$[x]^B = 0$ if and only if $x = 0$.*

4. *For every $\alpha_1, \ldots, \alpha_n \in F$ there exists a vector $x \in V$ for which*

$$[x]^B = \begin{pmatrix} \alpha_1 \\ \cdot \\ \cdot \\ \cdot \\ \alpha_n \end{pmatrix}.$$

Proof:

1. Let $x = \sum_{i=1}^{n} \alpha_i e_i$, $y = \sum_{i=1}^{n} \beta_i e_i$; then

$$x + y = \sum_{i=1}^{n} \alpha_i e_i + \sum_{i=1}^{n} \beta_i e_i = \sum_{i=1}^{n} (\alpha_i + \beta_i) e_i.$$

So

$$[x + y]^{\mathcal{B}} = \begin{pmatrix} \alpha_1 + \beta_1 \\ \cdot \\ \cdot \\ \cdot \\ \alpha_n + \beta_n \end{pmatrix} = \begin{pmatrix} \alpha_1 \\ \cdot \\ \cdot \\ \cdot \\ \alpha_n \end{pmatrix} + \begin{pmatrix} \beta_1 \\ \cdot \\ \cdot \\ \cdot \\ \beta_n \end{pmatrix} = [x]^{\mathcal{B}} + [y]^{\mathcal{B}}.$$

2. Let $\alpha \in F$. Then $\alpha x = \sum_{i=1}^{n} \alpha \alpha_i e_i$, and therefore

$$[\alpha x]^{\mathcal{B}} = \begin{pmatrix} \alpha \alpha_1 \\ \cdot \\ \cdot \\ \cdot \\ \alpha \alpha_n \end{pmatrix} = \alpha \begin{pmatrix} \alpha_1 \\ \cdot \\ \cdot \\ \cdot \\ \alpha_n \end{pmatrix} = \alpha [x]^{\mathcal{B}}.$$

3. If $x = 0$, then $x = \sum_{i=1}^{n} 0 e_i$, and

$$[x]^{\mathcal{B}} = \begin{pmatrix} 0 \\ \cdot \\ \cdot \\ \cdot \\ 0 \end{pmatrix}.$$

Conversely, if $[x]^{\mathcal{B}} = 0$, then $x = \sum_{i=1}^{n} 0 e_i = 0$.

4. Let $\alpha_1, \ldots, \alpha_n \in F$; then we define a vector $x \in V$ by $x = \sum_{i=1}^{n} \alpha_i e_i$. Then clearly,

$$[x]^{\mathcal{B}} = \begin{pmatrix} \alpha_1 \\ \cdot \\ \cdot \\ \cdot \\ \alpha_n \end{pmatrix}. \qquad \square$$

2.9   Change of Basis Transformations

Let V be a finite-dimensional linear space of dimension n, and let $\mathcal{B} = \{e_1, \ldots, e_n\}$ and $\mathcal{B}_1 = \{f_1, \ldots, f_n\}$ be two different bases in V. We will explore the connection between the coordinate vectors with respect to the two bases.

For each $x \in V$ there exist unique $\alpha_j, \beta_j \in F$ for which

$$x = \sum_{j=1}^{n} \alpha_j e_j = \sum_{j=1}^{n} \beta_j f_j. \qquad (2.5)$$

Since, for each $j = 1, \ldots, n$, the basis vector e_j has a unique expansion as a linear combination of the f_i, we set

$$e_j = \sum_{i=1}^{n} t_{ij} f_i.$$

These equations define n^2 numbers, which we arrange in an $n \times n$ matrix. We denote this matrix by $[I]_B^{B_1}$. We refer to this matrix as the **basis transformation matrix** from the basis B to the basis B_1. Substituting back into Eq. (2.5), we get

$$x = \sum_{i=1}^{n} \beta_i f_i = \sum_{j=1}^{n} \alpha_j e_j = \sum_{j=1}^{n} \alpha_j \sum_{i=1}^{n} t_{ij} f_i$$

$$= \sum_{j=1}^{n} \sum_{i=1}^{n} t_{ij} \alpha_j f_i = \sum_{i=1}^{n} \left(\sum_{j=1}^{n} t_{ij} \alpha_j \right) f_i.$$

Equating coefficients of the f_i, we obtain

$$\beta_i = \sum_{j=1}^{n} t_{ij} \alpha_j.$$

Thus we conclude that

$$[x]^{B_1} = [I]_B^{B_1} [x]^B.$$

So the basis transformation matrix transforms the coordinate vector with respect to the basis B to the coordinate vector with respect to the basis B_1.

Theorem 2.9.1 *Let V be a finite-dimensional linear space of dimension n, and let $B = \{e_1, \ldots, e_n\}$, $B_1 = \{f_1, \ldots, f_n\}$, and $B_2 = \{g_1, \ldots, g_n\}$ be three bases for V. Then we have*

$$[I]_B^{B_2} = [I]_{B_1}^{B_2} [I]_B^{B_1}.$$

Proof: Let $[I]_B^{B_2} = (r_{ij})$, $[I]_{B_1}^{B_2} = (s_{ij})$, and $[I]_B^{B_1} = (t_{ij})$. This means

$$e_j = \sum_{i=1}^{n} r_{ij} g_i,$$

$$f_k = \sum_{i=1}^{n} s_{ik} g_i,$$

$$e_j = \sum_{i=1}^{n} t_{kj} f_k.$$

Therefore, $r_{ij} = \sum_{i=1}^{n} s_{ik} t_{kj}$. $\qquad\qquad \square$

Corollary 2.9.1 *Let V be a finite-dimensional linear space with two bases $\mathcal{B}$ and $\mathcal{B}_1$; then*

$$[I]_{\mathcal{B}}^{\mathcal{B}_1} = ([I]_{\mathcal{B}_1}^{\mathcal{B}})^{-1}.$$

Proof: Clearly, $[I]_{\mathcal{B}}^{\mathcal{B}} = I$, where I denots the identity matrix. By Theorem 2.9.1 we get

$$I = [I]_{\mathcal{B}}^{\mathcal{B}} = [I]_{\mathcal{B}_1}^{\mathcal{B}} [I]_{\mathcal{B}}^{\mathcal{B}_1}. \qquad \square$$

Theorem 2.9.2 *The mapping in F^n defined by the matrix $[I]_{\mathcal{B}}^{\mathcal{B}_1}$ as*

$$[x]^{\mathcal{B}} \mapsto [I]_{\mathcal{B}}^{\mathcal{B}_1} [x]^{\mathcal{B}} = [x]^{\mathcal{B}_1}$$

has the following properties:

$$[I]_{\mathcal{B}}^{\mathcal{B}_1} ([x]^{\mathcal{B}} + [y]^{\mathcal{B}}) = [I]_{\mathcal{B}}^{\mathcal{B}_1} [x]^{\mathcal{B}} + [I]_{\mathcal{B}}^{\mathcal{B}_1} [y]^{\mathcal{B}}$$

$$[I]_{\mathcal{B}}^{\mathcal{B}_1} (\alpha[x]^{\mathcal{B}}) = \alpha[I]_{\mathcal{B}}^{\mathcal{B}_1} [x]^{\mathcal{B}}.$$

Proof: We compute

$$[I]_{\mathcal{B}}^{\mathcal{B}_1} ([x]^{\mathcal{B}} + [y]^{\mathcal{B}}) = [I]_{\mathcal{B}}^{\mathcal{B}_1} [x + y]^{\mathcal{B}} = [x + y]^{z\mathcal{B}_1}$$

$$= [x]^{\mathcal{B}_1} + [y]^{\mathcal{B}_1} = [I]_{\mathcal{B}}^{\mathcal{B}_1} [x]^{\mathcal{B}} + [I]_{\mathcal{B}}^{\mathcal{B}_1} [y]^{\mathcal{B}}.$$

Similarly,

$$[I]_{\mathcal{B}}^{\mathcal{B}_1} (\alpha[x]^{\mathcal{B}}) = [I]_{\mathcal{B}}^{\mathcal{B}_1} [\alpha x]^{\mathcal{B}} = [\alpha x]^{\mathcal{B}_1}$$

$$= \alpha[x]^{\mathcal{B}_1} = \alpha[I]_{\mathcal{B}}^{\mathcal{B}_1} [x]^{\mathcal{B}}. \qquad \square$$

These properties characterize linear maps. We will discuss those in details in the next chapter.

2.10   Lagrange Interpolation

Let $F_n[z] = \{p \in F[z] | \deg p < n - 1\}$. Clearly, $F_n[z]$ is an n-dimensional subspace of $F[z]$. In fact, $\{1, z, \ldots, z^{n-1}\}$ is a basis for $F_n[z]$, and we will refer to this as the **standard basis** of $F_n[z]$. Obviously, if $p(z) = \sum_{i=0}^{n-1} p_i z^i$, then

$$[p]^{st} = \begin{pmatrix} p_0 \\ \cdot \\ \cdot \\ \cdot \\ p_{n-1} \end{pmatrix}.$$

We exhibit next another important basis that is intimately related to polynomial interpolation. To this end we introduce:

The Lagrange interpolation problem. Given distinct numbers $\alpha_i \in F, i = 1, \ldots, n$, and another set of arbitrary numbers $c_i \in F, i = 1, \ldots, n$, find a polynomial $p \in F_n[z]$ such that

$$p(\alpha_i) = c_i, i = 1, \ldots, n.$$

We can replace this problem by a set of simpler

Special interpolation problems. Find polynomials $\pi_i \in F_n[z]$ such that, for all $i = 1, \ldots, n$,

$$\pi_i(\alpha_j) = \delta_{ij}, j = 1, \ldots, n.$$

If π_i are such polynomials, then the solution to the Lagrange interpolation problem is given by

$$p(z) = \sum_{i=1}^{n} c_i \pi_i(z).$$

This is seen by a direct evaluation at all α_i.

The existence and properties of the solution to the special interpolation problem are summarized in the following.

Proposition 2.10.1 *Let the polynomials $\pi_i(z)$ be defined by*

$$\pi_i(z) = \frac{\prod_{j \neq i}(z - \alpha_j)}{\prod_{j \neq i}(\alpha_i - \alpha_j)}.$$

Then

1. *The polynomials π_i are in $F_n[z]$ and*

$$\pi_i(\alpha_j) = \delta_{ij}, \qquad j = 1, \ldots, n.$$

2. *The set $\{\pi_1, \ldots, \pi_n\}$ forms a basis for $F_n[z]$.*

3. *For all $p \in F_n[z]$, we have*

$$p(z) = \sum_{i=1}^{n} p(\alpha_i)\pi_i(z). \tag{2.6}$$

Proof:

1. Clearly, π_i has degree $n - 1$; hence $\pi_i \in F_n[z]$. They are well-defined as, by the distinctness of the α_i, all of the denominators are different from zero. Moreover, by construction, we have

$$\pi_i(\alpha_j) = \delta_{ij}, j = 1, \ldots, n.$$

2. We show that the polynomials $\pi_1, \ldots, \pi_n$ are linearly independent. Assume that $\sum_{i=1}^n c_i \pi_i(z) = 0$. Evaluating at α_j, we get

$$0 = \sum_{i=1}^n c_i \pi_i(\alpha_j) = \sum_{i=1}^n c_i \delta_{ij} = c_j.$$

Since $\dim F_n[z] = n$, they actually form a basis.

3. As $\{\pi_1, \ldots, \pi_n\}$ forms a basis for $F_n[z]$, then for every $p \in F_n[z]$ there exist c_j such that

$$p(z) = \sum_{i=1}^n c_i \pi_i(z).$$

Evaluating at α_j, we get

$$p(\alpha_j) = \sum_{i=1}^n c_i \pi_i(\alpha_j) = \sum_{i=1}^n c_i \delta_{ij} = c_j. \qquad \square$$

Corollary 2.10.1 *Let $\alpha_1, \ldots, \alpha_n \in F$ be distinct. Let $\mathcal{B}_{in} = \{\pi_1, \ldots, \pi_n\}$ be the Lagrange interpolation basis and let $\mathcal{B}_{st}$ be the standard basis in $F_n[z]$. Then the change of basis transformation from the standard basis to the interpolation basis is given by*

$$[I]_{st}^{in} = \begin{pmatrix} 1 & \alpha_1 & \cdot & \cdot & \alpha_1^{n-1} \\ \cdot & \cdot & \cdot & \cdot & \cdot \\ \cdot & \cdot & \cdot & \cdot & \cdot \\ \cdot & \cdot & \cdot & \cdot & \cdot \\ 1 & \alpha_n & \cdot & \cdot & \alpha_n^{n-1} \end{pmatrix}. \tag{2.7}$$

Proof: From the equality (2.6) we get as special cases, for $i = 0, \ldots, n-1$,

$$z^j = \sum_{i=1}^n \alpha_i^j \pi_i(z),$$

and Eq. (2.7) follows. $\qquad \square$

The matrix in Eq. (2.6) is called the **Vandermonde matrix**.

2.11 Taylor Expansion

We consider the Taylor expansion of polynomials within the framework of change of basis transformations.

Let $p \in F[z]$, with $\deg p = n$, and let $\alpha \in F$. We would like to have a representation of p in the form

$$p(z) = \sum_{j=0}^{n} p_j (z - \alpha)^j.$$

We refer to this as the **Taylor expansion** of p at the point α. Naturally, the standard representation of a polynomial, $p(z) = \sum_{j=0}^{n} p_j z^j$, is the Taylor expansion at 0.

Proposition 2.11.1 *Given a positive integer n and $\alpha \in F$:*

1. *The set of polynomials $\mathcal{B}_\alpha = \{1, z - \alpha, \ldots, (z - \alpha)^{n-1}\}$ forms a basis for $F_n[z]$.*

2. *For every $p \in F[z]$ there exist unique numbers $p_{i,\alpha}$ such that*

$$p(z) = \sum_{j=0}^{n} p_{j,\alpha}(z - \alpha)^j.$$

3. *The change of basis transformation is given by*

$$[I]_\alpha^{st} = \begin{pmatrix} 1 & -\alpha & \alpha^2 & & & (-\alpha)^{n-1} \\ 0 & 1 & -2\alpha & & & \cdot \\ 0 & & 1 & & & \cdot \\ 0 & & & & & \cdot \\ 0 & & & \cdot & & -(n-1)\alpha \\ 0 & & & & & 1 \end{pmatrix},$$

that is, the i-th column entries come from the binomial expansion of $(z - \alpha)^{i-1}$.

Proof:

1. In $\mathcal{B}_\alpha$ we have one polynomial of each degree from 0 to $n - 1$; hence these polynomials are linearly independent. Since $\dim F_n[z] = n$, they form a basis.

2. Follows from the fact that $\mathcal{B}_\alpha$ is a basis.

3. We use the binomial expansion of $(z - \alpha)^{i-1}$. $\square$

2.12 Exercises

1. Let K, L, M be subspaces of a linear space. Show

$$\begin{aligned} K \cap (K \cap L + M) &= K \cap L + K \cap M \\ (K + L) \cap (K + M) &= K + (K + L) \cap M. \end{aligned}$$

2. Let M_i be subspaces of a finite-dimensional vector space X. Show that if $\dim(\sum_{i=1}^{k} M_i) = \sum_{i=1}^{k} \dim(M_i)$, then $M_1 + \cdots + M_k$ is a direct sum.

3. Let V be a finite-dimensional vector space over F. Let f be a bijective map on V. Define operations by

$$\left\{ \begin{array}{rcl} v_1 v_2 & = & f^{-1}(f(v_1) + f(v_2)) \\ \alpha \nabla v & = & f^{-1}(\alpha f(v)). \end{array} \right.$$

Show that with these operations V is a vector space over F.

4. Let $V = \{p_{n-1}x^{n-1} + \cdots + p_1 x + p_0 \in F[x] \mid p_{n-1} + \cdots + p_1 + p_0 = 0\}$. Show that V is a finite-dimensional subspace of $F[x]$ and find a basis for it.

5. Let q be a monic polynomial with distinct zeros $\lambda_1, \ldots, \lambda_n$. Let p be a polynomial of degree $n - 1$. Show that $\sum_{i=1}^{n} (g(\lambda_j))/(f'(\lambda_j)) = 1$.

6. Let $f, g \in F[z]$ with g nonzero. Then f has a unique representation of the form $f(z) = \sum_i a_i(z) g(z)^i$ with $\deg a_i < \deg g$.

7. Let $C_{\mathbf{R}}(X)$ be the space of all real-valued, continuous functions on a topological space X. Let $V_{\pm} = \{f \in C_{\mathbf{R}}(X) \mid f(x) = \pm f(-x)\}$. Show that $C_{\mathbf{R}}(X) = V_+ \oplus V_-$.

2.13   Notes and Remarks

Linear algebra is geometric in origin. It traces its development to the work of Fermat and Descartes on analytic geometry. In that context, points in the plane are identified with ordered pairs of real numbers. This has been extended in the work of Hamilton, Cayley, and others. The modern axiomatic approach seems due to Grassmann, and the first formal modern definition of a linear space appears in the works of Peano, who also treats linear transformations. Extensions of the axioms to include topological considerations are due to Banach, Wiener, and Von Neumann.

Matrices were introduced by Sylvester, but it was Cayley who introduced the modern notation. The theory of systems of linear equations is due to Kronecker.

3
Determinants

3.1 Basic Properties

Let R be a commutative ring with identity. Let X be a matrix, and let $x_1, \ldots, x_n$ be its columns.

Definition 3.1.1 *A **determinant** is a function $D : R^{n \times n} \longrightarrow R$ that as a function of the columns of a matrix in $R^{n \times n}$ satisfies the following:*

1. *$D(x_1, \ldots, x_n)$ is **multilinear**, that is, it is a linear function in each of its columns.*

2. *$D(x_1, \ldots, x_n)$ is **alternating**, by which we mean that it is zero whenever two adjacent columns coincide.*

3. *It is **normalized**, so that if $e_i, \ldots, e_n$ are the columns of the identity matrix, then*

$$D(e_1, \ldots, e_n) = 1.$$

Proposition 3.1.1 *If two adjacent columns in a matrix are interchanged, then the determinant changes signs.*

Proof: Let x and y be the the i-th and $i+1$-th columns, respectively. Then

$$0 = D(\cdots, x+y, x+y, \cdots)$$

$$= D(\cdots, x, x+y, \cdots) + D(\cdots, y, x+y, \cdots)$$

$$= D(\cdots, x, x, \cdots) + D(\cdots, x, y, \cdots) + D(\cdots, y, x, \cdots)$$
$$\quad + D(\cdots, y, y, \cdots)$$

$$= D(\cdots, x, y, \cdots) + D(\cdots, y, x, \cdots). \qquad \square$$

This property of determinants explains the usage of *alternating*.

Corollary 3.1.1 *If any two columns in a matrix are interchanged, then the determinant changes signs.*

Proof: Suppose that the i-th and j-th columns of the matrix A are interchanged. Without loss of generality, assume that $i < j$. By $j - i$ transpositions of adjacent columns, the j-th column can be brought to the i-th place. Now the i-th column is brought to the j-th place by $j - i - 1$ transpositions of adjacent columns. This changed the value of the determinant by only a factor of $(-1)^{2j-2i-1} = -1$. $\qquad \square$

Corollary 3.1.2 *If in a matrix any two columns are equal, then the determinant vanishes.*

Corollary 3.1.3 *If the j-th column of a matrix, multiplied by c, is added to the i-th column, the value of the dereminant does not change.*

Proof: We use the linearity in the i-th variable to compute

$$D(x_1, \cdots, x_i + cx_j, \cdots, x_j, \ldots, x_n)$$
$$= D(x_1, \cdots, x_i, \cdots, x_j, \ldots, x_n) + cD(x_1, \cdots, x_j, \cdots, x_j, \ldots, x_n).$$
$$= D(x_1, \cdots, x_i, \cdots, x_j, \ldots, x_n). \qquad \square$$

So far, we have proved elementary properties of the determinant function, but we do not know yet if such a function exists. So our aim now is to show the existence of a determinant function, which we will do by a direct, inductive, construction.

Definition 3.1.2 *Let A be an $n \times n$ matrix over R. We define the i, j-th* **minor** *of A, which we denote by M_{ij}, as the determinant of the matrix of order $n - 1$ obtained from A by eliminating the i-th row and j-th column. We define the i, j-***cofactor***, which we denote by A_{ij}, through the following:*

$$A_{ij} = (-1)^{i+j} M_{ij}.$$

Theorem 3.1.1 *For each integer $n \geq 1$, a determinant function exists and is given by the following formulas, to which we refer as the expansion by the i-th row:*

$$D(A) = \sum_{j=1}^{n} a_{ij} A_{ij}. \tag{3.1}$$

Proof: The construction is by an inductive process. For $n = 1$ we define $D(a) = a$, and this clearly satisfies the required properties of the determinant function.

Assume that we have constructed determinant functions for integers $\leq n - 1$. Now we define a determinant of A via the expansion by rows formula (3.1).

We show now that the properties of a determinant are satisfied.

- Multilinearity: Fix an index k. We will show that $D(A)$ is linear in the k-th column. For $j \neq k$, a_{ij} is not a function of the entries of the k-th column, but A_{ij} is linear in the entries of all columns appearing in it, including the k-th. For $j = k$, the cofactor A_{ik} is not a function of the elements of the k-th column, but the coefficient a_{ik} is a linear function. So $D(A)$ is linear as a function of the k-th column.

- Alternacy: Assume that the k and $k + 1$-th columns are equal. Then, for each index $j \neq k, k + 1$, we clearly have $A_{ij} = (-1)^{i+j} M_{ij}$ and M_{ij} contains two adjacent and equal columns; hence it is zero by the induction hypothesis. Therefore,

$$
\begin{aligned}
D(A) &= a_{ik} A_{ik} + a_{i(k+1)} A_{i(k+1)} \\
&= a_{ik}(-1)^{i+k} M_{ik} + a_{i(k+1)}(-1)^{i+k+1} M_{i(k+1)} \\
&= a_{ik}(-1)^{i+k} M_{ik} + a_{ik}(-1)^{i+k+1} M_{i(k+1)} = 0
\end{aligned}
$$

for $a_{ik} = a_{i(k+1)}$ and $M_{ik} = M_{i(k+1)}$.

- Normalization: We compute

$$
D(I) = \sum_{j=1}^{n} \delta_{ij}(-1)^{i+j} I_{ij} = (-1)^{2i} I_{ii} = 1
$$

for I_{ii} is the determinant of the identity matrix of order $n - 1$, and hence is equal to 1 by the induction hypothesis. $\qquad \square$

We use now the availability of the determinant function to discuss permutations. By a permutation σ of n elements we will mean a bijective map of the set $\{1, \ldots, n\}$ onto itself. The set of all such permutations S_n is a group and is called the **symmetric group**. We use alternatively the notation $(\sigma_1, \ldots, \sigma_n)$ for the permutation.

Assume now that $(\sigma_1, \ldots, \sigma_n)$ is a permutation. Let $e_1, \ldots, e_n$ be the standard unit vectors in R^n. Since a determinant function exists, $D(e_{\sigma_1}, \ldots, e_{\sigma_n}) = \pm 1$. We can bring the permutation matrix, whose columns are the e_{σ_i}, to the identity matrix in a finite number of column exchanges. The sign depends on the permutation only. Thus we define

$$
(-1)^{\sigma} = \text{sign}(\sigma) = D(e_{\sigma_1}, \ldots, e_{\sigma_n}). \tag{3.2}
$$

We will call $(-1)^\sigma$ the sign of the permutation. We say that the permutation is even if $(-1)^\sigma = 1$ and odd if $(-1)^\sigma = -1$.

Given a matrix A, we denote by $A^{(i)}$ its i-th column. Thus we have

$$A^{(j)} = \sum_{i=1}^{n} a_{ij} e_i.$$

Therefore we can compute

$$
\begin{aligned}
D(A) &= D(A^{(1)}, \ldots, A^{(n)}) \\
&= D\left(\sum_{i_1=1}^{n} a_{i_1 1} e_{i_1}, \ldots, \sum_{i_n=1}^{n} a_{i_n n} e_{i_n}\right) \\
&= \sum_{i_1=1}^{n} \cdots \sum_{i_n=1}^{n} D(a_{i_1 1} e_{i_1}, \ldots, a_{i_n n} e_{i_n}) \\
&= \sum_{\sigma \in S_n} a_{\sigma_1 1} \cdots a_{\sigma_n n} D(e_{\sigma_1}, \ldots, e_{\sigma_n}) \qquad (3.3) \\
&= \sum_{\sigma \in S_n} (-1)^\sigma a_{\sigma_1 1} \cdots a_{\sigma_n n} D(e_1, \ldots, e_n) \\
&= \sum_{\sigma \in S_n} (-1)^\sigma a_{\sigma_1 1} \cdots a_{\sigma_n n}.
\end{aligned}
$$

This gives us another explicit expression for the determinant function. In fact, this expression for the determinant function could be an alternative starting point for the development of the theory, but we choose to do it differently.

From now on we will use the notation $\det(A)$ for the determinant of A, that is, we have

$$\det(A) = \sum_{\sigma \in S_n} (-1)^\sigma a_{i_1 1} \cdots a_{i_n 1}.$$

The existence of this expression leads immediately to the following:

Theorem 3.1.2 *Let R be a commutative ring with an identity. Then, for any integer n, there exists a unique determinant function, and it is given by Eq. (3.2).*

Proof: Let D be any function defined on $R^{n \times n}$ and satisfying the determinantal axioms. Then clearly, using the computation in Eq. (3.3), we get $D(A) = \det(A)$. □

There was a basic asymmetry in our developments of determinants, as we focused on the columns. The next result is a step toward putting the theory of determinants into a more symmetric state.

Theorem 3.1.3 *Given an $n \times n$ matrix over the ring R, we have*

$$\det(\tilde{A}) = \det(A).$$

Proof: We use the fact that $\tilde{a}_{ij} = a_{ji}$, where $\tilde{a}_{ij}$ denotes the ij entry of the transposed matrix $\tilde{A}$. Thus we have

$$
\begin{aligned}
\det(\tilde{A}) &= \sum_{\sigma \in S_n} (-1)^\sigma \tilde{a}_{\sigma(1)1} \cdots \tilde{a}_{\sigma(n)n} \\
&= \sum_{\sigma \in S_n} (-1)^\sigma a_{1\sigma(1)} \cdots a_{n\sigma(n)} \\
&= \sum_{\sigma^{-1} \in S_n} (-1)^{\sigma^{-1}} a_{\sigma(1)^{-1}1} \cdots a_{\sigma(n)^{-1}n} \\
&= \det(A). \qquad \square
\end{aligned}
$$

Corollary 3.1.4 *For each integer $n \geq 1$, a determinant function exists, and it is given by the following formulas, to which we refer as the expansion by the j-th column:*

$$
D(A) = \sum_{i=1}^{n} a_{ij} A_{ij}. \tag{3.4}
$$

Proof: We use the expansion by the i-th column for the adjoint matrix. We use the fact that $\tilde{a}_{ji} = a_{ij}$ and $\tilde{A}_{ji} = A_{ij}$:

$$
\begin{aligned}
\det(A) &= D(\tilde{A}) = \sum_{j=1}^{n} \tilde{a}_{ji} \tilde{A}_{ji} \\
&= \sum_{j=1}^{n} a_{ij} A_{ij}. \qquad \square
\end{aligned}
$$

Corollary 3.1.5 *The determinant, as a function of the rows of a matrix, is multilinear, alternating, and satisfies $\det(I) = 1$.*

We now can prove the important multiplicative rule of determinants.

Theorem 3.1.4 *Let A and B be matrices in $R^{n \times n}$. Then*

$$
\det(AB) = \det(A)\det(B).
$$

Proof: Let $C = AB$. Let $A^{(i)}$ and $C^{(i)}$ be the j-th columns of A and C, respectively. Clearly, $C^{(j)} = \sum_{i=1}^{n} b_{ij} A^{(i)}$. Hence,

$$
\begin{aligned}
\det(C) &= \det(C^{(1)}, \ldots, C^{(n)}) \\
&= \det\left(\sum_{i_1=1}^{n} b_{i_1 j} A^{(i_1)}, \ldots, \sum_{i_n=1}^{n} b_{i_n j} A^{(i_n)} \right) \\
&= \sum_{i_1=1}^{n} \cdots \sum_{i_n=1}^{n} b_{i_1 1} \cdots b_{i_n n} \det(A^{(i_1)}, \ldots, A^{(i_n)}) \\
&= \sum_{\sigma \in S_n} b_{\sigma(1)1} \cdots b_{\sigma(n)n} (-1)^\sigma \det(A^{(1)}, \ldots, A^{(n)}) \\
&= \det(A) \sum_{\sigma \in S_n} (-1)^\sigma b_{\sigma(1)1} \cdots b_{\sigma(n)n} \\
&= \det(A)\det(B). \qquad \square
\end{aligned}
$$

We can use the expansion by rows and columns to obtain the following result.

Corollary 3.1.6 *Let A be an $n \times n$ matrix over R. Then*

1. $\displaystyle\sum_{k=1}^{n} a_{ik} A_{jk} = \delta_{ij} \det(A).$

2. $\displaystyle\sum_{k=1}^{n} a_{ki} A_{kj} = \delta_{ij} \det(A).$

Proof: The first equation is the expansion of the determinant by the j-th row if $i = j$, whereas if $i \neq j$, it is the expansion by the j-th row of a matrix derived from A by replacing the j-th row with the i-th. Thus it is the determinant of a matrix with two equal rows and hence its value is zero.

The other formula is proved analogously. $\qquad\square$

The determinant expansion rules can be written in a matrix form by defining a matrix $\text{adj}A$, called the **classical adjoint** of A, via

$$(\text{adj}A)_{ij} = A_{ji}.$$

Clearly, the expansion formulas given in Corollary 3.1.6 now can be written concisely as

$$(\text{adj}A)A = A(\text{adj}A) = \det A \cdot I.$$

This leads to an important corollary.

Corollary 3.1.7 *Let A be a square matrix over the field F. Then A is invertible if and only if $\det A \neq 0$.*

Proof: Assume that A is invertible. Then there exists a matrix B such that $AB = I$. By the multiplication rule of determinants we have

$$\det(AB) = \det(A) \cdot \det(B) = \det(I) = 1,$$

and hence $\det(A) \neq 0$.

Conversely, assume that $\det(A) \neq 0$. Then

$$A^{-1} = \frac{1}{\det A} \text{adj}A. \tag{3.5}$$

$\qquad\square$

3.2   Cramer's Rule

We can use determinants to give a closed-form representation of the solution of a system of linear equations, provided that the coefficient matrix of the system is nonsingular.

Theorem 3.2.1 *Given the system of linear equations $Ax = b$ with A a nonsingular $n \times n$ matrix, denote by $A^{(i)}$ the columns of the matrix A. Let*

$$x = \begin{pmatrix} x_1 \\ \cdot \\ \cdot \\ \cdot \\ x_n \end{pmatrix}$$

be the solution vector, that is, $b = \sum_{i=1}^{n} x_i A^{(i)}$. Then

$$x_i = \frac{\det(A^{(1)}, \ldots, b, \ldots, A^{(n)})}{\det(A)}$$

or, equivalently,

$$x_i = \frac{1}{\det A} \begin{vmatrix} a_{11} & \cdot & b_1 & \cdot & \cdot & a_{1n} \\ \cdot & \cdot & \cdot & \cdot & \cdot & \cdot \\ \cdot & \cdot & \cdot & \cdot & \cdot & \cdot \\ \cdot & \cdot & \cdot & \cdot & \cdot & \cdot \\ \cdot & \cdot & \cdot & \cdot & \cdot & \cdot \\ a_{n1} & \cdot & b_n & \cdot & \cdot & a_{nn} \end{vmatrix}.$$

Proof: We compute

$$
\begin{aligned}
\det(A^{(1)}, \ldots, b, \ldots, A^{(n)}) &= \det(A^{(1)}, \ldots, b, \sum_{j=1}^{n} x_j A^{(j)}, A^{(n)}) \\
&= \sum_{j=1}^{n} x_j \det(A^{(1)}, \ldots, b, A^{(j)}, A^{(n)}) \\
&= x_i \det(A^{(1)}, \ldots, b, A^{(j)}, A^{(n)}) \\
&= x_i \det(A).
\end{aligned}
$$

Another way to see this is to note that, in this case, the unique solution is given by $x = A^{-1}b$. We will compute the i-th component, x_i:

$$
\begin{aligned}
x_i &= (A^{-1}b)_i = \sum_{j=1}^{n} (A^{-1})_{ij} b_j \\
&= \frac{1}{\det A} \sum_{j=1}^{n} (\mathrm{adj} A)_{ij} b_j = \frac{1}{\det A} \sum_{j=1}^{n} b_j A_{ji}.
\end{aligned}
$$

But this is just the expansion by the i-th column of the matrix A, where the i-th column has been replaced by the elements of the vector b.   □

We now prove an elementary but useful result about the computation of determinants.

Proposition 3.2.1 *Let A and B be square matrices, not necessarily of the same size. Then*

$$\det \begin{pmatrix} A & C \\ 0 & B \end{pmatrix} = \det A \cdot \det B.$$

Proof:

$$\det \begin{pmatrix} A & C \\ 0 & B \end{pmatrix}$$

is multilinear and alternating in the columns of A and in the rows of B; therefore,

$$\det \begin{pmatrix} A & C \\ 0 & B \end{pmatrix} = \det A \det B \det \begin{pmatrix} I & C \\ 0 & I \end{pmatrix}.$$

But a consideration of elementary operations in rows and columns leads to

$$\det \begin{pmatrix} I & C \\ 0 & I \end{pmatrix} = \det \begin{pmatrix} I & 0 \\ 0 & I \end{pmatrix} = 1. \qquad \square$$

Lemma 3.2.1 *Let $A, B, C,$ and D be matrices of appropriate sizes such that C and D commute and D is invertible. Then*

$$\det \begin{pmatrix} A & B \\ C & D \end{pmatrix} = \det(AD - BC).$$

Proof: From the equality

$$\begin{pmatrix} A & B \\ C & D \end{pmatrix} = \begin{pmatrix} A - BD^{-1}C & B \\ 0 & D \end{pmatrix} \begin{pmatrix} I & 0 \\ D^{-1}C & I \end{pmatrix},$$

we conclude that

$$\begin{aligned} \det \begin{pmatrix} A & B \\ C & D \end{pmatrix} &= \det(A - BD^{-1}C) \cdot \det D & (3.6) \\ &= \det(A - BCD^{-1}) \cdot \det D \\ &= \det(AD - BC). & \square \end{aligned}$$

3.3   The Sylvester Resultant

We now give a determinantal condition for the coprimeness of two polynomials. Our approach is geometric in nature.

Let $p, q \in F[z]$. We know, from Corollary 1.3.6, that p and q are coprime if and only if the Bezout equation $ap + bq = 1$ has a polynomial solution. The following gives a matrix condition for coprimeness:

Theorem 3.3.1 *Let $p, q \in F[z]$ with $p(z) = p_0 + \cdots + p_m z^m$ and $q(z) = q_0 + q_1 z + \cdots + q_n z^n$. Then the following conditions are equivalent:*

1. *The polynomials p and q are coprime.*

2. $F_{m+n}[z] = pF_n[z] + qF_m[z].$

3. *The resultant matrix*

$$\text{Res}(p,q) = \begin{pmatrix} p_0 & & & & & q_0 & & & & \\ \cdot & \cdot & & & & \cdot & \cdot & & & \\ \cdot & \cdot & \cdot & & & \cdot & \cdot & \cdot & & \\ p_m & \cdot & \cdot & \cdot & & \cdot & \cdot & \cdot & q_0 \\ & \cdot & \cdot & \cdot & p_0 & q_n & \cdot & \cdot & \cdot \\ & & \cdot & \cdot & \cdot & \cdot & \cdot & \cdot & \cdot \\ & & & \cdot & \cdot & \cdot & & \cdot & \cdot \\ & & & & \cdot & \cdot & & & \cdot \\ & & & & p_m & & & & q_n \end{pmatrix} \tag{3.7}$$

is nonsingular.

4. *The resultant, defined as the determinant of the resultant matrix, is nonzero.*

Proof: (1) $\Rightarrow$ (2) Follows from the Bezout equation directly or from Proposition 2.6.2 as a special case.

(2) $\Rightarrow$ (1). Since $1 \in F_{m+n}[z]$, it follows that a solution exists to the Bezout equation $ap + bq = 1$; moreover, with the conditions $\deg a < n$ and $\deg b < m$ satisfied, it is unique.

(2) $\Rightarrow$ (3). By Proposition 2.6.2, the equality $F_{m+n}[z] = pF_n[z] + qF_m[z]$ is equivalent to this sum being a direct sum. Thus a union of bases for $pF_n[z]$ and $qF_m[z]$ is a basis for $f_{m+n}[z]$. Now $\{z^i p | i = 0, \ldots, n-1\}$ is a basis for $pF_n[z]$ and $\{z^i q | i = 0, \ldots, m-1\}$ is a basis for $qF_m[z]$. Thus $\mathcal{B}_{res} = \{p, zp, \ldots, z^{n-1}p, q, zq, \ldots, z^{m-1}q\}$ is a basis for $F_{m+n}[z]$. The last space also has the standard basis $\mathcal{B}_{st} = \{1, z, \ldots, z^{m+n-1}\}$. It is easy to check that $\text{Res}(p,q) = [I]_{res}^{st}$, that is, it is a change of basis transformation and hence is necessarily invertible.

(3) $\Rightarrow$ (2). If $\text{Res}(p,q)$ is nonsingular, this shows that $\mathcal{B}_{res}$ is a basis for $F_{m+n}[z]$. This implies the equality

$$F_{m+n}[z] = pF_n[z] + qF_m[z]. \qquad \square$$

We conclude this short chapter by computing the determinant of the Vandermonde matrix derived in Eq. (2.7).

Proposition 3.3.1 *Let*

$$V = \begin{pmatrix} 1 & \lambda_1 & \cdot & \cdot & \lambda_1^{n-1} \\ \cdot & \cdot & \cdot & \cdot & \cdot \\ \cdot & \cdot & \cdot & \cdot & \cdot \\ \cdot & \cdot & \cdot & \cdot & \cdot \\ 1 & \lambda_n & \cdot & \cdot & \lambda_n^{n-1} \end{pmatrix}$$

be the Vandermonde matrix. Then we have

$$\det V = \begin{vmatrix} 1 & \lambda_1 & . & . & \lambda_1^{n-1} \\ . & . & . & . & . \\ . & . & . & . & . \\ . & . & . & . & . \\ 1 & \lambda_n & . & . & \lambda_n^{n-1} \end{vmatrix} = \prod_{1 \le j < i \le n} (\lambda_i - \lambda_j). \qquad (3.8)$$

Proof: By induction on n. For $n = 1$, the equality is trivially satisfied. Assume that this holds for all integers $< n$. We consider the following determinant:

$$f(z) = \begin{vmatrix} 1 & \lambda_1 & . & . & \lambda_1^{n-1} \\ . & . & . & . & . \\ . & . & . & . & . \\ 1 & \lambda_{n-1} & . & . & \lambda_{n-1}^{n-1} \\ 1 & z & . & . & z^{n-1} \end{vmatrix}.$$

Obviously, f is a polynomial of degree $n - 1$ that vanishes at $\lambda_1, \ldots, \lambda_{n-1}$. Thus $f(z) = C_{n-1}(z - \lambda_1) \cdots (z - \lambda_{n-1})$. Expanding by the last row and using the induction hypothesis, we get

$$C_{n-1} = \begin{vmatrix} 1 & \lambda_1 & . & . & \lambda_1^{n-2} \\ . & . & . & . & . \\ . & . & . & . & . \\ . & . & . & . & . \\ 1 & \lambda_{n-1} & . & . & \lambda_{n-1}^{n-2} \end{vmatrix} = \prod_{1 \le j < i \le n-1} (\lambda_i - \lambda_j).$$

Evaluating f at $z = \lambda_n$ implies Eq. (3.8). $\qquad \square$

We note that for the previous proposition to hold we do not have to assume the λ_i to be distinct. However, clearly the determinant of a Vandermonde matrix is nonzero if and only if the λ_i are distinct.

3.4   Exercises

1. Let A, B be $n \times n$ matrices. For the classical adjoint show the following:

   (a) $\mathrm{adj}(AB) = \mathrm{adj}A \cdot \mathrm{adj}B$.
   (b) $\det \mathrm{adj}A = (\det A)^{n-1}$.
   (c) $\mathrm{adj}(\mathrm{adj}A) = (\det A)^{n-2} \cdot A$.
   (d) $\det \mathrm{adj}(\mathrm{adj}A) = (\det A)^{(n-1)^2}$.

2. Let a square $n \times n$ matrix A be defined by $a_{ij} = 1 - \delta_{ij}$. Show that $\det A = (n - 1)(-1)^{n-1}$.

3. Let A, B be square matrices. Show that

$$\det \begin{pmatrix} A & B \\ B & A \end{pmatrix} = \det(A + B) \cdot \det(A - B).$$

4. Let A be a square $n \times n$ matrix, $x, y \in F^n$, and $\alpha \in F$. Show that

$$\det \begin{pmatrix} A & x \\ \tilde{y} & \alpha \end{pmatrix}.$$

Show that

$$\mathrm{adj}(I - x\tilde{y}) = x\tilde{y} - (1 - \tilde{y}x)I.$$

5. Let p be a polynomial of degree $\leq n - 1$. Show that $p(z)$ and $z^n - 1$ are coprime if and only if

$$\det \begin{pmatrix} p_0 & p_{n-1} & \cdot & \cdot & p_1 \\ p_1 & p_0 & \cdot & \cdot & \cdot \\ \cdot & \cdot & \cdot & \cdot & \cdot \\ \cdot & \cdot & \cdot & \cdot & p_{n-1} \\ p_{n-1} & \cdot & \cdot & p_1 & p_0 \end{pmatrix}.$$

6. Let $V(\lambda_1, \ldots, \lambda_n)$ be the Vandermonde determinant. Show that

$$\begin{vmatrix} 1 & \lambda_1 & \cdot & \cdot & \lambda_1^{n-2} & \lambda_2 \cdots \lambda_n \\ \cdot & \cdot & \cdot & \cdot & \cdot & \cdot \\ \cdot & \cdot & \cdot & \cdot & \cdot & \cdot \\ \cdot & \cdot & \cdot & \cdot & \cdot & \cdot \\ \cdot & \cdot & \cdot & \cdot & \cdot & \cdot \\ 1 & \lambda_n & \cdot & \cdot & \lambda_n^{n-2} & \lambda_1 \cdots \lambda_{n-1} \end{vmatrix} = (-1)^{n-1} V(\lambda_1, \ldots, \lambda_n).$$

7. Let $V(\lambda_1, \ldots, \lambda_n)$ be the Vandermonde determinant. Show that

$$\begin{vmatrix} 1 & \lambda_1 & \cdot & \cdot & \lambda_1^{n-2} & (\lambda_2 + \cdots + \lambda_n)^{n-1} \\ \cdot & \cdot & \cdot & \cdot & \cdot & \cdot \\ \cdot & \cdot & \cdot & \cdot & \cdot & \cdot \\ \cdot & \cdot & \cdot & \cdot & \cdot & \cdot \\ \cdot & \cdot & \cdot & \cdot & \cdot & \cdot \\ 1 & \lambda_n & \cdot & \cdot & \lambda_n^{n-2} & (\lambda_1 + \cdots + \lambda_{n-1})^{n-1} \end{vmatrix} =$$

$$(-1)^{n-1} V(\lambda_1, \ldots, \lambda_n).$$

8. Given $\lambda_1, \ldots, \lambda_n$, define $s_k = p_1 \lambda_1^k + \cdots + p_n \lambda_n^k$. Set, for $i, j = 0, \ldots, n - 1$, $a_{ij} = s_{i+j}$. Show that

$$\det(a_{ij}) = \begin{vmatrix} s_0 & \cdot & \cdot & \cdot & s_{n-1} \\ \cdot & \cdot & \cdot & \cdot & \cdot \\ \cdot & \cdot & \cdot & \cdot & \cdot \\ \cdot & \cdot & \cdot & \cdot & \cdot \\ s_{n-1} & \cdot & \cdot & \cdot & s_{2n-2} \end{vmatrix} = p_1 \cdots p_n \prod_{i>j} (\lambda_i - \lambda_j)^2.$$

9. Show that

$$
\begin{vmatrix}
\dfrac{1}{x_1 + y_1} & \cdots & \dfrac{1}{x_1 + y_n} \\
\vdots & \ddots & \vdots \\
\dfrac{1}{x_n + y_1} & \cdots & \dfrac{1}{x_n + y_n}
\end{vmatrix}
= \frac{\displaystyle\prod_{i>j}(x_i - x_j)(y_i - y_j)}{\displaystyle\prod_{i,j}(x_i + y_j)}.
$$

This determinant is generally known as the **Cauchy determinant**.

3.5 Notes and Remarks

Determinants appear already in the work of Leibnitz on the solution of systems of linear equations. The axiomatic treatment presented here is due to Kronecker and Weierstrass.

4
Linear Transformations

4.1 Linear Transformations

Definition 4.1.1 *Let V and W be two vector spaces over the field F. A mapping T from V to W, denoted $T : V \longrightarrow W$, is a* **linear transformation** *if*

$$T(\alpha x + \beta y) = \alpha(Tx) + \beta(Ty)$$

for all $x, y \in V$ and all $\alpha, \beta \in F$.

For an arbitrary linear space V, we define the zero and identity transformations by $0x = 0$ and $Ix = x$ for all $x \in V$.

The property of linearity of a transformation T is equivalent to the following two properties:

- Additivity: $T(x + y) = Tx + Ty$.

- Homogeneity: $T(\alpha x) = \alpha(Tx)$.

The structure of linear transformations is very rigid. It suffices to know its action on basis vectors, and this determines the transformation uniquely.

Theorem 4.1.1 *Let V, W be two linear spaces over the same field F. Let $\{e_1, \ldots, e_n\}$ be a basis for V, and let $\{f_1, \ldots, f_n\}$ be arbitrary vectors in W. Then there exists a unique transformation $T : V \longrightarrow W$ for which*

$$Te_i = f_i \quad \text{for all } i = 1, \ldots, n.$$

Proof: Every $x \in V$ has a unique representation $x = \sum_{i=1}^{n} \alpha_i e_i$. We define $T : V \longrightarrow W$ by

$$Tx = \sum_{i=1}^{n} \alpha_i f_i.$$

Clearly, $Te_i = f_i$ for all $i = 1, \ldots, n$. Next we show that T is linear. Let $y \in V$ have the representation $y = \sum_{i=1}^{n} \beta_i e_i$. Then

$$
\begin{aligned}
T(\alpha x + \beta y) &= T \sum_{i=1}^{n} (\alpha \alpha_i + \beta \beta_i) e_i = \sum_{i=1}^{n} (\alpha \alpha_i + \beta \beta_i) f_i \\
&= \sum_{i=1}^{n} (\alpha \alpha_i) f_i + \sum_{i=1}^{n} (\beta \beta_i) f_i = \alpha \sum_{i=1}^{n} \alpha_i f_i + \beta \sum_{i=1}^{n} \beta_i f_i \\
&= \alpha(Tx) + \beta(Ty).
\end{aligned}
$$

To prove uniqueness, let $S : V \longrightarrow W$ be a linear transformation satisfying $Se_i = f_i$ for all i. Then

$$Sx = S \sum_{i=1}^{n} \alpha_i e_i = \sum_{i=1}^{n} \alpha_i Se_i = \sum_{i=1}^{n} \alpha_i f_i = Tx.$$

So $T = S$. $\square$

Given a linear transformation $T : V \longrightarrow W$, there are two important subspaces determined by it, namely,

1. $Ker\, T$, the **kernel** of T, defined by

$$Ker\, T = \{x \in V | Tx = 0\}.$$

2. $Im\, T$, the **image** of T, defined by

$$Im T = \{Tx | x \in V\}.$$

Theorem 4.1.2 *Let $T : V \longrightarrow W$ be a linear transformation. Then $Ker\, T$ and $Im\, T$ are subspaces of V and W, respectively.*

Proof:

1. It is clear that $Ker\, T \subset V$. Let $x, y \in V$ and $\alpha, \beta \in F$. Since

$$T(\alpha x + \beta y) = \alpha(Tx) + \beta(Ty) = 0,$$

it follows that $\alpha x + \beta y \in Ker\, T$.

2. By definition, $Im\, T \subset W$. Let $y_1, y_2 \in Im T$. Then there exist vectors $x_1, x_2 \in V$ for which $y_i = Tx_i$. Let $\alpha_1, \alpha_2 \in F$. Then

$$\alpha_1 y_1 + \alpha_2 y_2 = \alpha_1 T x_1 + \alpha_2 T x_2 = T(\alpha_1 x_1 + \alpha_2 x_2)$$

or $\alpha_1 x_1 + \alpha_2 x_2$. $\qquad\square$

We define the **rank** of a linear transformation T, denoted by $\mathrm{rank}(T)$, by

$$\mathrm{rank}(T) = \dim \mathrm{Im} T$$

and the **nullity** of T, denoted null (T), by

$$\mathrm{null}(T) = \dim Ker\, T.$$

The rank and nullity of a linear transformation are connected via the following theorem:

Theorem 4.1.3 *Let* $T : V \longrightarrow W$ *be a linear transformation. Then*

$$\mathrm{rank}(T) + \mathrm{null}(T) = \dim V$$

or, equivalently,

$$\dim \mathrm{Im} T + \dim Ker\, T = \dim V.$$

Proof: Let $\{e_1, \ldots, e_n\}$ be a basis of V for which $\{e_1, \ldots, e_p\}$ is a basis of $Ker\, T$. Let $x \in V$. Then x has a unique representation of the form $x = \sum_{i=1}^n \alpha_i e_i$. As T is linear, we have

$$Tx = T \sum_{i=1}^n \alpha_i e_i = \sum_{i=1}^n \alpha_i T e_i = \sum_{i=p+1}^n \alpha_i T e_i$$

as $T e_i = 0$ for $i = 1, \ldots, p$. Therefore it follows that $\{T e_{p+1}, \ldots, T e_n\}$ is a spanning set of $\mathrm{Im} T$. We show that they are also linearly independent. Assume that there exist $\{\alpha_{p+1}, \ldots, \alpha_n\}$ such that $\sum_{i=p+1}^n \alpha_i(T e_i) = 0$. Now $\sum_{i=p+1}^n \alpha_i(T e_i) = T \sum_{i=p+1}^n \alpha_i(e_i)$, which implies that $\sum_{i=p+1}^n \alpha_i(e_i) \in Ker\, T$. Therefore, it can be represented as a linear combination of the form $\sum_{i=1}^p \alpha_i e_i$; we get

$$\sum_{i=1}^p \alpha_i e_i - \sum_{i=p+1}^n \alpha_i e_i = 0.$$

Using the fact that $\{e_1, \ldots, e_n\}$ is a basis of V, we conclude that $\alpha_i = 0$ for all $i = 1, \ldots, n$. Thus we proved that $\{T e_{p+1}, \ldots, T e_n\}$ is a basis of $\mathrm{Im} T$. So $\dim Ker\, T = p$, $\dim \mathrm{Im} T = n - p$, and $\dim Ker\, T + \dim \mathrm{Im} T = p + (n - p) = n = \dim V$. $\qquad\square$

Let U and V be linear spaces over the field F. We denote by $L(U, V)$ the space of all linear transformations from U to V. We define the sum of two transformations by

$$(T + S)x = Tx + Sx$$

and the product with a scalar $\alpha \in F$ by

$$(\alpha T)x = \alpha(Tx).$$

Theorem 4.1.4 *From the previous definitions, $L(U,V)$ is a linear space.*

Proof: We show that $T+S$ and αT are indeed linear transformations. Let $x, y \in V$ and $\alpha, \beta \in F$. We have

$$
\begin{aligned}
(T+S)(x+y) &= T(x+y) + S(x+y) = (Tx + Ty) + (Sx + Sy) \\
&= (Tx + Sx) + (Ty + Sy) = (T+S)x + (T+S)y
\end{aligned}
$$

and

$$
\begin{aligned}
(T+S)(\alpha x) &= T(\alpha x) + S(\alpha x) \\
&= \alpha(Tx) + \alpha(Sx) = \alpha(Tx + Sx) \\
&= \alpha((T+S)x).
\end{aligned}
$$

This shows that $T+S$ is a linear transformation. The proof for αT goes along similar lines. We leave it to the reader to verify that all of the axioms of a linear space hold for $L(U,V)$. $\qquad\square$

It is clear that the dimension of $L(U,V)$ is determined by the spaces U and V. The next theorem characterizes the dimension of $L(U,V)$.

Theorem 4.1.5 *Let V and W be linear spaces over the field F. Let $\dim V = n$ and $\dim W = m$. Then*

$$
\dim L(U,V) = \dim U \cdot \dim V.
$$

Proof: Let $\{e_1, \ldots, e_n\}$ be a basis for V and $\{f_1, \ldots, f_m\}$ be a basis for W. We proceed to construct a basis for $L(V,W)$. By Theorem 4.1.1, a linear transformation is determined by its values on basis elements. We define a set of mn linear transformations E_{ij}, $i = 1, \ldots, m, j = 1, \ldots, n$ by letting

$$
E_{ij}e_k = \delta_{jk}f_i.
$$

Here δ_{ij} is the Kronecker delta function, which is defined by

$$
\delta_{ij} = \begin{cases} 1 & i = j \\ 0 & i \neq j. \end{cases}
$$

On the vector $x = \sum_{k=1}^{n} \alpha_k e_k$ the transformation E_{ij} acts as

$$
E_{ij}x = E_{ij}\sum_{k=1}^{n} \alpha_k e_k = \sum_{k=1}^{n} \alpha_k E_{ij}e_k = \sum_{k=1}^{n} \alpha_k \delta_{jk}f_i = \alpha_j f_i
$$

or

$$
E_{ij}x = \alpha_j f_i.
$$

We now show that the set of transformations E_{ij} forms a basis for $L(V,W)$. To prove this, we have to show that this set is linearly independent and spans $L(V,W)$.

We begin by showing linear independence. Assume that there exist $\alpha_{ij} \in F$ for which

$$\sum_{i=1}^{m} \sum_{j=1}^{n} \alpha_{ij} E_{ij} = 0.$$

We operate with the zero transformation, in both of its representations, on an arbitrary basis element e_k in V:

$$0 = 0 \cdot e_k = \sum_{i=1}^{m} \sum_{j=1}^{n} \alpha_{ij} E_{ij} e_k = \sum_{i=1}^{m} \sum_{j=1}^{n} \alpha_{ij} \delta_{jk} f_i = \sum_{i=1}^{m} \alpha_{ik} f_i.$$

Since the f_i are linearly independent, we get $\alpha_{ik} = 0$ for all $i = 1, \ldots, m$. The index k was chosen arbitrarily, hence we conclude that all of the α_{ij} are zero. This completes the proof of linear independence.

Let T be an arbitrary transformation in $L(V, W)$. Since $Te_k \in W$, it has a unique representation of the form

$$Te_k = \sum_{i=1}^{m} b_{ik} f_i, \qquad k = 1, \ldots, n.$$

We now show that

$$T = \sum_{i=1}^{m} \sum_{j=1}^{n} b_{ij} E_{ij}.$$

Of course it suffices, by Theorem 4.1.1, to show equality on the basis elements in V. We compute

$$\begin{aligned}
\left(\sum_{i=1}^{m} \sum_{j=1}^{n} b_{ij} E_{ij} \right) e_k &= \sum_{i=1}^{m} \sum_{j=1}^{n} b_{ij} (E_{ij} e_k) = \sum_{i=1}^{m} \sum_{j=1}^{n} b_{ij} \delta_{jk} f_i \\
&= \sum_{i=1}^{m} b_{ik} f_i = Te_k.
\end{aligned}$$

Since k can be chosen arbitrarily, the proof is complete. $\square$

In some cases, linear transformations can be composed, or multiplied. Let U, V, W be linear spaces over F. Let $S \in L(V, W)$ and $T \in L(U, V)$, that is,

$$U \xrightarrow{\ T\ } V \xrightarrow{\ S\ } W.$$

We define the transformation $ST : U \longrightarrow W$ by

$$(ST)x = S(Tx), \qquad \forall x \in U.$$

We call the transformation ST the **composition** or **product** of S and T.

Theorem 4.1.6 *The transformation ST is linear, that is, $ST \in L(U, W)$.*

Proof: Let $x, y \in U$ and $\alpha, \beta \in F$. By the linearity of S and T we get

$$\begin{aligned}
(ST)(\alpha x + \beta y) &= S(T(\alpha x + \beta y)) \\
&= S(\alpha(Tx) + \beta(Ty)) = \alpha S(Tx) + \beta S(Ty) \\
&= \alpha(ST)x + \beta(ST)y.
\end{aligned}$$
$\square$

Composition of transformations is associative; that is, if

$$U \xrightarrow{T} V \xrightarrow{S} W \xrightarrow{R} Y,$$

then

$$(RS)T = R(ST).$$

We already saw the existence of the identity map I in each linear space V. Clearly we have $IT = TI = T$ for every transformation T.

Definition 4.1.2 *Let $T \in L(V, W)$. We say that T is **right invertible** if there exists a transformation $S \in L(W, V)$ for which*

$$TS = I_W.$$

*Similarly, we say that T is **left invertible** if there exists a transformation $S \in L(W, V)$ for which*

$$ST = I_V.$$

*We say that T is **invertible** if it is both right and left invertible.*

Proposition 4.1.1 *If T is both right and left invertible, then the right inverse and the left inverse are equal and will be denoted by T^{-1}.*

Proof: Let $TS = I_W$ and $RT = I_V$. Then $R(TS) = (RT)S$ implies

$$R = RI_W = I_V S = S.$$
$\square$

Left and right invertibility are intrinsic properties of the transformation T, as follows from the next theorem.

Theorem 4.1.7 *Let $T \in L(V, W)$. Then*

1. *T is right invertible if and only if T is surjective.*

2. *T is left invertible if and only if T is injective.*

Proof:

1. Assume that T is right invertible. Thus there exists $S \in L(W, V)$ for which $TS = I_W$. Therefore, for all $x \in W$,

$$T(Sx) = (TS)x = I_W x = x$$

or $x \in \mathrm{Im}T$, so $\mathrm{Im}T = W$ and T is surjective.

Conversely, assume that T is surjective. Let $\{f_1, \ldots, f_n\}$ be a basis in W, and let $\{e_1, \ldots, e_n\}$ be vectors in V satisfying

$$Te_i = f_i, \qquad i = 1, \ldots, n.$$

Clearly the vectors $\{e_1, \ldots, e_n\}$ are linearly independent. We define a linear transformation $S \in L(W, V)$ by

$$Sf_i = e_i, \qquad i = 1, \ldots, n.$$

Then, for all i, $(TS)f_i = T(Sf_i) = Te_i = f_i$, that is, $TS = I_W$.

2. Assume that T is left invertible. Let $S \in L(W, V)$ be a left inverse, that is, $ST = I_V$. Let $x \in Ker\, T$. Then

$$x = I_V x = (ST)x = S(Tx) = S(0) = 0,$$

that is, $Ker\, T = \{0\}$.

Conversely, assume that $Ker\, T = \{0\}$. Let $\{e_1, \ldots, e_n\}$ be a basis in V. It follows from our assumption that $\{Te_1, \ldots, Te_n\}$ are linearly independent vectors in W. We next complete this set of vectors to a basis $\{Te_1, \ldots, Te_n, f_{m+1}, \ldots, f_n\}$. We now define a linear transformation $S \in L(W, V)$ by

$$S \begin{cases} Te_i = e_i & i = 1, \ldots, m \\ Tf_i = 0 & i = m+1, \ldots, n. \end{cases}$$

Obviously, $STe_i = e_i$ for all i, and hence $ST = I_V$. $\qquad\square$

It should be noted that the left inverse is generally not unique. In our construction, we had the freedom to define S differently on the vectors f_i, but our choice was for the simplest definition.

The previous characterizations are special cases of the more general result.

Theorem 4.1.8 *Let V, V_1, V_2 be linear spaces over F.*

1. *Assume that $A \in L(V_1, V)$ and $C \in L(V_1, V_2)$. Then there exists a $B \in L(V_2, V)$ such that*

$$A = BC$$

if and only if

$$Ker\, A \supset Ker\, C. \tag{4.1}$$

2. *Assume that $A \in L(V_1, V)$ and $B \in L(V_2, V)$. Then there exists a $C \in L(V_1, V_2)$ such that*

$$A = BC$$

if and only if

$$Im A \subset Im B. \tag{4.2}$$

Proof:

1. If $A = BC$, then for $x \in Ker\, C$ we have

$$Ax = B(Cx) = B0 = 0,$$

that is, $x \in Ker\, A$ and Eq. (4.1) follows.

Conversely, assume that $Ker\, A \supset Ker\, C$. Let $\{e_1, \ldots, e_r, e_{r+1}, \ldots, e_p, e_{p+1}, \ldots, e_n\}$ be a basis in V such that $\{e_1, \ldots, e_p\}$ is a basis for $Ker\, A$ and $\{e_1, \ldots, e_r\}$ is a basis for $Ker\, C$. The vectors $\{Ce_{r+1}, \ldots, Ce_n\}$ are linearly independent. We complete them to a basis of V_2 and define a linear transformation $B : V_2 \longrightarrow V$ by

$$BCe_i = \begin{cases} Ae_i & i = p+1, \ldots, n \\ 0 & i = r+1, \ldots, p \end{cases}$$

and arbitrarily on the other basis elements of V_2. Then, for every $x = \sum_{i=1}^{n} \alpha_i e_i \in V$, we have

$$
\begin{aligned}
Ax &= A\sum_{i=1}^{n} \alpha_i e_i = \sum_{i=1}^{n} \alpha_i A e_i = \sum_{i=p+1}^{n} \alpha_i (A e_i) \\
&= \sum_{i=p+1}^{n} \alpha_i (BC) e_i = B \sum_{i=p+1}^{n} \alpha_i C e_i = BC \sum_{i=p+1}^{n} \alpha_i e_i
\end{aligned}
$$

or $A = BC$.

2. If $A = BC$, then for every $x \in V_1$ we have

$$Ax = (BC)x = B(Cx)$$

or $\mathrm{Im}A \subset \mathrm{Im}B$.

Conversely, assume that $\mathrm{Im}A \subset \mathrm{Im}B$. Then let $\{e_1, \ldots, e_r, e_{r+1}, \ldots, e_n\}$ be a basis for V_1 such that $\{e_{r+1}, \ldots, e_n\}$ is a basis for $Ker\, A$. Then $\{Ae_1, \ldots, Ae_r\}$ are linearly independent vectors in V_2. By our assumption $\mathrm{Im}A \subset \mathrm{Im}B$, there exist vectors vectors $\{g_1, \ldots, g_r\}$, necessarily linearly independent, such that

$$Ae_i = Bg_i, \qquad i = 1, \ldots, r.$$

We complete these to a basis $\{g_1, \ldots, g_r, g_{r+1}, \ldots, g_m\}$ of V_2. We now define $C \in L(V_1, V_2)$ by

$$Ce_i = \begin{cases} g_i & i = 1, \ldots, r \\ 0 & i = r+1, \ldots, n. \end{cases}$$

Then we compute

$$
\begin{aligned}
A\sum_{i=1}^{n} \alpha_i e_i &= \sum_{i=1}^{r} \alpha_i A e_i = \sum_{i=1}^{r} \alpha_i B g_i \\
&= B\sum_{i=1}^{r} \alpha_i g_i = B\sum_{i=1}^{r} \alpha_i C e_i = BC\sum_{i=1}^{r} \alpha_i e_i
\end{aligned}
$$

or $A = BC$. $\square$

Theorem 4.1.9 *Let $T \in L(U, V)$ be invertible. Then $T^{-1} \in L(V, U)$.*

Proof: Let $y_1, y_2 \in V$ and $\alpha_1, \alpha_2 \in F$. There exist unique vectors $x_1, x_2 \in U$ such that $Tx_i = y_i$ or, equivalently, $x_i = T^{-1}y_i$. Since

$$T(\alpha_1 x_1 + \alpha_2 x_2) = \alpha_1 Tx_1 + \alpha_2 Tx_2 = \alpha_1 y_1 + \alpha_2 y_2,$$

therefore

$$T^{-1}(\alpha_1 y_1 + \alpha_2 y_2) = \alpha_1 x_1 + \alpha_2 x_2 = \alpha_1 T^{-1}y_1 + \alpha_2 T^{-1}y_2.$$ $\square$

A linear transformation $T : V \longrightarrow W$ is called an **isomorphism** if it is invertible. In this case, we say that V and W are **isomorphic spaces**. Clearly, isomorphism of linear spaces is an equivalence relation.

Theorem 4.1.10 *Every n-dimensional linear space over F is isomorphic to F^n.*

Proof: Let $\mathcal{B} = \{e_1, \ldots, e_n\}$ be a basis for V . The map $x \mapsto [x]^{\mathcal{B}}$ is an isomorphism. $\square$

Theorem 4.1.11 *Two finite-dimensional linear spaces V and W over F are isomorphic if and only if they have the same dimension.*

Proof: If U and V have the same dimension, then we can construct a map of the basis elements in U onto the basis elements in V. This map is necessarily an isomorphism.

Conversely, if U and V are isomorphic, the image of a basis in U is necessarily a basis in V. Thus, the dimensions are equal. $\square$

4.2   Matrix Representations

Let V, W be linear vector spaces over a field F. Let $B = \{e_1, \ldots, e_n\}$ and $B_1 = \{f_1, \ldots, f_n\}$ be bases of V and W, respectively. We saw that the set of linear transformations $\{E_{ij}\}$ defined by

$$E_{ij}e_k = \delta_{jk}f_i$$

is a basis for $L(V, W)$. We denote this basis by $\mathcal{B}_1 \times \mathcal{B}_2$.

A natural problem presents itself, namely, finding the coordinate vector of a linear transformation $T \in L(V, W)$ with respect to this basis.

We observed before that T can be written as

$$T = \sum_{i=1}^{m} \sum_{j=1}^{n} t_{ij} E_{ij}.$$

Hence,

$$Te_k = \sum_{i=1}^{m}\sum_{j=1}^{n} t_{ij}E_{ij}e_k = \sum_{i=1}^{m}\sum_{j=1}^{n} t_{ij}\delta_{jk}f_i = \sum_{i=1}^{m} t_{ik}f_i,$$

that is,

$$Te_k = \sum_{i=1}^{m} t_{ik}f_i.$$

Therefore, the coordinate vector of T with respect to the basis $\mathcal{B}_1 \times \mathcal{B}_2$ will have the entries t_{ik} arranged in the order of the basis elements.

Contrary to the case of an abstract linear space, we arrange the coordinates in this case in an $m \times n$ matrix (t_{ij}) and call this the **matrix representation** of T with respect to the bases $\mathcal{B}$ in V and $\mathcal{B}_1$ in W. We use the notation

$$[T]_{\mathcal{B}}^{\mathcal{B}_1} = (t_{ik}).$$

The importance of the matrix representation stems froms the following theorem. This theorem reduces the application of an arbitrary linear transformation to matrix multiplication.

Theorem 4.2.1 *Let $T : V \longrightarrow W$ be a linear transformation, and let $\mathcal{B}_1$ and $\mathcal{B}_2$ be bases of V and W, respectively. Then the following diagram is commutative:*

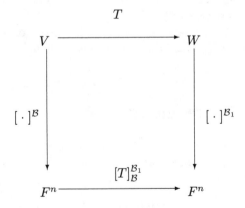

In other words,

$$[Tx]^{\mathcal{B}_1} = [T]_{\mathcal{B}}^{\mathcal{B}_1}[x]^{\mathcal{B}}.$$

Proof: Assume that $x = \sum_{j=1}^{n} \alpha_j e_j$ and $Te_j = \sum_{i=1}^{m} t_{ij}f_i$. Then

$$Tx = T\sum j = 1^n \alpha_j e_j = \sum_{j=1}^{n}\alpha_j\sum_{i=1}^{m} t_{ij}f_i = \sum_{i=1}^{m}\left(\sum_{j=1}^{n} t_{ij}\alpha_j\right)f_i.$$

So

$$[Tx]^{\mathcal{B}_1} = \begin{pmatrix} \sum_{j=1}^{n} t_{1j}\alpha_j \\ \vdots \\ \sum_{j=1}^{n} t_{mj}\alpha_j \end{pmatrix} = [T]_{\mathcal{B}}^{\mathcal{B}_1}[x]^{\mathcal{B}}.$$

□

The previous theorem shows that, by choosing bases, we can pass from an abstract representation of spaces and transformations to a concrete one in the form of column vectors and matrices. In the later form, computations are easily mechanized.

The next theorem deals with the matrix representation of the product of two linear transformations.

Theorem 4.2.2 *Let* U, V, W *be linear spaces over a field* F. *Let* $T \in L(U, V)$ *and* $S \in L(V, W)$. *Let* $\mathcal{B} = \{e_1, \ldots, e_n\}$, $\mathcal{B}_1 = \{f_1, \ldots, f_m\}$, *and* $\mathcal{B}_2 = \{g_1, \ldots, g_p\}$ *be bases in* U, V, *and* W, *respectively. Then*

$$[ST]_{\mathcal{B}}^{\mathcal{B}_2} = [S]_{\mathcal{B}_1}^{\mathcal{B}_2}[T]_{\mathcal{B}}^{\mathcal{B}_1}.$$

Proof: Let

$$Te_j = \sum_{k=1}^{m} t_{kj}f_k$$

$$Sf_k = \sum_{i=1}^{p} s_{ik}g_i$$

$$(ST)e_j = \sum_{i=1}^{p} r_{ij}g_i.$$

Therefore,

$$\begin{aligned}
(ST)e_j &= S(Te_j) = S\sum_{k=1}^{m} t_{kj}f_k \\
&= \sum_{k=1}^{m} t_{kj}Sf_k = \sum_{k=1}^{m} t_{kj}\sum_{i=1}^{p} s_{ik}g_i \\
&= \sum_{i=1}^{p} \left[\sum_{k=1}^{m} s_{ik}t_{kj}\right] g_i = \sum_{i=1}^{p} r_{ij}g_i.
\end{aligned}$$

Since $\{g_1, \ldots, g_p\}$ is a basis for W, we get

$$r_{ij} = \sum_{k=1}^{m} s_{ik}t_{kj},$$

or

$$[ST]_{\mathcal{B}}^{\mathcal{B}_2} = [S]_{\mathcal{B}_1}^{\mathcal{B}_2}[T]_{\mathcal{B}}^{\mathcal{B}_1}$$

follows.

□

Clearly we proved the commutativity of the following diagram:

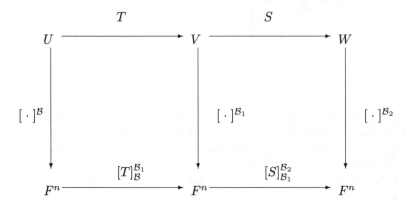

Corollary 4.2.1 *Let the linear space V have the two bases $\mathcal{B}$ and $\mathcal{B}_1$. Then we have*

$$[I]_{\mathcal{B}_1}^{\mathcal{B}}[I]_{\mathcal{B}}^{\mathcal{B}_1} = [I]_{\mathcal{B}}^{\mathcal{B}} = I$$

or, alternatively,

$$[I]_{\mathcal{B}}^{\mathcal{B}} = ([I]_{\mathcal{B}}^{\mathcal{B}})^{-1}.$$

An important special case of the previous result is that of the relation between matrix representations of a linear transformation with respect to two different bases:

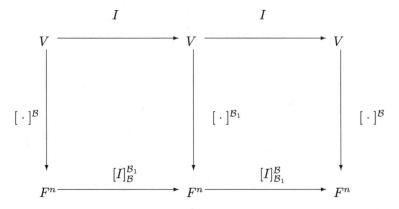

Theorem 4.2.3 *Let T be a linear transformation in a linear space V. Let $\mathcal{B}$ and $\mathcal{B}_1$ be two bases of V. Then we have*

$$[T]_{\mathcal{B}_1}^{\mathcal{B}_1} = [I]_{\mathcal{B}}^{\mathcal{B}_1}[T]_{\mathcal{B}}^{\mathcal{B}}[I]_{\mathcal{B}_1}^{\mathcal{B}}$$

or

$$[T]_{\mathcal{B}_1}^{\mathcal{B}_1} = ([I]_{\mathcal{B}_1}^{\mathcal{B}})^{-1}[T]_{\mathcal{B}}^{\mathcal{B}}[I]_{\mathcal{B}_1}^{\mathcal{B}}.$$

Proof: We observe the following commutative diagram, from which the result immediately follows:

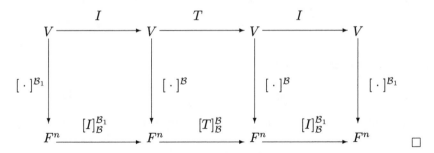

The previous theorem plays a key role in the study of linear transformations. The structure of a linear transformation is found through the search for simple matrix representations, which is equivalent to finding appropriate bases. The previous theorem's content is that through a change of basis the matrix representation of a given linear transformation undergoes a similarity transformation.

4.3 Linear Functionals and Duality

Given a linear space V over a field F, a linear transformation from V to F is called a **linear functional**. The set of all linear functionals on V, that is, $L(V, F)$, is called the **dual space** of V and will be denoted by V^*. Thus,

$$V^* = L(V, F).$$

Examples:

- Let $V = F^n$. Then, for $\alpha_1, \ldots, \alpha_n \in F$, $f : F^n \longrightarrow F$ defined by

$$f\begin{pmatrix} x_1 \\ \cdot \\ \cdot \\ \cdot \\ x_n \end{pmatrix} = \sum_{i=1}^{n} \alpha_i x_i$$

  defines a linear functional.

- Let $F^{n \times n}$ be the space of all square matrices of order n, and let $X, A \in F^{n \times n}$. Then f defined by

$$f(X) = \text{tr}(AX)$$

  is a linear functional.

Let V be an n-dimensional linear space, and let $\mathcal{B} = \{e_1, \ldots, e_n\}$ be a basis for V. Thus each vector $x \in V$ has a unique representation of the form

$$x = \sum_{j=1}^{n} \alpha_j e_j.$$

We define linear functionals f_i by

$$f_i(x) = \alpha_i, \qquad i = 1, \ldots, n.$$

It is easily checked that the f_i are linear functionals. To see this, let $y = \sum_{j=1}^{n} \beta_j e_j$. Then

$$
\begin{aligned}
f_i(x + y) &= f_i \left(\sum_{j=1}^{n} \alpha_j e_j + \sum_{j=1}^{n} \beta_j e_j \right) \\
&= f_i \left(\sum_{j=1}^{n} (\alpha_j + \beta_j) e_j \right) \\
&= \alpha_i + \beta_i = f_i(x) + f_i(y).
\end{aligned}
$$

In the same way,

$$
\begin{aligned}
f_i(\alpha x) &= f_i \left(\alpha \sum_{j=1}^{n} \alpha_j e_j \right) \\
&= f_i \left(\sum_{j=1}^{n} (\alpha \alpha_j) e_j \right) \\
&= \alpha \alpha_i = \alpha f_i(x).
\end{aligned}
$$

Since clearly

$$e_i = \sum_{j=1}^{n} \delta_{ij} e_j,$$

we have that

$$f_i(e_j) = \delta_{ij}, \qquad 1 \le i, j \le n.$$

Theorem 4.3.1 *Let V be an n-dimensional linear space, and let $\mathcal{B} = \{e_1, \ldots, e_n\}$ be a basis for V. Then:*

1. We have

$$\dim V^* = n.$$

2. *The set of linear functionals $\{f_1, \ldots, f_n\}$ defined, through linear extensions, by $f_i(e_j) = \delta_{ij}$ is a basis for V^*.*

Proof:

1. The proof follows from the second part. Alternatively, we can use Theorem 4.1.1, that is, the fact that $\dim L(U, V) = \dim U \dim V$ and the fact that $\dim F = 1$.

2. The functionals $\{f_1, \ldots, f_n\}$ are linearly independent. Let $\sum_{i=1}^{n} \alpha_i f_i = 0$. Then

$$0 = 0 \cdot e_j = \left(\sum_{i=1}^{n} \alpha_i f_i \right) e_j = \sum_{i=1}^{n} \alpha_i f_i(e_j) = \sum_{i=1}^{n} \alpha_i \delta_{ij} = \alpha_j.$$

Therefore, $\alpha_j = 0$ for all j, and linear independence is proved.

   Now let f be an arbitrary functional in V^*. Let $f(e_i) = \alpha_i$. We will show that $f = \sum_{i=1}^{n} \alpha_i f_i$. It suffices to show that the two functionals agree on basis elements. Indeed,

$$\left(\sum_{i=1}^{n} \alpha_i f_i \right) (e_j) = \sum_{i=1}^{n} \alpha_i f_i(e_j) = \sum_{i=1}^{n} \alpha_i \delta_{ij} = \alpha_j = f(e_j). \qquad \square$$

Definition 4.3.1 *Let V be an n-dimensional linear space, and let $\mathcal{B} = \{e_1, \ldots, e_n\}$ be a basis for V. Then the basis $\{f_1, \ldots, f_n\}$ defined by $f_i(e_j) = \delta_{ij}$ is called the **dual basis** to $\mathcal{B}$ and will be denoted by $\mathcal{B}^*$.*

Definition 4.3.2 *Let S be a subset of a linear space V. We denote by $S^\perp$ the subset of V^* defined by*

$$S^\perp = \{f \in V^* | f(s) = 0, \text{ for all } s \in S\}.$$

*The set $S^\perp$ is called the **annihilator** of S.*

Proposition 4.3.1 *Let S be a subset of the linear space V. Then:*

- *The set $S^\perp$ is a subspace of V^*.*

- *If $Q \subset S$, then $S^\perp \subset Q^\perp$.*

- *We have $S^\perp = (L(S))^\perp$.*

Proof:

- Let $f, g \in M^\perp$ and $\alpha, \beta \in F$. Then for an arbitrary $x \in M$

$$(\alpha f + \beta g)x = \alpha f(x) + \beta g(x) = \alpha \cdot 0 + \beta \cdot 0 = 0,$$

   that is, $\alpha f + \beta g \in M^\perp$.

- Let $f \in N^{\perp}$. Then $f(x) = 0$ for all $x \in N$ and particularly, by the inclusion $M \subset N$, for all $x \in M$. So $f \in M^{\perp}$.

- It is clear that $M \subset L(M)$, and therefore $L(M)^{\perp} \subset M^{\perp}$. On the other hand, if $f \in M^{\perp}$, $x_i \in M$, and $\alpha_i \in F$, then

$$f\left(\sum_{i=1}^{n} \alpha_i x_i\right) = \sum_{i=1}^{n} \alpha_i f(x_i) = \sum_{i=1}^{n} \alpha_i \cdot 0 = 0,$$

which means that f annihilates all linear combinations of elements of M. So $f \in L(M)^{\perp}$ or $M^{\perp} \subset L(M)^{\perp}$.

From these two inclusions the equality $M^{\perp} = L(M)^{\perp}$ follows.   □

We will find the following proposition useful in the study of duality:

Proposition 4.3.2 *Let V be a linear space and M a subspace. Then the dual space to the quotient space V/M is isomorphic to $M^{\perp}$.*

Proof: Let $\phi \in M^{\perp}$. Then ϕ induces a linear functional Φ on V/M by $\Phi([x]) = \phi(x)$. The functional Φ is well defined as $[x_1] = [x_2]$ if and only if $x_1 - x_2 \in M$, and this implies $\phi(x_1) - \phi(x_2) = \phi(x_1 - x_2) = 0$. The linearity of Φ follows from the linearity of ϕ.

Conversely, given $\Phi \in (V/M)^*$, we define ϕ by $\phi(x) = \Phi([x])$. Clearly, $\phi(x) = 0$ for $x \in M$, that is, $\phi \in M^{\perp}$.   □

Theorem 4.3.2 *Let V be an n-dimensional linear space. Let M be a subspace of V and $M^{\perp}$ its annihilator. Then*

$$\dim V = \dim M + \dim M^{\perp}.$$

Proof: Let $\{e_1, \ldots, e_n\}$ be a basis for V, with $\{e_1, \ldots, e_m\}$ a basis for M. Let $\{f_1, \ldots, f_n\}$ be the dual basis. We show that $\{f_{m+1}, \ldots, f_n\}$ is a basis for $M^{\perp}$. The linear independence of the vectors $\{f_{m+1}, \ldots, f_n\}$ is clear as they are a subset of a basis. Moreover, they are clearly in $M^{\perp}$. It remains to show that they actually span $M^{\perp}$. So let $f \in M^{\perp}$; it can be written as

$$f = \sum_{j=1}^{n} \alpha_j f_j.$$

But as $f \in M^{\perp}$, we have $f(e_1) = \cdots = f(e_m) = 0$; so

$$f(e_i) = \sum_{j=1}^{n} \alpha_j f_j(e_i) = \sum_{j=1}^{n} \alpha_j \delta_{ji} = \alpha_i.$$

So $\alpha_1 = \cdots = \alpha_m = 0$, and therefore,

$$f = \sum_{j=m+1}^{n} \alpha_j f_j.$$

The proof is complete as $\dim M = m$ and $\dim M^\perp = n - m$. $\square$

Let U be a linear space over F. We define a **hyperspace** M to be a maximal nontrivial subspace of U.

Proposition 4.3.3 *Let $\dim U = n$. The following statements are equivalent:*

1. *M is a hyperspace.*

2. *We have $\dim M = n - 1$.*

3. *M is the kernel of a nonzero functional ϕ.*

Proof: (1) $\Rightarrow$ (2). Assume that M is a hyperplane. Fix a nonzero vector $x \in M$. Then $L(x, M) = \{\alpha x + m \mid \alpha \in F, m \in M\}$ is a subspace that properly contains M. Thus, necessarily, $L(x, M) = U$. Let $\{e_1, \ldots, e_k\}$ be a basis for M. Then $\{e_1, \ldots, e_k, x\}$ are linearly independent and span M, and so they are a basis for U. Necessarily, $\dim M = n - 1$.

(2) $\Rightarrow$ (3). Let M be an $n - 1$ dimensional subspace of U. Let $x \notin M$. Define a linear functional ϕ by setting

$$\begin{cases} \phi(x) = 1 \\ \phi|M = 0. \end{cases}$$

For each $x \in U$ there exist unique $\gamma \in F$ and $m \in M$ such that $x = \gamma f + m$. It follows that

$$\phi(x) = \phi(\gamma f + m) = \gamma \phi(f) + \phi(m) = \gamma.$$

Therefore, $x \in Ker\,\phi$ if and only if $x \in M$.

(3) $\Rightarrow$ (2). Assume that $M = Ker\,\phi$ with ϕ a nontrivial functional. Fix $f \notin M$, which exists by the nontriviality of ϕ. We now show that every $x \in U$ has a unique representation of the form $x = \gamma f + m$ with $m \in Ker\,\phi$. In fact, it suffices to take $\gamma = (\phi(x))/(\phi(f))$. Thus, M is a hyperspace. $\square$

Recalling Definition 2.7.1, where the codimension of a space was introduced, we can extend Proposition 4.3.3 in the following way:

Proposition 4.3.4 *Let U be an n-dimensional linear space over F, and let M be a subspace. Then the following statements are equivalent:*

1. *M has codimension k.*

2. *We have $\dim M = n - k$.*

3. *M is the intersection of the kernels of k linearly independent functionals.*

Proof: (1) $\Rightarrow$ (2). Let $f_1, \ldots, f_k$ be linearly independent vectors over M, and let $L(M, f_1, \ldots, f_k) = U$. Let $e_1, \ldots, e_p$ be a basis for M. We claim that $\{e_1, \ldots, e_p, f_1, \ldots, f_k\}$ is a basis for U. The spanning property is obvious. To see linear independence, assume that there exist $\alpha_i, \beta_i \in F$ such that $\sum_{i=1}^{p} \alpha_i e_i + \sum_{i=1}^{k} \beta_i f_i = 0$. This implies that $\sum_{i=1}^{k} \beta_i f_i = -\sum_{i=1}^{p} \alpha_i e_i \in M$. This in turn implies that all β_i are zero. From this we conclude that all α_i are zero. Thus, we must have that $p + k = n$ or $\dim M = n - k$.

(2) $\Rightarrow$ (3). Assume that $\dim M = n - k$. Let $\{e_1, \ldots, e_{n-k}\}$ be a basis for M. We extend it to a basis $\{e_1, \ldots, e_n\}$. Using linear extensions, define k linear functionals $\phi_1, \ldots, \phi_k$ such that

$$\phi_i(e_j) = \delta_{ij}, \qquad j = n - k + 1, \ldots, n.$$

Obviously, $\phi_1, \ldots, \phi_k$ are linearly independent and $\cap_i Ker\, \Phi_i = M$.

(3) $\Rightarrow$ (1). Assume that $\phi_1, \ldots, \phi_k$ are linearly independent and $\cap_i Ker\, \Phi_i = M$. We choose a sequence of vectors x_i inductively. We choose a vector x_1 such that $\phi_1(x_1) = 1$. Suppose that we already have chosen $x_1, \ldots, x_{i-1}$. Then we choose x_i so that $x_i \in \cap_{j=1}^{i-1} Ker\, \Phi_j$ and $\phi_i(x_i) = 1$. To see that this is possible, we note that if this is not the case, then

$$Ker\, \phi_i \subset \cap_{j=1}^{i-1} Ker\, \Phi_j = Ker \begin{pmatrix} \phi_1 \\ \cdot \\ \cdot \\ \cdot \\ \phi_{i-1} \end{pmatrix}.$$

Using Theorem 4.1.8, it follows that there exist α_j such that $\phi_i = \sum_{j=1}^{i-1} \alpha_j \phi_j$, contrary to the assumption of linear independence. Clearly the vectors, $x_1, \ldots, x_k$ are linearly independent. Moreover, it is easy to check that $x - \sum_{j=1}^{k} \alpha_j x_j \in \cap_{j=1}^{k} Ker\, \Phi_j$. This implies $\dim M = n - k$. $\qquad\square$

Let V be a linear space over F, and let V^* be its dual space. Thus, given $x \in V$ and $f \in V^*$, we have $f(x) \in F$. But, fixing a vector $x \in F$, we can view the expression $f(x)$ as a numerical-valued function defined on the dual space V^*. We denote this function by $\hat{x}$, and it is defined by

$$\hat{x}(f) = f(x).$$

Theorem 4.3.3 *The function* $\hat{x} : V^* \longrightarrow F$ *is a linear map.*

Proof: Let $f, g \in V^*$ and $\alpha, \beta \in F$. Then

$$\begin{aligned} \hat{x}(\alpha f + \beta g) &= (\alpha f + \beta g)(x) = (\alpha f)(x) + (\beta g)(x) \\ &= \alpha f(x) + \beta g(x) = \alpha \hat{x}(f) + \beta \hat{x}(g). \end{aligned}$$ $\qquad\square$

Together with the function $\hat{x} \in L(V^*, F) = V^{**}$, we have the function $\phi : V \longrightarrow V^{**}$ defined by

$$\phi(x) = \hat{x}.$$

The function ϕ is referred to as the **canonical embedding** of V in V^{**}.

Theorem 4.3.4 *The function* $\phi : V \longrightarrow V^{**}$ *defined by*

$$\phi(x) = \hat{x}$$

is injective and surjective, that is, it is an invertible linear map.

Proof: We begin by showing the linearity of ϕ. Let $x, y \in V$, $\alpha, \beta \in F$, and $f \in V^*$. Then

$$
\begin{aligned}
(\phi(\alpha x + \beta y))f &= f(\alpha x + \beta y) = \alpha f(x) + \beta f(y) \\
&= \alpha \hat{x}(f) + \beta \hat{y}(f) = (\alpha \hat{x} + \beta \hat{y})(f) \\
&= (\alpha \phi(x) + \beta \phi(y))f.
\end{aligned}
$$

Since this is true for all $f \in V^*$, we get

$$\phi(\alpha x + \beta y) = \alpha \phi(x) + \beta \phi(y),$$

that is, the canonical map ϕ is linear.

Since $\dim V^* = \dim V$, we also get $\dim V^{**} = \dim V$. So to show that ϕ is invertible, it suffices to show injectivity. To this end, let $x \in Ker\,\phi$. Then for each $f \in V^*$ we have

$$(\phi(x))(f) = 0 = \hat{x}(f) = f(x).$$

This implies that $x = 0$. If $x \neq 0$, there exists a functional f for which $f(x) \neq 0$. The easiest way to see this is to complete x to a basis and take the dual basis. $\square$

Corollary 4.3.1 *Let* $L \in V^{**}$. *Then there exists an* $x \in V$ *such that* $L = \hat{x}$.

Corollary 4.3.2 *Let* V *be an* n-*dimensional linear space, and let* V^* *be its dual space. Let* $\{f_1, \ldots, f_n\}$ *be a basis for* V^*. *Then there exists a basis* $\{e_1, \ldots, e_n\}$ *for* V *for which*

$$f_i(e_j) = \delta_{ij},$$

that is, every basis in V^* *is the dual of a basis in* V.

Proof: Let $\{E_1, \ldots, E_n\}$ be the basis in V^{**} that is dual to the basis $\{f_1, \ldots, f_n\}$. Now, there exist unique vectors $\{e_1, \ldots, e_n\}$ in V for which $E_i = \hat{e}_i$. So, for each $f \in V^*$, we have

$$E_i(f) = \hat{e}_i(f) = f(e_i)$$

and in particular

$$f_j(e_i) = E_i(f_j) = \delta_{ij}. \qquad \square$$

Now let $M \subset V^*$ be a subspace. We define

$$^{\perp}M = \bigcap_{f \in M} Ker\,f = \{x \in V | f(x) = 0, \forall f \in M\}.$$

Theorem 4.3.5 *We have*

$$\phi(^\perp M) = M^\perp.$$

Proof: We have $x \in \cap_{f \in M} Ker\, f$ if and only if for each $f \in M$ we have $\hat{x}(f) = f(x) = 0$. However the last condition is equivalent to $\hat{x} \in M^\perp$. □

Corollary 4.3.3 *Let V be an n-dimensional linear space and M an $n - k$ dimensional subspace of V^*. Let $\phi(^\perp M) = \cap_{f \in M} Ker\, f$. Then $\dim \phi(^\perp M) = k$.*

Proof: We have

$$\begin{aligned} \dim V^* &= n = \dim M + \dim M^\perp \\ &= n - k + \dim M^\perp, \end{aligned}$$

that is, $\dim M^\perp = k$. We conclude by observing that $\dim \phi(^\perp M) = \dim M^\perp$. □

Theorem 4.3.6 *Let $f, g_1, \ldots, g_p \in V^*$. Then $f \in \cap_{i=1}^p Ker\, g_i$ if and only if $f \in L(g_1, \ldots, g_n)$.*

Proof: Define the map $g \,:\, V \longrightarrow F^p$ by

$$g(x) = \begin{pmatrix} g_1(x) \\ \cdot \\ \cdot \\ \cdot \\ g_p(x) \end{pmatrix}.$$

Obviously, we have the equality $Ker\, g = \cap_{i=1}^p Ker\, g_i$. We apply Theorem 4.1.8 to conclude the proof. □

4.4　The Adjoint Transformation

Let V and W be two vector spaces over the field F. Assume that $T \in L(V, W)$ and $f \in W^*$. Let us consider the composition of maps

$$V \xrightarrow{T} W \xrightarrow{f} F.$$

It is clear that the product, or composition, of f and T, that is, fT, is a linear transformation from U to F. This means that $fT \in V^*$. We denote this functional by $T^* f$. Therefore, $T^* : V^* \longrightarrow V^*$ is defined by

$$T^* f = fT$$

or

$$(T^* f)x = f(Tx) \qquad x \in U.$$

The transformation T^* is called the **adjoint transformation** of T.

Theorem 4.4.1 *The transformation T^* is linear, that is, $T^* \in L(V^*, U^*)$.*

Proof: Let $f, g \in V^*$ and $\alpha, \beta \in F$. Then, for every $x \in U$,

$$
\begin{aligned}
(T^*(\alpha f + \beta g))x &= (\alpha f + \beta g))(Tx) \\
&= \alpha f(Tx) + \beta g(Tx) = \alpha(T^*f)x + \beta(T^*g)x \\
&= (\alpha T^*f + \beta T^*g)x.
\end{aligned}
$$

Therefore,

$$
T^*(\alpha f + \beta g) = \alpha T^*f + \beta T^*g. \qquad \square
$$

Theorem 4.4.2 *Let $T \in L(U, V)$. Then*

$$
(\mathrm{Im}\,T)^{\perp} = Ker\,T^*.
$$

Proof: For every $x \in U$ and $f \in V^*$ we have

$$
(T^*f)x = f(Tx).
$$

So $f \in (\mathrm{Im}\,T)^{\perp}$ if and only if for all $x \in U$ we have $0 = f(Tx) = (T^*f)x$, which is equivalent to $f \in Ker\,T^*$. $\qquad \square$

Corollary 4.4.1 *Let $T \in L(U, V)$. Then*

$$
\mathrm{rank}(T) = \mathrm{rank}(T^*).
$$

Proof: Let $\dim U = n$ and $\dim V = m$. Assume that $\mathrm{rank}T = \dim \mathrm{Im}T = p$. Now

$$
\begin{aligned}
m &= \dim V = \dim V^* = \dim \mathrm{Im}T^* + \dim Ker\,T^* \\
&= \mathrm{rank}T^* + \dim Ker\,T^*.
\end{aligned}
$$

Since we have

$$
\begin{aligned}
\dim Ker\,T^* &= \dim(\mathrm{Im}T)^{\perp} = m - \dim(\mathrm{Im}T) \\
&= m - \mathrm{rank}T = m - r,
\end{aligned}
$$

$\mathrm{rank}T^* = p$. $\qquad \square$

Theorem 4.4.3 *Let $T \in L(U, V)$, and let $\mathcal{B} = \{e_1, \ldots, e_n\}$ be a basis in U and $\mathcal{B}_1 = \{f_1, \ldots, f_m\}$ be a basis in V. Let $\mathcal{B}^* = \{\phi_1, \ldots, \phi_n\}$ and $\mathcal{B}_1^* = \{\psi_1, \ldots, \psi_m\}$ be the dual bases in U^* and V^*, respectively. Then*

$$
[T^*]_{\mathcal{B}_1^*}^{\mathcal{B}^*} = \widetilde{[T]_{\mathcal{B}}^{\mathcal{B}_1}}.
$$

Proof: We recall that

$$
([T]_{\mathcal{B}}^{\mathcal{B}_1})_{ij} = \psi_i(Te_j).
$$

So, in order to compute $([T^*]_{\mathcal{B}_1^*}^{\mathcal{B}^*})_{ij}$, we have to find the dual basis to $\mathcal{B}^*$. Now we know that $B^{**} = \{\hat{e}_1, \ldots, \hat{e}_n\}$, so

$$
([T^*]_{\mathcal{B}_1^*}^{\mathcal{B}^*})_{ij} = \hat{e}_i(T^*\psi_j) = (T^*\psi_j)e_i = \psi_j(Te_i) = ([T]_{\mathcal{B}}^{\mathcal{B}_1})_{ji}. \qquad \square
$$

Corollary 4.4.2 *Let A be an $m \times n$ matrix. Then the row and column ranks of A are equal.*

Proof: Follows from $\text{rank} A = \text{rank} \tilde{A}$.   □

We now consider a special class of linear transformations T that satisfy $T^* = T$. Assuming that $T \in L(U, V)$, it follows that $T^* \in L(V^*, U^*)$. For the equality $T^* = T$ to hold, it is therefore necessary that $V = U^*$ and $V^* = U$. Of course, $V = U^*$ implies $V^* = U^{**}$, and therefore $V^* = U$ would hold only if we identify U^{**} with U, which we can do by using canonical embedding.

Now let $x, y \in U$. Then we have

$$(Tx)(y) = \hat{y}(Tx) = (T^*\hat{y})(x) = \hat{x}(T^*\hat{y}). \tag{4.3}$$

If we rewrite the action of a functional x^* on a vector x by

$$x^*(x) = <x^*, x>, \tag{4.4}$$

we can rewrite Eq. (4.3) as

$$<Tx, y> = <x, T^*y>. \tag{4.5}$$

We say that a linear transformation $T \in L(U, U^*)$ is **self-dual** if $T^* = T$ or, equivalently, for all $x, y \in U$ we have $<Tx, y> = <x, Ty>$.

Proposition 4.4.1 *Let U be a linear space and U^* its dual, $\mathcal{B}$ a basis in U and $\mathcal{B}^*$ its dual basis. Let $T \in L(U, U^*)$ be a self-dual linear transformation. Then $[T]_{\mathcal{B}}^{\mathcal{B}^*}$ is a symmetric matrix.*

Proof: We use the fact that $U^{**} = U$ and $\mathcal{B}^{**} = \mathcal{B}$. Then

$$[T]_{\mathcal{B}}^{\mathcal{B}^*} = [T^*]_{\mathcal{B}^{**}}^{\mathcal{B}^*} = \widetilde{[T]_{\mathcal{B}}^{\mathcal{B}^*}}.   □$$

If $\mathcal{B} = \{e_1, \ldots, e_n\}$, then every vector $x \in U$ has an expansion $x = \sum_{i=1}^n \xi_i e_i$. This leads to $<Tx, x> = \sum_{i=1}^n \sum_{j=1}^n T_{ij} \xi_i \xi_j$, where $(T_{ij} = [T]_{\mathcal{B}}^{\mathcal{B}^*}$. The expression $\sum_{i=1}^n \sum_{j=1}^n T_{ij} \xi_i \xi_j$ is called a **quadratic form**. We will return to this topic in much greater detail in Chapter 8.

4.5 Polynomial Module Structure on Vector Spaces

Let U be a vector space over the field F. Given a linear transformation T in U, there is in U a naturally induced module structure over the ring $F[z]$. This module structure will be central to our study of linear transformations.

Given a polynomial $p \in F[z]$, $p(z) = \sum_{j=0}^{k} p_j z^j$, and a linear transformation T in a vector space U over F, we define

$$p(T) = \sum_{j=0}^{k} p_j T^j, \qquad (4.6)$$

with $T^0 = I$. The action of a polynomial p on a vector $x \in U$ is defined by

$$p \cdot x = p(T)x. \qquad (4.7)$$

Proposition 4.5.1 *Given a linear transformation T in a vector space V over F, the map $p \mapsto p(T)$ is an algebra homomorphism of $F[z]$ into $L(V)$.*

Proof: Clearly, $(\alpha p + \beta q)(T) = \alpha p(T) + \beta q(T)$ and $(pq)(T) = p(T)q(T)$. □

Given two spaces U_1, U_2 and linear transformations $T_i \in L(U_i)$, a linear transformation $X : U_1 \longrightarrow U_2$ is said to **intertwine** T_1 and T_2, if

$$XT_1 = T_2 X. \qquad (4.8)$$

Obviously, Eq. (4.8) implies, for an arbitrary polynomial p, that $Xp(T_1) = p(T_2)X$. Thus, for $x \in U_1$, we have $X(p \cdot x) = p \cdot Xx$. This shows that intertwining maps are $F[z]$ module homomorphisms from U_1 to U_2. Two operators T_i are said to be **similar** if there exists an invertible map intertwining them. A natural strategy for studying similarity is to characterize intertwining maps and find conditions guarranteeing their invertibility.

Definition 4.5.1 *Let U be an n-dimensional linear space over the field F and $T : U \longrightarrow U$ a linear transformation. Then:*

1. *A subspace $M \subset U$ is called an* **invariant subspace** *of T if for all $x \in M$ we also have $Tx \in M$.*

2. *A subspace $M \subset U$ is called a* **reducing subspace** *of T if it is an invariant subspace of T and there exists another invariant subspace $N \subset U$ such that*

$$U = M \oplus N.$$

We note that, given a linear transformation T in U, T-invariant subspaces are $F[z]$-submodules of U relative to the module structure defined in U by Eq. (4.7). Similarly, T-reducing subspaces are equivalent to module direct summands of the module U.

Proposition 4.5.2

1. *Let M be a subspace of U invariant under T. Let $\mathcal{B} = \{e_1, \ldots, e_n\}$ be a basis for U such that $\mathcal{B}_1 = \{e_1, \ldots, e_m\}$ is a basis for M. Then, with respect to this basis, T has the block triangular matrix representation*

$$[T]_{\mathcal{B}}^{\mathcal{B}} = \begin{pmatrix} T_{11} & T_{12} \\ 0 & T_{22} \end{pmatrix}.$$

Moreover, $T_{11} = [T|_M]_{\mathcal{B}_1}^{\mathcal{B}_1}$.

2. *Let M be a reducing subspace for T, and let N be a complementary invariant subspace. Let $\mathcal{B} = \{e_1, \ldots, e_n\}$ be a basis for U such that $\mathcal{B}_1 = \{e_1, \ldots, e_m\}$ is a basis for M and $\mathcal{B}_2 = \{e_{m+1}, \ldots, e_n\}$ is a basis for N. Then the matrix representation of T with respect to this basis is block diagonal. Specifically,*

$$[T]_{\mathcal{B}}^{\mathcal{B}} = \begin{pmatrix} T_1 & 0 \\ 0 & T_2 \end{pmatrix},$$

where $T_1 = [T|_M]_{\mathcal{B}_1}^{\mathcal{B}_1}$ and $T_2 = [T|_N]_{\mathcal{B}_2}^{\mathcal{B}_2}$.

Corollary 4.5.1

1. *Let T be a linear transformation in U. Let $\{0\} \subset M_1 \subset \cdots, M_k \subset U$ be invariant subspaces. Let $\mathcal{B} = \{e_1, \ldots, e_n\}$ be a basis for U such that $\mathcal{B}_i = \{e_{n_1 + \cdots + n_{i-1} + 1}, \ldots, e_{n_1 + \cdots + n_i}\}$ is a basis for M_i. Then*

$$[T]_{\mathcal{B}}^{\mathcal{B}} = \begin{pmatrix} T_{11} & T_{12} & . & . & T_{1n} \\ 0 & T_{22} & & & . \\ . & . & . & & . \\ . & & . & . & . \\ . & & . & 0 & T_{kk} \end{pmatrix}.$$

2. *Let $U = M_1 \oplus \cdots \oplus M_k$, with M_i T-invariant subspaces. Let $\mathcal{B}_i$ be a basis for M_i, and let B be the union of the bases $\mathcal{B}_i$. Then*

$$[T]_{\mathcal{B}}^{\mathcal{B}} = \begin{pmatrix} T_{11} & & & & \\ & T_{22} & & & \\ & & . & & \\ & & & . & \\ & & & & T_{kk} \end{pmatrix}.$$

An invariant subspace of a linear transformation T in T induces two other transformations, namely, the restriction of T to the invariant subspace and the induced transformation in the quotient space U/M.

Proposition 4.5.3 *Let $M \subset U$ be a T-invariant subspace. Let π be the canonical projection of U on the quotient space U/M. Then:*

1. *There exists a unique linear transformation $\bar{T} : M \longrightarrow U/M$ that makes the following diagram commutative:*

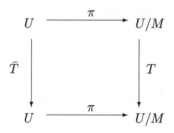

that is,

$$\bar{T}\pi x = \pi T x. \tag{4.9}$$

2. *Let $\mathcal{B} = \{e_1, \ldots, e_n\}$ be a basis for U such that $\mathcal{B}_1 = \{e_1, \ldots, e_m\}$ is a basis for M. Then $\overline{\mathcal{B}} = \{\pi e_{m+1}, \ldots, \pi e_n\}$ is a basis for U/M. If*

$$[T]_{\mathcal{B}}^{\mathcal{B}} = \begin{pmatrix} T_{11} & T_{12} \\ 0 & T_{22} \end{pmatrix},$$

then

$$[\bar{T}]_{\overline{\mathcal{B}}}^{\overline{\mathcal{B}}} = T_{22}.$$

Proof:

1. In terms of equivalence classes modulo M, we have $\bar{T}[x] = [Tx]$. For $\bar{T}$ to be well defined, we have to show that $[x] = [y]$ implies $[Tx] = [Ty]$. This is the direct consequence of the invariance of M under T.

2. It suffices to show that $\pi e_{m+1}, \ldots, \pi e_n$ are linearly independent, as they obviously span $U/$. Assume therefore that $\sum_{i=m+1}^{n} \alpha_i \pi e_i = \pi \sum_{i=m+1}^{n} \alpha_i e_i = 0$, that is, $\sum_{i=m+1}^{n} \alpha_i e_i \in M$. Thus, it can be expressed in terms of the basis of M as $\sum_{i=m+1}^{n} \alpha_i e_i = \sum_{i=1}^{m} \beta_i e_i$. Since $E - 1, \ldots, e_n$ is a basis of U, all of the coefficients are zero. Thus, the linear dependence of $\pi e_{m+1}, \ldots, \pi e_n$ is proved. $\square$

Definition 4.5.2

1. *Let $M \subset U$ be a T-invariant subspace. The **restriction** of T to M is the unique linear transformation $T|_M : M \longrightarrow M$ defined by*

$$T|_M x = x, \qquad x \in M.$$

2. *The **induced map**, namely, the map induced by T on the quotient space U/M, is the unique map defined by Eq. (4.9). We use the notation $T|_{U/M}$ for the induced map.*

Invariance and reducibility of linear transformations are conveniently described in terms of projection operators. We turn to them now.

Assume that the vector space U admits a direct sum decomposition $U = M \oplus N$. Thus every vector $x \in U$ can be written, in a unique way, as $x = m + n$ with $m \in M$ and $n \in N$. We define a transformation $P_N : U \longrightarrow U$ by $P_N x = m$. We call P_N the **projection** on M in the direction of N. Clearly, P_N is a linear transformation and satisfies $Ker\, P_N = N$ and $Im P_N = M$.

Proposition 4.5.4 *A linear transformation P is a projection if and only if $P^2 = P$.*

Proof: Assume that P_N is the projection on M in the direction of N. If $x = m + n$, it is clear that $P_N m = m$, so $P_N^2 x = P_N m = m = P_N x$.

Conversely, assume that $P^2 = P$. Let $M = Im P$ and $N = Ker\, P$. Since, for every $x \in U$, we have $x = Px + (I - P)x$, with $Px \in Im P$ and $(I - P)x \in Ker\, P$, it follows that $U = M + N$. To show that this is a direct sum decomposition, assume that $x \in M \cap N$. This implies that $x = Px = (I - P)y$. This implies in turn that $x = Px = P(I - P)y = 0$. $\square$

Proposition 4.5.5 *A linear transformation P in U is a projection if and only if $I - P$ is a projection.*

Proof: It suffices to show that $(I - P)^2 = I - P$, and this is immediate. $\square$

If $U = M \oplus N$ and P_N is the projection on M in the direction of N, then $P_M = I - P_N$ is the projection on M in the direction of M.

The next proposition is central. It expresses the geometric conditions of inariance and reducibility in terms of arithmetic conditions involving projection operators.

Proposition 4.5.6 *Let P be a projection operator in a linear space X, and let T be a linear transformation in X. Then:*

1. *A subspace $M \subset X$ is invariant under T if and only if for any projection P on M we have*

$$TP = PTP. \tag{4.10}$$

2. *Let $X = M \oplus N$ be a direct sum decomposition of X, and let P be the projection on M in the direction of N. Then the direct sum decomposition reduces T if and only if*

$$TP = PT. \tag{4.11}$$

Proof:

1. Assume that Eq. (4.10) holds, and $Im P = M$. Then, for every $m \in M$, we have

$$Tm = TPm = PTPm = PTm \in M.$$

Thus, M is invariant, and the converse is immediate.

2. The direct sum reduces T if and only if both M and N are invariant subspaces. By part 1, this is equivalent to the two conditions $TP = PTP$ and $T(I - P) = (I - P)T(I - P)$. These two conditions are equivalent to Eq. (4.11).   □

We clearly have, for a projection P, that $I = P + (I - P)$, which corresponds to the direct sum decomposition $X = \text{Im}P \oplus \text{Im}(I - P)$. This generalizes in the following way:

Theorem 4.5.1 *Given the direct sum decomposition $X = M_1 \oplus \cdots \oplus M_k$, there exist k projections P_i on X such that*

1. $P_i P_j = \delta_{ij} P_j$.

2. $I = P_1 + \cdots + P_k$.

3. $\text{Im}P_i = M_i$, $i = 1, \ldots, k$.

Conversely, given k operators P_i satisfying conditions 1 to 3, with $M_i = \text{Im}P_i$, we have $X = M_1 \oplus \cdots \oplus M_k$.

Proof: Assume that we are given the direct sum decomposition $X = M_1 \oplus \cdots \oplus M_k$. Thus, each $x \in X$ has a unique representation in the form $x = m_1 + \cdots + m_k$ with $m_i \in M_i$. We define $P_i x = m_i$. The operators P_i are clearly linear projections with $\text{Im}P_i = M_i$, and satisfy conditions 1 to 3. Moreover, we have $Ker P_i = \sum_{j \neq i} M_j$.

Conversely, assume that P_i satisfy conditions 1 to 3, and let $M_i = \text{Im}P_i$. From $I = P_1 + \cdots + P_k$ it follows that $x = P_1 x + \cdots + P_k x$ and hence $X = M_1 + \cdots + M_k$. This representation of x is unique, for if $x = m_1 + \cdots + m_k$, with $m_i \in M_i$, is another such representation, then, since $m_j \in M_j$, we have $m_j = P_j y_j$ for some y_j. Therefore,

$$P_i x = P_i \sum_{j=1}^{k} m_j = \sum_{j=1}^{k} P_i m_j = \sum_{j=1}^{k} P_i P_j y_j = \sum_{j=1}^{k} \delta_{ij} P_j y_j = P_i y_i = m_i.$$

Thus, the sum is a direct sum.   □

In our search for nontrivial invariant subspaces, we begin by looking for those that are one-dimensional. If M is a one-dimensional invariant subspace and x a nonzero vector in M, then every other vector in M is of the form αx. Thus, the invariance condition is $Tx = \alpha x$.

Definition 4.5.3

1. *Let T be a linear transformation in a vector space U over the field F. A nonzero vector $x \in U$ will be called an* **eigenvector,** *or* **characteristic vector,** *of T if there exists an $\alpha \in F$ such that*

$$Tx = \alpha x.$$

   Such an α will be called an **eigenvalue,** *or* **characteristic value,** *of T.*

2. *The* **characteristic polynomial** *of T, $d_T(z)$, is defined by*

$$d_T(z) = \det(zI - T).$$

   Clearly,

$$\deg d_T = \dim U.$$

Proposition 4.5.7 *A number $\alpha \in F$ is an eigenvalue of T if and only if it is a zero of the characteristic polynomial of T.*

Proof: The homogeneous system $(\alpha I - T)x = 0$ has a nontrivial solution if and only if $\alpha I - T$ is singular, that is, if $\det(\alpha I - T) = 0$. □

Let T be a linear transformation in an n-dimensional vector space U. We say that a polynomial $p \in F[z]$ annihilates T if $p(T) = 0$, where $p(T)$ is defined by Eq. (4.6). Clearly, every linear transformation is annihilated by the zero polynomial. A priori, it is not clear that an arbitrary linear transformation has a nontrivial annihilator. This is proved next.

Proposition 4.5.8 *For every linear transformation T in an n-dimensional vector space U there exists a nontrivial annihilating polynomial.*

Proof: $L(U)$, the space of all linear transformations in U, is n^2-dimensional. Therefore, the set of linear transformations $\{I, T, \ldots, T^{n^2}\}$ is linearly dependent. Hence, there exist numbers, not all zero, $p_i, i = 0, \ldots, n^2$ for which $\sum_{i=0}^{n^2} p_i T^i = 0$ or $p(T) = 0$, where $\sum_{i=0}^{n^2} p_i z^i$. □

Theorem 4.5.2 *Let T be a linear transformation in an n-dimensional vector space U. Then there exists a unique monic polynomial of minimal degree that annihilates T.*

Proof: Let

$$J = \{p \in F[z] \mid p(T) = 0\}.$$

Clearly, J is a nontrivial ideal in $F[z]$. As $F[z]$ is a principal ideal domain, there exists a unique monic polynomial m for which $J = mF[z]$. Obviously, if $0 \neq p \in J$, then $\deg p \geq \deg m$. □

The polynomial m whose existence is established in the previous theorem is called the **minimal polynomial** of T.

Both the characteristic and the minimal polynomials are similarity invariants.

Proposition 4.5.9 *Let A and B be similar linear transformations. Let d_A, d_B be their characteristic polynomials and m_A, m_B be their minimal polynomials. Then $d_A = d_B$ and $m_A = m_B$.*

Assume that we are given a linear transformation A in a finite-dimensional vector space V over the field F. Without loss of generality, by choosing a matrix representation, we may as well assume that $V = F^n$ and that A is a square matrix.

By $F^n[z]$ we denote the space of vector polynomials with the coefficients in F^n. We freely use the isomorphism between $F^n[z]$ and $F[z]^n$, and we identify the two spaces. With the linear polynomial matrix $zI - A$, we associate a map $\pi_{zI-A} : F^n[z] \longrightarrow F^n$ given by

$$\pi_{zI-A} \sum_{j=0}^{k} \xi_j z^j = \sum_{j=0}^{k} A^j \xi_j. \tag{4.12}$$

The operation defined above can be considered as taking the remainder of a polynomial vector after left division by the polynomial matrix $zI - A$.

Proposition 4.5.10 *For the map π_{zI-A} defined by Eq. (4.12), we have:*

1. *π_{zI-A} is surjective.*

2. *$Ker \, \pi_{zI-A} = (zI - A)F^n[z].$* $\tag{4.13}$

3. *For the map $S_+ : F^n[z] \longrightarrow F^n[z]$ defined by*

$$(S_+ f)(z) = z f(z),$$

the following diagram is commutative:

$$
\begin{array}{ccc}
F^n[z] & \xrightarrow{\ \pi_{zI-A}\ } & F^n \\
\Big\downarrow{\scriptstyle S_+} & & \Big\downarrow{\scriptstyle A} \\
F^n[z] & \xrightarrow{\ \pi_{zI-A}\ } & F^n
\end{array}
$$

This implies that

$$Ax = \pi_{zI-A} z \cdot x. \tag{4.14}$$

4. *Given a polynomial $p \in F[z]$, we have*

$$p(A)x = \pi_{zI-A}p(z)x. \tag{4.15}$$

Proof:

1. For each constant polynomial $x \in F^n[z]$, we have $\pi_{zI-A}x = x$. The surjectivity follows.

2. Assume that $f(z) = (zI - A)g(z)$ with $g(z) = \sum_{i=0}^k g_i z^i$. Then

$$\begin{aligned}
f(z) &= (zI - A)\sum_{i=0}^k g_i z^i \\
&= \sum_{i=0}^k g_i z^{i+1} - \sum_{i=0}^k A g_i z^i.
\end{aligned}$$

Therefore,

$$\pi_{zI-A}f = \sum_{i=0}^k A^{i+1} g_i - \sum_{i=0}^k A^i A g_i = 0,$$

that is, $(zI - A)F^n[z] \subset Ker\,\pi_{zI-A}$.

Conversely, assume that $f(z) = \sum_{i=0}^k f_i z^i \in Ker\,\pi_{zI-A}$, that is, $\sum_{i=0}^k A^i f_i = 0$. Recalling that $z^i I - A^i = (zI - A)\sum_{j=0}^{i-1} z^{i-1-j}A^j$, we compute

$$\begin{aligned}
f(z) &= \sum_{i=0}^k f_i z^i = \sum_{i=0}^k f_i z^i - \sum_{i=0}^k A^i f_i \\
&= \sum_{i=0}^k (z^i I - A^i)f_i = \sum_{i=0}^k (zI - A)\left(\sum_{j=0}^{i-1} z^{i-1-j}A^j\right)f_i \\
&= (zI - A)\sum_{i=0}^k \left(\sum_{j=0}^{i-1} z^{i-1-j}A^j\right)f_i = (zI - A)g.
\end{aligned}$$

So $Ker\,\pi_{zI-A} \subset (zI - A)F^n[z]$, and hence, equality (4.13) follows.

3. To prove the commutativity of the diagram, we compute, with $f(z) = \sum_{i=0}^k f_i z^i$,

$$\begin{aligned}
\pi_{zI-A}S_+f &= \pi_{zI-A}z\sum_{i=0}^k f_i z^i = \pi_{zI-A}\sum_{i=0}^k f_i z^{i+1} \\
&= \sum_{i=0}^k A^{i+1} f_i = A\sum_{i=0}^k A^i f_i \\
&= A\pi_{zI-A}f.
\end{aligned}$$

4. By linearity it suffices to prove this for polynomilas of the form z^k. We do this by induction. For $k = 1$, this holds by Eq. (4.14). Assume that it holds up to $k - 1$. Using the fact that

$$z\mathit{Ker}\,(zI - A) = z(zI - A)F^n[z] \subset (zI - A)F^n[z] = \mathit{Ker}\,(zI - A),$$

we compute

$$\pi_{zI-A}z^k x = \pi_{zI-A}z\pi_{zI-A}z^{k-1}x = \pi_{zI-A}zA^{k-1}x = A^{k-1}x = A^k x.$$
$\square$

Clearly, the equality $f(z) = (zI - A)g(z) + \pi_{zI-A}f$ can be interpreted as $\pi_{zI-A}f$ being the remainder of f after division by $zI - A$.

As a corollary, we obtain the celebrated Cayley–Hamilton theorem.

Theorem 4.5.3 (Cayley–Hamilton) *Let A be a linear transformation in an n-dimensional vector space U over F, and let d_A be its characteristic polynomial. Then*

$$d_A(A) = 0.$$

Proof: By Cramer's rule, we have $d_A(z)I = (zI - A)\mathrm{adj}(zI - A)$; hence we have the inclusion

$$d_A(z)F^n[z] \subset (zI - A)F^n[z].$$

This implies, for each $x \in U$, that $d_A(A)x = \pi_{zI-A}d(z)x = 0$, so $d_A(A) = 0$.
$\square$

Corollary 4.5.2 *Let A be a linear transformation in an n-dimensional vector space U over F. Then its minimal polynomial m_A divides its characteristic polynomial d_A.*

4.6 Exercises

1. Let T be an $m \times n$ matrix over the field F. Define the **determinant rank** of T to be the order of the largest nonvanishing minor of T. Show that the rank of T is equal to its determinant rank.

2. Let M be a subspace of a finite-dimensional vector space V. Show that $M^* \simeq V^*/M^\perp$.

3. Let X be an n-dimensional complex vector space. Let $T : X \longrightarrow X$ be such that, for every $x \in X$, the vectors $x, Tx, \ldots, T^m x$ are linearly dependent. Show that $I, T, \ldots, T^m$ are linearly dependent.

4. Let A, B, C be linear transformations. Show that there exists a linear transformation Z such that $C = AZB$ if and only if

$$\begin{cases} \mathrm{Im}\,C & \subset & \mathrm{Im}\,A \\ \mathit{Ker}\,C & \supset & \mathit{Ker}\,B. \end{cases}$$

5. Let V be a finite-dimensional vector space and let $A_i \in L(V)$, $i = 1, \ldots, s$. Show that if $M = \sum_{i=1}^{s} \operatorname{Im} A_i$ $(M = \cap_{i=1}^{s} Ker A_i)$, then there exist $B_i \in L(V)$ such that $\operatorname{Im} \sum_{i=1}^{s} A_i B_i = M$ $(Ker \sum_{i=1}^{s} B_i A_i = M)$.

6. Let V be a finite-dimensional vector space. Show that there is a bijective correspondence between the left (right) ideals in $L(V)$ and subspaces of V. The correspondence is given by $J \leftrightarrow \cap_{A \in J} Ker A$ $(J \leftrightarrow \sum_{A \in J} \operatorname{Im} A)$.

7. Show that $Ker A^2 \supset Ker A$. Also show that $Ker A^2 = Ker A$ implies $Ker A^p = Ker A$ for all $p > 0$.

8. Let T be an injective linear transformation on a not necessarily finite-dimensional vector space V. Show that if, for some integer k, we have $T^k = T$, then T is also surjective.

9. Let A be an $n \times n$ complex matrix with eigenvalues $\lambda_1, \ldots, \lambda_n$. Show that $\det A = \prod_{i=1}^{n} \lambda_i$. Also show that, given a polynomial p, the eigenvalues of $p(A)$ are $p(\lambda_1), \ldots, p(\lambda_n)$ (spectral mapping theorem).

10. Let A be an $n \times n$ complex matrix with eigenvalues $\lambda_1, \ldots, \lambda_n$. Show that the eigenvalues of adj A are $\prod_{j \neq 1} \lambda_j, \ldots, \prod_{j \neq n} \lambda_j$.

11. Let A be invertible and let $d(z)$ be its characteristic polynomial. Show that the characteristic polynomial of A^{-1} is $d^{\sharp}(z) = d(0)^{-1} z^n d(z^{-1})$.

12. Let the minimal polynomial of a linear transformation A be $\Pi(z - \lambda_j)^{\nu_j}$. Show that the minimal polynomial of $\begin{pmatrix} A & I \\ 0 & A \end{pmatrix}$ is $\Pi(z - \lambda_j)^{\nu_j + 1}$.

13. Let A be a linear transformation in a finite-dimensional vector space V over the field F. Let $m(z)$ be its minimal polynomial. Prove that, given a polynomial p, $p(A)$ is invertible if and only if p and m are coprime. Show that the minimal polynomial can be replaced by the characteristic polynomial and the result still will hold.

4.7   Notes and Remarks

The material in this chapter is mostly standard. Definitions 4.12 and 4.15 have far-reaching implications. They can be generalized by replacing $zI - A$ by an arbitrary nonsingular polynomial matrix, thus leading to the theory of polynomial models, which was initiated in Fuhrmann [1976]. This is the algebraic counterpart of the functional models used so effectively in operator theory, for example, Sz.-Nagy and Foias [1970].

An excellent source for linear algebra is Hoffmann and Kunze [1961]. The classic treatise of Gantmacher [1959] is still a rich source for results and ideas, and is highly recommended as a general reference.

5

The Shift Operator

5.1 Basic Properties

We turn our attention now to the study of a special class of cyclic transformations, namely, shift operators. These turn out later to serve as models for all cyclic transformations, in the sense that every cyclic transformation is similar to a shift operator.

We now introduce an extremely important class of linear transformations that will play a central role in the analysis of the structure of linear transformations. Recall that, for a nonzero polynomial q, we denote by $\pi_q f$ the remainder of the polynomial f after division by q. $\pi_q f$ is a projection operator in $F[z]$.

Definition 5.1.1 *Let q be a monic polynomial in $F[z]$.*

1. We define the set X_q by

$$X_q = \operatorname{Im}\pi_q = \{\pi_q f \mid f \in F[x]\}. \tag{5.1}$$

2. We define a linear transformation $S_q : X_q \longrightarrow X_q$ by

$$S_q f = \pi_q z f. \tag{5.2}$$

*We call S_q the **shift operator** in X_q and X_q, with the $F[z]$-module structure induced by S_q a **polynomial model**.*

Since π_q is a projection, so is $I - \pi_q$. The identity $I = \pi_q + (I - \pi_q)$, taken together with $\operatorname{Ker}\pi_q = qF[z]$, implies the direct sum

$$F[z] = X_q \oplus qF[z].$$

Proposition 5.1.1 *Let* $q(z) = z^n + q_{n-1}z^{n-1} + \cdots + q_0$. *Then:*

1. *We have* $\dim X_q = \deg q = n$.

2. *The following sets are bases for* X_q:

   (a) *The* **standard basis***, namely,* $\mathcal{B}_{st} = \{1, z, \ldots, z^{n-1}\}$.

   (b) *The* **control basis***, namely,* $\mathcal{B}_{co} = \{e_1, \ldots, e_n\}$, *where*

   $$e_i(z) = z^{n-i} + q_{n-1}z^{n-i-1} + \cdots + q_i.$$

   (c) *In case* $\alpha_1, \ldots, \alpha_n$, *the zeros of* q *are distinct. Then the polynomials* $p_i(z) = \Pi_{j \neq i}(z - \alpha_j), i = 1, \ldots, n$ *form a basis for* X_q. *We refer to this as the* **spectral basis** $\mathcal{B}_{sp}$ *of* X_q.

   (d) *Under the same assumption, the Lagrange interpolation polynomials, given by* $\pi_i(z) = (p_i(z))/(p_i(\alpha_i))$, *are a basis naturally called the* **interpolation basis***.*

Proof:

1. Clearly the elements of X_q are all polynomials of degree $< \deg q = n$. This is obviously an n-dimensional space.

2. Each of the sets has n linearly independent elements and hence is a basis for X_q. The linear independence of the Lagrange interpolation polynomials has been proved in Chapter 2, and the polynomials p_i are, up to a multiplicative constant, equal to the Lagrange interpolation polynomials. $\qquad\square$

We proceed by studying the matrix representations of S_q with respect to these bases of X_q.

Proposition 5.1.2 *Let* $S_q : F_n[z] \longrightarrow F_n[z]$ *be defined by Eq. (5.2).*

1. *With respect to the standard basis,* S_q *has the matrix representation*

$$C_q^{\#} = [S_q]_{st}^{st} = \begin{pmatrix} 0 & & & -q_0 \\ 1 & & & \cdot \\ & \cdot & & \cdot \\ & & \cdot & \\ & & 1 & -q_{n-1} \end{pmatrix}. \qquad (5.3)$$

2. *With respect to the control basis,* S_q *has the diagonal matrix representation*

$$C_q^b = [S_q]_{co}^{co} = \begin{pmatrix} 0 & 1 & & \\ & \cdot & & \cdot \\ & & \cdot & \\ & & & 1 \\ -q_0 & \cdot & \cdot & -q_{n-1} \end{pmatrix}. \qquad (5.4)$$

3. With respect to the spectral basis, S_q has the matrix representation

$$[S_q]^{sp}_{sp} = \begin{pmatrix} \alpha_1 & & & \\ & \cdot & & \\ & & \cdot & \\ & & & \cdot \\ & & & & \alpha_n \end{pmatrix}.$$

Proof:

1. Clearly we have

$$S_q z^i = \begin{cases} z^{i+1} & i = 0, \ldots, n-2 \\ \\ -\displaystyle\sum_{i=0}^{n-1} q_i z^i & i = n-1. \end{cases}$$

2. We compute, defining $e_0(z) = 0$,

$$\begin{aligned} S_q e_i &= \pi_q z e_i(z) = \pi_q z(z^{n-i} + q_{n-1} z^{n-i-1} + \cdots + q_i) \\ &= \pi_q(z^{n-i+1} + q_{n-1} z^{n-i} + \cdots + q_i z) \\ &= \pi_q(z^{n-i+1} + q_{n-1} z^{n-i} + \cdots + q_i z + q_{i-1}) - q_{i-1} e_n(z). \end{aligned}$$

So we get

$$S_q e_i = e_{i-1} - q_{i-1} e_n. \tag{5.5}$$

3. Noting that $q(z) = (z - \alpha_i) p_i(z)$, we compute

$$\begin{aligned} S_q p_i(z) &= \pi_q(z - \alpha_i + \alpha_i) p_i = \pi_q(q + \alpha_i p_i) \\ &= \alpha_i p_i. \end{aligned} \qquad \square$$

The matrices $C_q^\sharp, C_q^\flat$ are called the **companion matrices** of the polynomial q. This particularly suggestive notation was introduced by Kalman.

We note that the change of basis transformation from the control to the standard basis has a particularly nice form. In fact, we have

$$[I]^{st}_{co} = \begin{pmatrix} q_1 & \cdot & \cdot & q_{n-1} & 1 \\ \cdot & & & \cdot & \\ \cdot & & \cdot & & \\ q_{n-1} & \cdot & & & \\ 1 & & & & \end{pmatrix}.$$

Corollary 5.1.1 *For the shift operator defined by Eq. (5.2), we have*

$$S_q^k f = \pi_q z^k f.$$

We now relate the invariant subspaces of the shift operators S_q, or equivalently, the submodules of X_q, to the factorization of the polynomial q.

Theorem 5.1.1 *Given a monic polynomial q, $M \subset X_q$ is an S_q-invariant subspace if and only if*

$$M = q_1 X_{q_2}$$

for some factorization

$$q = q_1 q_2.$$

Proof: Assume that $q = q_1 q_2$ and $M = q_1 X_{q_2}$. Thus, $f \in M$ implies $f = q_1 f_1$ with $\deg f_1 < \deg q_2$. Using Lemma 1.3.2, we compute

$$S_q f = \pi_q z f = \pi_{q_1 q_2} z q_1 f_1 = q_1 \pi_{q_2} z f_1 = q_1 S_{q_2} f_1 \in M.$$

Conversely, let M be an S_q-invariant subspace. Now, for each $f \in X_q$ there exists a scalar α that depends on f for which

$$S_q f = zf - \alpha q. \tag{5.6}$$

Consider now the set $N = M + qF[z]$. Obviously, N is closed under additions, and, using Eq. (5.6), $z\{M + qF[z]\} \subset \{M + qF[z]\}$. Thus N is an ideal in $F[z]$ and hence is of the form $q_1 F[z]$. As, obviously, $qF[z] \subset q_1 F[z]$, it follows from Proposition 1.3.4 that q_1 is a divisor of q, that is, $q = q_1 q_2$. It is clear that

$$M = \pi_q \{M + qF[z]\} = \pi_q q_1 F[z] = \pi_{q_1 q_2} q_1 F[z] = q_1 X_{q_2}. \qquad \square$$

The following sums up the basic arithmetic properties of invariant subspaces of the shift operator. This can be viewed as the counterpart of Proposition 1.3.4.

Proposition 5.1.3 *Given a monic polynomial $q \in F[z]$:*

1. *Let $q = q_1 q_2 = p_1 p_2$ be two factorizations. Then we have the inclusion*

$$q_1 X_{q_2} \subset p_1 X_{p_2} \tag{5.7}$$

   if and only if $p_1 | q_1$, or equivalently, $q_2 | p_2$.

2. *Given factorizations $q = p_i q_i, i = 1, \ldots, s$, then $\cap_{i=1}^{s} p_i X_{q_i} = p X_q$ with p the l.c.m. of the p_i and q the g.c.d. of the q_i.*

3. *Given factorizations $q = p_i q_i, i = 1, \ldots, s$, then $\sum_{i=1}^{s} p_i X_{q_i} = p X_q$ with q the l.c.m. of the q_i and p the g.c.d. of the p_i.*

Proof:

1. Assume that $p_1|q_1$, that is, $q_1 = p_1 r$ for some polynomial r. Then $q = q_1 q_2 = (p_1 r)q_2 = p_1(rq_2) = p_1 p_2$ and, in particular, $p_2 = rq_2$. This implies that $q_1 X_{q_2} = p_1 r X_{q_2} \subset p_1 X_{rq_2} = p_1 X_{p_2}$.

   Conversely, assume that the inclusion (5.7) holds. From this we have

   $$q_1 X_{q_2} + qF[z] = q_1 X_{q_2} + q_1 q_2 F[z] = q_1 [X_{q_2} + q_2 F[z]] = q_1 F[z].$$

   So inclusion (5.7) implies the inclusion $q_1 F[z] \subset p_1 F[z]$. By Proposition 1.3.4, it follows that $p_1|q_1$.

2. Let $t = p_i q_i, i = 1, \ldots, s$. Since $\cap_{i=1}^s p_i X_{q_i}$ is a submodule of X_q, it is of the form pX_q with $t = pq$. Now the inclusion $pX_q \subset p_i X_{q_i}$ implies that $p_i|p$ and $q|q_i$, so p is a common multiple of the p_i and q a common divisor of the q_i. Now let q' be any common divisor of the q_i. Since necessarily $q'|t$, we can write $t = p'q'$. Now, applying Proposition 1.3.4, we have $p' X_{q'} \subset p_i X_{q_i}$ and hence $p' X_{q'} \subset \cap_{i=1}^s p_i X_{q_i} = pX_q$. This implies that $q'|q$ and hence q is a g.c.d. of the q_i. By the same token, we conclude that p is the l.c.m. of the p_i.

3. Since $p_1 X_{q_1} + \cdots + p_s X_{q_s}$ is an invariant subspace of X_t, it is of the form pX_q with $t = pq$. Now the inclusions $p_i X_{q_i} \subset pX_q$ imply the division relations $q_i|q$ and $p|p_i$. So q is a common multiple of the q_i and p a common divisor of the p_i. Let p' be any other common divisor of the p_i. Then $p_i = p' e_i$ for some polynomials e_i. Now $t = p_i q_i = p' e_i q_i = p'q'$, so $e_i q_i = q'$ and q' is a common multiple of the q_i. Now $p_i = p' e_i$ implies $p_i X_{q_i} = p' e_i X_{q_i} \subset p' X_{e_i q_i} = p' X_{q'}$ and hence $pX_q = p_1 X_{q_1} + \cdots + p_s X_{q_s} = p' X_{q'}$. This shows that $q|q'$, and so q is the l.c.m. of the q_i. □

Corollary 5.1.2 *Given the factorizations $q = p_i q_i, i = 1, \ldots, s$. Then:*

1.
$$X_q = p_1 X_{q_1} + \cdots + p_s X_{q_s}$$

   if and only if the p_i are coprime.

2. *The sum $p_1 X_{q_1} + \cdots + p_s X_{q_s}$ is a direct sum if and only if $q_1, \ldots, q_s$ are pairwise coprime.*

3. *We have the direct sum decomposition*

   $$X_q = p_1 X_{q_1} \oplus \cdots \oplus p_s X_{q_s}$$

   if and only if the p_i are coprime and the q_i are pairwise coprime.

4. *We have the direct sum decomposition*

$$X_q = p_1 X_{q_1} \oplus \cdots \oplus p_s X_{q_s}$$

if and only if the q_i are pairwise coprime and $q = q_1 \cdots q_s$. In this case, $p_i = \Pi_{j \neq i} q_j$.

Proof:

1. Let the invariant subspace $p_1 X_{q_1} + \cdots + p_s X_{q_s}$ have the representation $p_\nu X_{q_\nu}$ with p_ν the g.c.d. of the p_i and q_ν the l.c.m. of the q_i. Therefore, $p_\nu X_{q_\nu} = X_q$ if and only if $p_\nu = 1$, or equivalently, $q_\nu = q$.

2. The sum $p_1 X_{q_1} + \cdots + p_s X_{q_s}$ is a direct sum if and only if, for each index i, we have

$$p_i X_{q_i} \cap \sum_{j \neq i} p_j X_{q_j} = \{0\}.$$

   Now $\sum_{j \neq i} p_j X_{q_j}$ is an invariant subspace and hence of the form $\pi_i X_{\sigma_i}$ for some factorization $q = \pi_i \sigma_i$. Here π_i is the g.c.d. of the $p_j, j \neq i$, and σ_i is the l.c.m. of the $q_j, j \neq i$. Now $p_i X_{q_i} \cap \pi_i X_{\sigma_i} = \{0\}$ if and only if σ_i and q_i are coprime. This, however, is equivalent to q_i being coprime with each of the $q_j, j \neq i$, that is, to the pairwise coprimeness of the $q_i, i = 1, \ldots, s$.

3. Follows from the two previous parts.

4. Clearly the p_i are coprime.   □

Corollary 5.1.3 *Let $p = p_1^{\nu_1} \cdots p_k^{\nu_k}$ be the primary decomposition of p. Define $\pi_i = \Pi_{j \neq i} p_j^{\nu_j}$. Then*

$$X_p = \pi_1 X_{p_1^{\nu_1}} \oplus \cdots \oplus \pi_k X_{p_k^{\nu_k}}. \tag{5.8}$$

Proof: Clearly the g.c.d. of the π_i is 1, whereas the l.c.m. of the $p_i^{\nu_i}$ is p.   □

The structure of the shift operator restricted to an invariant subspace can be deduced easily from the corresponding factorization.

Proposition 5.1.4 *Let $q = q_1 q_2$. Then we have the similarity*

$$S_q | q_1 X_{q_2} \simeq S_{q_2}. \tag{5.9}$$

Proof: Let $\phi : X_{q_2} \longrightarrow q_1 X_{q_2}$ be the map defined by

$$\phi(f) = q_1 f.$$

This is clearly an isomorphism of the two spaces. Next we compute, for $f \in X_{q_2}$,

$$\phi S_{q_2} f = q_1 S_{q_2} f = q \pi_q q_1 z f = \pi_q z q_1 f = S_q \phi f.$$

Therefore, the following diagram is commutative:

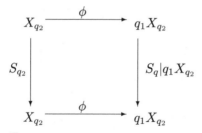

This is equivalent to Eq. (5.9).   □

Since eigenvectors span one-dimensional invariant subspaces, we expect a characterization of eigenvectors of the shift S_q in terms of the polynomial q.

Proposition 5.1.5 *Let q be a nonzero polynomial.*

1. *The eigenvalues of S_q coincide with the zeros of q.*

2. *$f \in X_q$ is an eigenvector of S_q corresponding to the eigenvalue α if and only if*

$$f(z) = \frac{cq(z)}{z - \alpha}. \qquad (5.10)$$

Proof: Let f be an eigenvector of S_q corresponding to the eigenvalue α, that is, $S_q f = \alpha f$. Now, by Eq. (5.6), there exists a scalar c for which $(S_q f)(z) = z f(z) - cq(z)$. Thus $z f(z) - cq(z) = \alpha f(z)$, which implies Eq. (5.10). Since f is a polynomial, we must have $q(\alpha) = 0$.

Conversely, if $q(\alpha) = 0$, q is divisible by $z - \alpha$ and hence f defined by Eq. (5.10) is in X_q. We compute

$$(S_q - \alpha I)f = \pi_q(z - \alpha)\frac{cq(z)}{z - \alpha} = \pi_q cq(z) = 0. \qquad □$$

The previous proposition suggests that the characteristic polynomial of S_q is q itself. This indeed is true and is proved next.

Proposition 5.1.6 *Let $q(z) = z^n + q_{n-1}z^{n-1} + \cdots + q_0$ be a monic polynomial of degree n, and let S_q be the shift operator defined by Eq. (5.2). Then the characteristic polynomial of S_q is q.*

Proof: It suffices to compute $\det(zI - C)$ for an arbitrary matrix representation of S_q. We find it convenient to do the computation in the standard basis. In this case, the matrix representation is given by the companion matrix $C_q^\sharp$ in Eq. (5.3). Thus, we compute, expanding the determinant by the first row,

$$\det(zI - C_q^\sharp) = \begin{vmatrix} z & & & q_0 \\ -1 & & & \cdot \\ & \ddots & & \cdot \\ & & -1 & z+q_{n-1} \end{vmatrix}$$

$$= z \begin{vmatrix} z & & & q_1 \\ -1 & & & \cdot \\ & \ddots & & \cdot \\ & & -1 & z+q_{n-1} \end{vmatrix}$$

$$+ (-1)^{n+1} q_0 \begin{vmatrix} -1 & z & & \\ & \ddots & \ddots & \\ & & & z \\ & & & -1 \end{vmatrix}$$

$$= z(z^{n-1} + q_{n-1}z^{n-2} + \cdots + q_1) + (-1)^{n+1}q_0(-1)^{n-1}$$

$$= q(z). \qquad \qquad \square$$

Lemma 5.1.1

1. *Given a nonzero polynomial q and $f \in X_q$, the smallest S_q-invariant subspace of X_q containing f is $q_1 X_{q_2}$, where $q_1 = q \wedge f$ and $q = q_1 q_2$.*

2. *S_q is a cyclic transformation in X_q.*

3. *A polynomial $f \in X_q$ is a cyclic vector of S_q if and only if f and q are coprime.*

Proof:

1. Let M be the subspace of X_q spanned by the vectors $\{S_q^i f | i \geq 0\}$. This is the smallest S_q-invariant subspace containing f. Therefore, it has the representation $M = q_1 X_{q_2}$ for a factorization $q = q_1 q_2$. As $f \in M$, there exists a polynomial $f_1 \in X_{q_2}$ for which $f = q_1 f_1$. This shows that q_1 is a common divisor of q and f.

   To show that it is the greatest common divisor, let us assume that q' is an arbitrary common divisor of q and f. Thus, we have $q = q'q''$ and $f = q'f'$. Using Lemma 1.3.3, we compute

$$S_q^k f = \pi_q x^k f = \pi_q x^k q' f'$$
$$= q' \pi_{q''} x^k f' = q' S_{q''}^k f'.$$

Thus, we have $M \subset q'X_{q''}$ or, equivalently, $q_1 X_{q_2} \subset q'X_{q''}$. This implies that $q'|q_1$, and hence q_1 is the g.c.d. of q and f.

2. Obviously, $1 \in X_q$ and $1 \wedge q = 1$. So 1 is a cyclic vector for S_q.

3. Obviously, since $\dim q_1 X_{q_2} = \dim X_{q_2} = \deg q_2$, $X = q_1 X_{q_2}$ if and only if $\deg q_1 = 0$, that is, f and q are coprime. $\square$

The availability of eigenvectors of the shift allows us to study under what conditions the shift S_q is diagonalizable, that is, has a diagonal matrix representation.

Proposition 5.1.7 *Let q be a monic polynomial of degree n. Then S_q is diagonalizable if and only if q splits into the product of n distinct linear factors, or equivalently, it has n distinct zeros.*

Proof: Assume that $\alpha_1, \ldots, \alpha_n$ are distinct zeros of q. So $q(z) = \Pi_{i=1}^n (z - \alpha_i)$. Let $p_i(z) = (q(z))/(z - \alpha_i) = \Pi_{j \neq i}(z - \alpha_j)$. Then $\mathcal{B}_{sp} = \{p_1, \ldots, p_n\}$ is the spectral basis for X_q, differing from the Lagrange interpolation basis by constant factors only. It is easily checked that $(S_q - \alpha_i)p_i = 0$. So

$$[S_q]_{sp}^{sp} = \begin{pmatrix} \alpha_1 & & & \\ & \cdot & & \\ & & \cdot & \\ & & & \cdot \\ & & & & \alpha_n \end{pmatrix}, \tag{5.11}$$

and S_q is diagonalizable.

Conversely, assume that S_q is diagonalizable. Then with respect to some basis it has the representation of Eq. (5.11). Since S_q is cyclic, its minimal and characteristic polynomials coincide. Necessarily, all of the α_i are distinct. $\square$

Proposition 5.1.8 *Let q be a monic polynomial and S_q the shift operator in X_q defined by Eq. (5.2). Then*

$$p(S_q)f = \pi_q(pf), \qquad f \in X_q. \tag{5.12}$$

Proof: Using linearity, it suffices to show that

$$S_q^k f = \pi_q z^k f, \qquad f \in X_q.$$

We prove this by induction. For $k = 1$, this is the definition. Assume that we proved it up to an integer k. Now, using the fact that $z \text{Ker} \, \pi_q \subset \text{Ker} \, \pi_q$, we compute

$$S_q^{k+1} f = S_q S_q^k f = \pi_q z \pi_q z^k f = \pi_q z^{k+1} f. \qquad \square$$

Clearly, the operators $p(S_q)$ all commute with the shift S_q. We proceed to state the simplest version of the commutant lifting theorem. It characterizes operators commuting with the shift in X_q via operators commuting with the shift S_+ in $F[z]$. The last class of operators is multiplication operators by polynomials.

Theorem 5.1.2 *Let q be a monic polynomial and S_q the shift operator in X_q defined by Eq. (5.2). Let X be any operator in X_q that commutes with S_q. Then there exists an operator $\bar{X}$ that commutes with S_+ and is such that*

$$X = \pi_q \bar{X}|_{X_q}. \tag{5.13}$$

Equivalently, the following diagram is commutative:

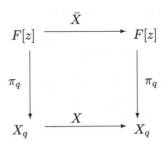

Proof: By Proposition 6.1.2, there exists a polynomial p for which $X = p(S_q)$. We define $\bar{X} : F[z] \longrightarrow F[z]$ by $\bar{X}f = pf$. It is easily checked that Eq. (5.13) holds.   □

Proposition 5.1.9 *Given the polynomials p, q, with q monic. Let r be the g.c.d. of p and q and let $q = rs$. Then*

$$Ker\, p(S_q) = sX_r \tag{5.14}$$

and

$$Im\, p(S_q) = rX_s. \tag{5.15}$$

We proceed to study the invertibility properties of transformations of the form $p(S_q)$. Naturally we expect to be able to give such a characterization in terms of the polynomials p and q. This is the content of the next theorem, which has far-reaching implications.

Theorem 5.1.3 *Let $p, q \in F[z]$, with q nonzero. Then the linear transformation $p(S_q)$ is invertible if and only if p and q are coprime. Moreover, we have*

$$p(S_q)^{-1} = a(S_q),$$

where the polynomial a arises out of any solution of the Bezout equation $a(z)p(z) + b(z)q(z) = 1$.

Proof: Assume that p and q are coprime. Then there exist polynomials a, b that solve the Bezout equation $ap + bq = 1$. Applying the functional calculus to the shift S_q, and noting that $q(S_q) = 0$, we get

$$a(S_q)p(S_q) + b(S_q)q(S_q) = a(S_q)p(S_q) = I$$

or

$$p(S_q)^{-1} = a(S_q).$$

To prove the converse, we assume that p and q have a nontrivial g.c.d. r. Let $q = q'r$ and $p = p'r$. Obviously, $q'X_r$ is a nontrivial invariant subspace of X_q. We will show that it is included, actually it is equal, to $Ker\, p(S_q)$. Let $f \in q'X_r$, that is, $f = q'f'$ with $f' \in X_r$. We now compute

$$\begin{aligned} p(S_q)f &= \pi_q pf = \pi_q(p'r)(q'f') \\ &= \pi_q(p'q')(rf') = \pi_q q(rf') = 0. \end{aligned} \qquad \square$$

5.2 Circulant Matrices

In this section we give a short account of a special class of structured matrices called circulant matrices. We do this for its own sake, as well as to illustrate the power of polynomial algebra.

Definition 5.2.1 *An $n \times n$ matrix over a field F is called a* **circulant matrix** *if it has the form*

$$C = \mathrm{circ}(c_0, \ldots, c_{n-1}) = \begin{pmatrix} c_0 & c_{n-1} & \cdot & \cdot & c_1 \\ c_1 & c_0 & \cdot & \cdot & \cdot \\ \cdot & \cdot & \cdot & \cdot & \cdot \\ \cdot & \cdot & \cdot & \cdot & c_{n-1} \\ c_{n-1} & \cdot & \cdot & c_1 & c_0 \end{pmatrix},$$

that is, $c_{ij} = c_{(i-j \bmod n)}$. We define a polynomial $c(z)$ by $c(z) = c_0 + c_1 z + \cdots + c_{n-1}z^{n-1}$. The polynomial c is called the **representer** *of $\mathrm{circ}(c_0, \ldots, c_{n-1})$.*

Theorem 5.2.1 *For circulant matrices the following properties hold:*

1. *The circulant matrix $\mathrm{circ}(c_0, \ldots, c_{n-1})$ is the matrix representation of $c(S_{z^n-1})$ with respect to the standard basis of X_{z^n-1}.*

2. *The sum of circulant matrices is a circulant matrix. Specifically,*

$$\mathrm{circ}(a_1, \ldots, a_n) + \mathrm{circ}(b_1, \ldots, b_n) = \mathrm{circ}(a_1 + b_1, \ldots, a_n + b_n).$$

3. *We have*

$$\alpha\, \mathrm{circ}(a_1, \ldots, a_n) = \mathrm{circ}(\alpha a_1, \ldots, \alpha a_n).$$

4. *Define the special circulant matrix Π by*

$$\Pi = \mathrm{circ}(0,1,0,\ldots,0) = \begin{pmatrix} 0 & & & 1 \\ 1 & \cdot & & \cdot \\ & \cdot & \cdot & \cdot \\ & & \cdot & \cdot \\ & & 1 & 0 \end{pmatrix}.$$

Then $C \in F^{n \times n}$ is a circulant if and only if $C\Pi = \Pi C$.

5. *The product of circulants is commutative.*

6. *The product of circulants is a circulant.*

7. *The inverse of a circulant is a circulant. Moreover, the inverse of a circulant with representer $c(z)$ is a circulant with representer $a(z)$, where a comes from a Bezout identity $a(z)c(z) + b(z)(z^n - 1) = 1$.*

8. *Over an algebraically closed field, circulants are diagonalizable.*

Proof:

1. We compute

$$[S_{z^n-1}]_{st}^{st} = \begin{pmatrix} 0 & & & 1 \\ 1 & & & \cdot \\ & \cdot & & \cdot \\ & & \cdot & \cdot \\ & & 1 & 0 \end{pmatrix}.$$

This is a special case of the companion matrix in Eq. (5.3). Now

$$S_{z^n-1}^i z^j = \begin{cases} z^{i+j} & i+j \leq n-1 \\ z^{i+j-n} & i+j \geq n. \end{cases}$$

So

$$
\begin{aligned}
c(S_{z^n-1})z^j &= \sum_{i=0}^{n-1} c_i S_{z^n-1}^i z^j \\
&= \sum_{i=0}^{n-1-j} c_i z^{i+j} + \sum_{i=n-j}^{n-1} c_i z^{i+j-n},
\end{aligned}
$$

and this implies the equality

$$[c(S_{z^n-1})]_{st}^{st} = \mathrm{circ}(c_0, \ldots, c_{n-1}).$$

2. Given polynomials $a, b \in F[z]$, we have

$$(a+b)(S_{z^n-1}) = a(S_{z^n-1}) + b(S_{z^n-1})$$

and hence

$$\begin{aligned} \mathrm{circ}(a_0 + b_0, \ldots, a_{n-1} + b_{n-1}) &= [(a+b)(S_{z^n-1})]_{st}^{st} \\ &= [a(S_{z^n-1})]_{st}^{st} + [b(S_{z^n-1})]_{st}^{st} \\ &= \mathrm{circ}(a_0, \ldots, a_{n-1}) + \mathrm{circ}(b_0, \ldots, b_{n-1}). \end{aligned}$$

3. We compute

$$\begin{aligned} \mathrm{circ}(\alpha a_0, \ldots, \alpha a_{n-1}) &= [\alpha a(S_{z^n-1})]_{st}^{st} \\ &= \alpha[a(S_{z^n-1})]_{st}^{st} = \alpha \mathrm{circ}(a_0, \ldots, a_{n-1}). \end{aligned}$$

4. Clearly, $\Pi = \mathrm{circ}(0,1,0,..,0) = [S_{z^n-1}]_{st}^{st}$, and obviously, S_{z^n-1} is cyclic. Hence a linear transformation K commutes with S_{z^n-1} if and only if $K = c(S_{z^n-1})$ for some polynomial c. Thus, assume that $C\Pi = \Pi C$. Then there exists a linear transformation K in X_{z^n-1} satisfying $[K]_{st}^{st} = C$ and K commutes with S_{z^n-1}. Therefore, $K = c(S_{z^n-1})$ and $C = \mathrm{circ}(c_0, \ldots, c_{n-1})$.

Conversely, if $C = \mathrm{circ}(c_0, \ldots, c_{n-1})$, we have

$$\begin{aligned} C\Pi &= [c(S_{z^n-1})]_{st}^{st}[S_{z^n-1}]_{st}^{st} \\ &= [S_{z^n-1}]_{st}^{st}[c(S_{z^n-1})]_{st}^{st} = \Pi C. \end{aligned}$$

5. Follows from

$$c(S_{z^n-1})d(S_{z^n-1}) = d(S_{z^n-1})c(S_{z^n-1}).$$

6. Follows from

$$c(S_{z^n-1})d(S_{z^n-1}) = (cd)(S_{z^n-1}).$$

7. Let $C = \mathrm{circ}(c_0, \ldots, c_{n-1}) = c(S_{z^n-1})$, where $c(z) = c_0 + c_1 z + \cdots + c_{n-1}z^{n-1}$. By Theorem 5.1.3, $c(S_{z^n-1})$ is invertible if and only if $c(z)$ and $z^n - 1$ are coprime. In this case, there exist polynomials a, b satisfying the Bezout identity $a(z)c(z) + b(z)(z^n - 1) = 1$. We may assume without loss of generality that $\deg a < n$. From the Bezout identity we conclude that $c(S_{z^n-1})c(S_{z^n-1}) = I$ and hence

$$\mathrm{circ}(c_0, \ldots, c_{n-1})^{-1} = [a(S_{z^n-1})]_{st}^{st} = \mathrm{circ}(a_0, \ldots, a_{n-1}).$$

8. The polynomial $z^n - 1$ has a multiple zero if and only if $z^n - 1$ and nz^{n-1} have a common zero. Clearly this cannot occur. As all the roots of $z^n - 1$ are distinct, it follows from Proposition 5.1.7 that S_{z^n-1} is diagonalizable. This implies the diagonalizability of $c(S_{z^n-1})$.   □

5.3   Rational Models

Given a field F, we saw that the ring of polynomials $F[z]$ is an entire ring. Hence, by Theorem 1.3.11, it is embeddable in its field of quotients. We call the field of quotients of $F[z]$ the field of **rational functions** and denote it by $F(z)$. Strictly speaking, the elements of $F(z)$ are equivalence classes of pairs of polynomials (p, q), with q nonzero. However, in each nonzero equivalence class, there is a unique pair with p, q coprime and q monic. The corresponding equivalence class will be denoted by $(p(z))/(q(z))$. Given such a pair of polynomials $p(z), q(z)$, there is a unique representation of p in the form $p(z) = a(z)q(z) + r(z)$, with $\deg r < \deg q$. This allows us to write

$$\frac{p(z)}{q(z)} = a(z) + \frac{r(z)}{q(z)}. \tag{5.16}$$

A rational function r/q with $\deg r \leq \deg q$ will be called **proper**, and if $\deg r < \deg q$ is satisfied, **strictly proper**. Thus, any rational function $g = p/q$ has a unique representation as a sum of a polynomial and a strictly proper rational function. We denote by $F_-(z)$ the space of strictly proper rational functions and observe that it is an infinite-dimensional linear space. Equation (5.16) means that for $F(z)$ we have the following direct sum decomposition:

$$F(z) = F[z] \oplus F_-(z). \tag{5.17}$$

With this direct sum decomposition we associate two projection operators in $F(z)$, π_+ and π_-, with images $F[z]$ and $F_-(z)$, respectively. To be precise, given the representation (5.16), we have

$$\begin{cases} \pi_+(\dfrac{p}{q}) = a \\[2mm] \pi_-(\dfrac{p}{q}) = \dfrac{r}{q}. \end{cases} \tag{5.18}$$

We note that the projection operator π_q now can be wriiten as

$$\pi_q f = q\pi_- q^{-1} f. \tag{5.19}$$

We say that a rational function g is **proper** if $g = p/q$ with $\deg p \leq \deg q$. The set of all proper rational functions will be denoted by $F_{pr}(z)$.

Proper rational functions have an expansion as formal power series in the variable z^{-1}, that is, in the form $\sum_{i=0}^{\infty} g_i / z^i$. Assume that $p(z) = \sum_{k=0}^{n} p_k z^k$, $q(z) = \sum_{i=0}^{n} q_i z^i$, with $q_n \neq 0$. We compute

$$\sum_{k=0}^{n} p_k z^k = \sum_{i=0}^{n} q_i z^i \sum_{j=0}^{\infty} \frac{g_j}{z^j} = \sum_{k=-\infty}^{n} \left\{ \sum_{i=k}^{n} q_i h_{i-k} \right\} z^k.$$

By comparing coefficients we get the infinite system of linear equations

$$p_k = \sum_{i=k}^{n} q_i h_{i-k}, \quad -\infty < k \le n.$$

This system has a unique solution that can be found by solving it recursively, starting with $k = n$. An alternative way of finding this expansion is via the process of long division of p by q.

The space $F_-(z)$ of strictly proper rational functions can be viewed as a space that is isomorphic to $F(z)/F[z]$. In fact, the projection $\pi_- : F(z) \longrightarrow F[z]$ is a surjective linear map with kernel equal to $F[z]$. However, both $F(z)$ and $F[z]$ also carry a natural $F[z]$-module structure, with polynomials acting by multiplication. This structure induces an $F[z]$-module structure in the quotient space. This module structure is transferred to $F_-(z)$ by defining

$$p \cdot g = \pi_-(pg), \quad g \in F_-(z).$$

In particular, we define the **backward shift operator** S_- acting in $F_-(z)$ by

$$S_- g = \pi_-(zg), \quad g \in F_-(z). \tag{5.20}$$

In terms of the expansion of $g \in F_-(z)$ around infinity, that is, in terms of the representation $g(z) = \sum_{i=1}^{\infty} g_i/z^i$, we have

$$S_- \sum_{i=1}^{\infty} \frac{g_i}{z^i} = \sum_{i=1}^{\infty} \frac{g_{i+1}}{z^i}.$$

This explains the usage of a backward shift for this operator.

It is easy to construct many finite-dimensional S_--invariant subspaces of $F_-(z)$. In fact, given any nonzero polynomial d, we let

$$X^d = \left\{ \frac{r}{d} \mid \deg r < \deg d \right\}.$$

It is easily checked that X^d is indeed an S_--invariant subspace and its dimension equals the degree of d.

It is natural to consider the restriction of the operator S_- to X^d. Thus, we define a linear transformation $S^d : X^d \longrightarrow X^d$ by $S^d = S_-|X^d$ or, equivalently, for $h \in X^d$,

$$S^d h = S_- h = \pi_- zh.$$

The modules X_d and X^d have the same dimensions and are defined by the same polynomial. It is natural to conjecture that they must be isomorphic, and this is indeed the case.

Theorem 5.3.1 *Let d be a nonzero polynomial. Then the operators S_d and S^d are isomorphic. The isomorphism is given by the map $\rho_d : X^d \longrightarrow X_d$ defined by*

$$\rho_d h = dh. \tag{5.21}$$

Proof: We compute

$$\rho_d S^d h = \rho_d \pi_- z h = d\pi_- z h = d\pi_- d^{-1} dz h = \pi_d z(dh) = S_d(\rho_d h). \qquad \square$$

The polynomial and rational models that have been introduced are isomorphic. Yet, they represent two fundamentally different points of view. In the case of polynomial models, all spaces of the form X_q, with $\deg q = n$, are the same, but the shifts S_q act differently. On the other hand, the operators S^q in the spaces X^q act in the same way, as they are all restrictions of the backward shift S_-; however, the spaces are different. Thus, the polynomial models represent an arithmetic perspective, whereas the rational models represent a geometric one.

Our next result is the characterization of all finite-dimensional S_--invariant subspaces.

Proposition 5.3.1 *A subset M of $F_-(z)$ is a finite-dimensional S_--invariant subspace if and only if, for some nonzero polynomial d, $M = X^d$.*

Proof: Assume that, for some d, $M = X^d$. Then

$$\pi_- z \frac{r}{d} = d^{-1} d\pi_- d^{-1} z r = d^{-1} \pi_d r \in M.$$

Conversely, let M be a finite-dimensional S_--invariant subspace. By Theorem 4.5.2, there exists a nonzero polynomial p of minimal degree such that $\pi_- p h = 0$ for all $h \in M$. Thus we get $M \subset X^p$. It follows that pM is a submodule of X_p and hence is of the form $p_1 X_{p_2}$ for some factorization $p = p_1 p_2$. We conclude that $M = X^{p_2}$. The minimality of p implies that $p = p_2$, up to a constant nonzero factor. $\qquad \square$

The spaces of rational functions of the form X^d will be referred to as **rational models**. Note that in the theory of differential equations, these spaces appear as the Laplace transforms of the spaces of solutions of a homogeneous linear differential equation with constant coefficients.

The following sums up the basic arithmetic properties of rational models. It is the counterpart of Proposition 5.1.3:

Proposition 5.3.2

1. *Given polynomials $p, q \in F[z]$, then we have the inclusion $X^p \subset X^q$ if and only if $p | q$.*

2. *Given polynomials $p_i \in F[z]$, $i = 1, \ldots, s$, then $\cap_{i=1}^s X^{p_i} = X^p$ with p the g.c.d. of the p_i.*

3. *Given polynomials $p_i \in F[z]$, then $\sum_{i=1}^s X^{p_i} = X^q$ with q the l.c.m. of the p_i.*

Proof: Follows from Proposition 5.1.3, using the isomorphism of polynomial and rational models given by Theorem 5.3.1.   □

The primary decomposition theorem and the direct sum representation (5.8) have a direct implication toward the partial fraction decomposition of rational functions.

Theorem 5.3.2 *Let* $p = \Pi_{i=1}^{s} p_i^{\nu_i}$ *be the primary decomposition of the nonzero polynomial* p. *Then:*

1. *We have*

$$X^p = X^{p_1^{\nu_1}} \oplus \cdots \oplus X^{p_s^{\nu_s}}. \tag{5.22}$$

2. *Each rational function* $g \in X^p$ *has a unique representation of the form*

$$g = \frac{r}{p} = \sum_{i=1}^{s} \sum_{j=1}^{\nu_i} \frac{r_{ij}}{p_i^j}$$

with $\deg r_{ij} < \deg p_i, j = 1, \ldots, \nu_i.$

Proof:

1. Given the primary decomposition of p, we define $\pi_i(z) = \Pi_{j \neq i} p_i^{\nu_i}$. Clearly, by Corollary 5.1.3, we have

$$X_p = \pi_1 X_{p_1^{\nu_1}} \oplus \cdots \oplus \pi_s X_{p_s^{\nu_s}}. \tag{5.23}$$

We use the isomorphism of the modules X_p and X^p and the fact that $p = \pi_i p_i^{\nu_i}$, which implies that $p^{-1} \pi_i X_{p_i^{\nu_i}} = p_i^{-\nu_i} \pi_i - 1 \pi_i X_{p_i^{\nu_i}} = X^{p_i^{\nu_i}}$, to get the direct sum decomposition (5.22).

2. For any $r_i \in X_{p_i^{\nu_i}}$ we have $r_i = \sum_{j=0}^{\nu_i - 1} r_{i(\nu_i - j)} p_i^j$. With $r/p = \sum_{i=1}^{s} (r_i/p_i^{\nu_i})$, Using Eq. (5.23) and taking $r_i \in X_{p_i^{\nu_i}}$,

$$\frac{r_i}{p_i^{\nu_i}} = \sum_{j=1}^{\nu_i} \frac{r_{ij}}{p_i^j}. \tag{5.24}$$
□

5.4   The Chinese Remainder Theorem

Theorem 5.4.1 (Chinese remainder theorem) *Let* $q_i \in F[z], i = 1, \ldots, s$, *be pairwise coprime polynomials, and let* $q = q_1 \cdots q_s$. *Then, given polynomials* a_i *such that* $\deg a_i < \deg q_i$, *there exists a unique polynomial* f, *satisfying* $\deg f < \deg q$ *and* $\pi_{q_i} f = a_i$, *for* $i = 1, \ldots, s$.

Proof: The interesting thing about the proof is its use of coprimeness in two distinct ways. Let us define $d_j = \prod_{i \neq j} q_i$. Then the pairwise coprimeness of the q_i implies the direct sum decomposition

$$X_q = d_1 X_{q_1} \oplus \cdots \oplus d_s X_{q_s}.$$

The condition $\deg f < \deg q$ is equivalent to $f \in X_q$. Let $f = \sum_{j=1}^s d_j f_j$ with $f_j \in X_{q_j}$. Since, for $i \neq j$, $q_i | d_j$, it follows that in this case $\pi_{q_i} d_j f_j = 0$. Hence,

$$\pi_{q_i} f = \pi_{q_i} \sum_{j=1}^s d_j f_j = \pi_{q_i} d_i f_i = d_i(S_{q_i}) f_i.$$

Now the module homomorphism $d_i(S_{q_i})$ in X_{q_i} is actually an isomorphism, by the coprimeness of d_i and q_i. Hence there exists a unique f_i in X_{q_i} such that $a_i = d_i(S_{q_i}) f_i$ and $f_i = d_i(S_{q_i})^{-1} a_i$. So $f = \sum_{j=1}^s d_j d_i(S_{q_i})^{-1} a_i$ is the required polynomial. Note that the inversion of $d_i(S_{q_i})$ can be done easily, by Theorem 5.1.3, using the Euclidean algorithm.

The uniqueness of f, under the condition $f \in X_q$, follows from the fact that it is a direct sum representation. This completes the proof.   □

5.5   Hermite Interpolation

We apply now the Chinese remainder theorem to the problem of Hermite interpolation. In Hermite interpolation, which generalizes the Lagrange interpolation, we prescribe not only the value of the interpolating polynomial at given points, but also the value of a certain number of derivatives; the number may differ from point to point.

Specifying the first ν derivatives, counting from zero, of a polynomial p at a point α means that we are given a representation

$$f(z) = \sum_{i=0}^{\nu-1} f_{i,\alpha}(z - \alpha_i)^i + (z - \alpha)^\nu g(z).$$

Of course, as $\deg \sum_{i=0}^{\nu-1} f_{i,\alpha}(z - \alpha_i)^i < \nu$, this means that

$$\deg \sum_{i=0}^{\nu-1} f_{i,\alpha}(z - \alpha)^i = \pi_{(z-\alpha)^\nu} f.$$

Hence, we can state the following:

The Hermite interpolation problem: Given distinct $\alpha_1, \ldots, \alpha_k \in F$, positive integers $\nu_1, \ldots, \nu_k$, and polynomials $f_i(z) = \sum_{j=0}^{\nu_i-1} f_{j,\alpha_i}(z - \alpha_i)^j$. Find polynomials f such that

$$\pi_{(z-\alpha_i)^{\nu_i}} f = f_i, \qquad i = 1, \ldots, k. \tag{5.25}$$

Proposition 5.5.1 *There exists a unique solution f, of degree $< n = \sum_{i=1}^{k} \nu_i$, to the Hermite interpolation problem. Any other solution of the Hermite interpolation problem is of the form $f + pg$, where g is an arbitrary polynomial and p is given by*

$$p(z) = \prod_{i=1}^{k} (z - \alpha_i)^{\nu_i}. \tag{5.26}$$

Proof: We apply the Chinese remainder theorem. Obviously, the polynomials $(z - \alpha_i)^{\nu_i}$, $i = 1, \ldots, k$, are pairwise coprime. Then, with p defined by Eq. (5.26), there exists a unique f with $\deg f < n$ for which Eq. (5.25) holds.

If $\hat{f}$ is any other solution, then $h = (f - \hat{f})$ satisfies $\pi_{(z-\alpha_i)^{\nu_i}} h = 0$, that is, h is divisible by $(z - \alpha_i)^{\nu_i}$. As these polynomials are pairwise coprime, it follows that $p|h$ or $h = pg$ for some polynomial g. $\qquad\square$

5.6   Duality

The availability of both polynomial and rational models allows us to proceed with a deeper study of duality. Our aim is to obtain an identification of the dual space to a polynomial model in terms of a polynomial model.

On $F(z)$ we introduce a bilinear form, given, for $f(z) = \sum_{j=-\infty}^{n_f} f_j z^j$ and $g(z) = \sum_{j=-\infty}^{n_g} g_j z^j$, by

$$[f, g] = \sum_{j=-\infty}^{\infty} f_j g_{-j-1}. \tag{5.27}$$

Clearly, the sum in Eq. (5.27) is well defined, as only a finite number of summands are nonzero. Given a subspace $M \subset F(z)$, we let $M^{\perp} = \{f \in F(z) | [m, f] = 0, \forall m \in M\}$. It is easy to check that $F[z]^{\perp} = F[z]$.

We will need the following simple computational rule:

Proposition 5.6.1 *Let ϕ, f, g be rational functions. Then*

$$[\phi f, g] = [f, \phi g]. \tag{5.28}$$

Proof: With the obvious notation we compute

$$\begin{aligned}
[\phi f, g] &= \sum_{j=-\infty}^{\infty} (\phi f)_j h_{-j-1} \\
&= \sum_{j=-\infty}^{\infty} \left(\sum_{i=-\infty}^{\infty} \phi_i f_{j-i} \right) h_{-j-1} \\
&= \sum_{i=-\infty}^{\infty} f_{j-i} \sum_{j=-\infty}^{\infty} \phi_i h_{-j-1}
\end{aligned}$$

$$= \sum_{k=-\infty}^{\infty} f_k \sum_{j=-\infty}^{\infty} \phi_{j-k} h_{-j-1}$$

$$= \sum_{k=-\infty}^{\infty} f_k \sum_{i=-\infty}^{\infty} \phi_i h_{-i-k-1}$$

$$= \sum_{k=-\infty}^{\infty} f_k (\phi h)_{-k-1} = [f, \phi h]. \qquad \square$$

Multiplication operators in $F(z)$, of the form $L_\phi h = \phi h$, are called **Laurent operators**. The function ϕ is called the **symbol** of the Laurent operator.

Before getting the characterization of the dual space to a polynomial model, we give a characterization of the dual space to $F[z]$.

Theorem 5.6.1 *The dual space of $F[z]$ is $z^{-1}F[[z^{-1}]]$.*

Proof: Clearly, every element $h \in z^{-1}F[[z^{-1}]]$ defines, by way of the pairing (5.27), a linear functional on $F[z]$. Conversely, let Φ be a linear functional on $F[z]$. It induces linear functionals ϕ_i on F by defining, for $\xi \in F$,

$$\phi_i(\xi) = [z^i \xi, \phi] = \Phi(z^i \xi),$$

and an element $h \in z^{-1}F[[z^{-1}]]$ is defined by letting $h(z) = \sum_{j=0}^{\infty} \phi_j z^{-j-1}$. It follows that $\Phi(f) = [f, h]$. $\qquad \square$

Point evaluations are clearly linear functionals in $F[z]$. It is easy to identify the representing functions. This in fact is an algebraic version of Cauchy's theorem.

Proposition 5.6.2 *Let $\alpha \in F$ and $f \in F[z]$. Then*

$$f(\alpha) = [f, (z - \alpha)^{-1}].$$

Proof: We have $(z-\alpha)^{-1} = \sum_{i=1}^{\infty} \alpha^{i-1} z^{-i}$. So this follows from Eq. (5.27), as

$$[f, \frac{1}{z-\alpha}] = \sum_{i=0}^{n} f_i \alpha^i = f(\alpha). \qquad \square$$

Theorem 5.6.2 *Let $M = dF[z]$ with $d \in F[z]$. Then $M^{\perp} = X^d$.*

Proof: Let $f \in F[z]$ and $h \in M$. Then

$$0 = [df, h] = [f, dh] = [f, \pi_- dh].$$

But this implies that $dh \in X_d$ or $h \in X^d$. $\qquad \square$

Next we compute the adjoint of the projection $\pi_d : F[z] \longrightarrow F[z]$. Clearly, π_d^* is a transformation acting in $z^{-1}F[[z^{-1}]]$.

Theorem 5.6.3 *The adjoint of π_d is π^d.*

Proof: Let $f \in F[z]$ and $h \in z^{-1}F[[z^{-1}]]$. Then

$$
\begin{aligned}
[\pi_d f, h] &= [d\pi_- d^{-1}f, h] = [\pi_- d^{-1}f, \tilde{d}h] = [d^{-1}f, \pi_+ \tilde{d}h] \\
&= [f, \tilde{d}^{-1}\pi_+ \tilde{d}h] = [\pi_+ f, \tilde{d}^{-1}\pi_+ \tilde{d}h] = [f, \pi_- \tilde{d}^{-1}\pi_+ \tilde{d}h] \\
&= [f, \pi^{\tilde{d}}h].
\end{aligned}
$$
$\square$

Not only are we interested in the study of duality on the level of $F[z]$ and its dual space $z^{-1}F[[z^{-1}]]$, but we also would like to study it on the level of the modules X_d and X^d. The key to this study is the fact that, if X is a linear space and M a subspace, then $(X/M)^* \simeq M^\perp$.

Theorem 5.6.4 *Let $d \in F[z]$ be nonsingular. Then X_d^* is isomorphic to X^d and $S_d^* = S^d$.*

Proof: Since X_d is isomorphic to $F[z]/dF[z]$, then X_d^* is isomorphic to $(F[z]/dF[z])^*$, which in turn is isomorphic to $(dF[z])^\perp$. But this last module is X^d. It is clear that, under the duality pairing we introduced, we actually have $X_d^* = X^d$. Finally, let $f \in X_d$ and let $h \in X^d$. Then

$$
[S_d f, h] = [\pi_d z f, h] = \left[z f, \pi^{\tilde{d}}h \right]
$$

$$
[z f, h] = [f, zh] = [\pi_+ f, zh]
$$

$$
[f, \pi_- zh] = [f, S_- h] = \left[f, S^{\tilde{d}}h \right].
$$

Now the $F[z]$–module $X^{\tilde{d}}$ is isomorphic to $X_{\tilde{d}}$ by the isomorphism given by the map $R_{\tilde{d}} : X^{\tilde{d}} \longrightarrow X_{\tilde{d}}$ defined by $R_{\tilde{d}}h = \tilde{d}h$. Hence we can identify $X_{\tilde{d}}$ with X_d^* by defining a new pairing,

$$
< f, g > = [d^{-1}f, g] = [f, d^{-1}g], \tag{5.29}
$$

for all $f, g \in X_d$. $\square$

As a direct corollary of Theorem 5.6.4 we have the following:

Theorem 5.6.5 *The dual space of X_d under the pairing $< , >$ introduced in Eq. (5.29) is X_d, and, moreover $S_d^* = S_d$.* $\square$

With the identification of the polynomial model X_d and its dual space, we can identify some pairs of dual bases.

Proposition 5.6.3

1. *Let $d(z) = z^n + d_{n-1}z^{n-1} + \cdots + d_0$, $\mathcal{B}_{st} = \{1, z, \ldots, z^{n-1}\}$ be the standard, and $\mathcal{B}_{co} = \{e_1(z), \ldots, e_n(z)\}$ the control bases, respectively, of X_d. Then $\mathcal{B}_{co} = \mathcal{B}_{st}^*$, that is, the control and standard bases, are dual to each other.*

2. Let $d(z) = \Pi_{i=1}^n (z - \lambda_i)$, with the λ_i distinct. Let $\mathcal{B}_{in} = \{\pi_1(z), \ldots, \pi_n(z)\}$, with π_i the Lagrange interpolation polynomials, be the interpolation basis in X_d. Let $\mathcal{B}_{sp} = \{p_1(z), \ldots, p_n(z)\}$ be the spectral basis, with $p_i(z) = \Pi_{j \neq i}(z - \lambda_j)$. Then $\mathcal{B}_{sp}^* = \mathcal{B}_{in}$.

Proof:

1. We use Eq. (5.29) to compute

$$
\begin{aligned}
< z^{i-1}, e_j > &= [d^{-1}z^{i-1}, \pi_+ z^{-j}d] = [d^{-1}z^{i-1}, z^{-j}d] \\
&= [z^{i-j-1}, 1] = \delta_{ij}.
\end{aligned}
$$

2. We use Proposition 5.6.2 and note that for every $f \in X_d$ we have

$$
< f, p_i >= [d^{-1}f, p_i] = [f, \frac{p_i}{d}] = \left[f, \frac{1}{z - \lambda_i} \right] = f(\lambda_i).
$$

In particular, $< \pi_i, p_j >= \pi_i(\lambda_j) = \delta_{ij}$. $\square$

This result explains the connection between the two companion matrices given in Proposition 5.1.2. Indeed,

$$
C_q^\sharp = [S_q]_{st}^{st} = \widetilde{[S_q^*]_{co}^{co}} = \widetilde{[S_q]_{co}^{co}} = \widetilde{C_q^\flat}.
$$

Next we compute the change of basis transformations.

Proposition 5.6.4

1. Let $q(z) = z^n + q_{n-1}z^{n-1} + \cdots + q_0$. Then

$$
[I]_{co}^{st} = \begin{pmatrix} q_1 & \cdot & \cdot & q_{n-1} & 1 \\ \cdot & & & \cdot & \\ \cdot & & \cdot & & \\ q_{n-1} & \cdot & & & \\ 1 & & & & \end{pmatrix} \tag{5.30}
$$

and

$$
[I]_{st}^{co} = \begin{pmatrix} & & & & 1 \\ & & \cdot & \psi_1 & \\ & & \cdot & \cdot & \cdot \\ & \cdot & & \cdot & \cdot \\ 1 & \psi_1 & \cdot & \cdot & \psi_{n-1} \end{pmatrix}, \tag{5.31}
$$

where, for $q^\sharp(z) = z^n q(z^{-1})$, $\psi(z) = \psi_0 + \cdots + \psi_{n-1}z^{n-1}$ is the unique solution, of degree $< n$, of the Bezout equation

$$
q^\sharp(z)\psi(z) + z^n \sigma(z) = 1. \tag{5.32}
$$

2. *The matrices in Eqs. (5.30) and (5.31) are inverses of each other.*

3. *Let $d(z) = \Pi_{j=1}^n (z - \alpha_j)$ with $\alpha_1, \ldots, \alpha_n$ distinct, and let $\mathcal{B}_{sp}, \mathcal{B}_{in}$ be the corresponding spectral and interpolation bases of X_d. Then we have the following change of basis transformations:*

$$[I]_{st}^{in} = \begin{pmatrix} 1 & \alpha_1 & . & . & \alpha_1^{n-1} \\ . & . & . & . & . \\ . & . & . & . & . \\ . & . & . & . & . \\ 1 & \alpha_n & . & . & \alpha_n^{n-1} \end{pmatrix} \tag{5.33}$$

$$[I]_{sp}^{co} = \begin{pmatrix} 1 & . & . & . & 1 \\ \alpha_1 & . & . & . & \alpha_n \\ . & . & . & . & . \\ . & . & . & . & . \\ \alpha_1^{n-1} & . & . & . & \alpha_n^{n-1} \end{pmatrix} \tag{5.34}$$

$$[I]_{sp}^{in} = \begin{pmatrix} p_1(\alpha_1) & & & \\ & . & & \\ & & . & \\ & & & p_n(\alpha_n) \end{pmatrix}. \tag{5.35}$$

Proof:

1. Note that, if ψ is a solution of Eq. (5.32), then necessarily $\psi_0 = 1$. With J the transposition matrix defined in Eq. (8.20), we compute

$$J[I]_{st}^{co} = ([I]_{co}^{st} J)^{-1} = (q^\#(S_{z^n}))^{-1}.$$

However, if we consider the map S_{z^n}, then

$$\begin{pmatrix} 1 & q_{n-1} & . & . & . & q_1 \\ . & & . & . & . & . \\ & & . & . & . & . \\ & & & . & . & . \\ & & & & . & q_{n-1} \\ & & & & & 1 \end{pmatrix} = q^\#(S_{z^n}),$$

and its inverse is given by $\psi(S_{z^n})$, where ψ solves the Bezout equation 5.3.2). This completes the proof.

2. Follows from the fact that $[I]_{co}^{st}[I]_{st}^{co} = I$.

3. The matrix representation for $[I]_{st}^{in}$ has been derived in Corollary 2.10.1.

The matrix representation for $[I]_{sp}^{co}$ follows from Eq. (5.33) by applying duality theory, in particular Theorem 4.4.3. We can also derive this matrix representation directly, which we proceed to do. For this we define polynomials $s_1, \ldots, s_n$ by

$$s_i(z) = e_1(z) + \alpha_i e_2(z) + \cdots + \alpha_i^{n-1} e_n(z).$$

We claim that $s_i(z)$ are eigenfunctions of S_d corresponding to the eigenvalues α_i. Indeed, using Eq. (5.5) and the fact that $0 = q(\alpha_i) = q_0 + q_1 \alpha_i + \cdots + \alpha_i^n$, we have

$$
\begin{aligned}
S_q s_i &= -q_0 e_n + \alpha_i(e_1 - q_1 e_n) + \cdots + \alpha_i(e_{n-1} - q_{n-1} e_n) \\
&= \alpha_i e_1 + \cdots + \alpha_i^{n-1} e_{n-1} - (q_0 + \cdots + q_{n-1}\alpha_i^{n-1})e_n \\
&= \alpha_i e_1 + \cdots + \alpha_i^n e_n = \alpha_i s_i(z).
\end{aligned}
$$

This implies that there exist constants γ_i such that $s_i(z) = \gamma_i p_i(z)$. Since both s_i and e_i are obviously monic, it necessarily follows that $\gamma_i = 1$ and $s_i = p_i$. The equations

$$p_i(z) = e_1(z) + \alpha_i e_2(z) + \cdots + \alpha_i^{n-1} e_n(z)$$

imply the matrix representation (5.34).

Finally the matrix representation in (5.35) follows from the trivial identities

$$p_i(z) = p_i(\alpha_i)\pi_i(z). \qquad \square$$

5.7 Reproducing Kernels

With the identification $X_q^* = X_q$, given in terms of the pairing (5.29), we can introduce reproducing kernels for polynomial models. To this end, we consider the ring $F[z, w]$ of polynomials in the two variables z, w. Now let q be a monic polynomial of degree n. Assume that $q(z) = z^n + q_{n-1} z^{n-1} + \cdots + q_0$. Given a polynomial $K(z, w) \in F[z, w]$ of degree less than n in each variable, it can be written as $K(z, w) = \sum_{i=1}^n \sum_{j=1}^n k_{ij} z^{i-1} w^{j-1}$. Such a polynomial K induces a linear transformation in X_d, which we also denote by K, defined by

$$(Kf)(z) = <K(z, \cdot), f> = \sum_{i=1}^n \sum_{j=1}^n k_{ij} z^{i-1} < w^{j-1}, f>. \qquad (5.36)$$

Now, any $f \in X_d$ has an expansion $f(w) = \sum_{k=1}^n f_k e_k(w)$, where $\{e_1, \ldots, e_n\}$ is the control basis in X_q. Thus, we compute

$$\begin{aligned}
(Kf)(z) &= \sum_{i=1}^{n}\sum_{j=1}^{n} k_{ij} z^{i-1} < w^{j-1}, f > \\
&= \sum_{i=1}^{n}\sum_{j=1}^{n}\sum_{k=1}^{n} k_{ij} z^{i-1} f_k < w^{j-1}, e_k > \\
&= \sum_{i=1}^{n}\sum_{j=1}^{n}\sum_{k=1}^{n} k_{ij} f_k z^{i-1} \delta_{jk} \\
&= \sum_{i=1}^{n}\sum_{j=1}^{n} k_{ij} f_j z^{i-1}.
\end{aligned}$$

This implies that $[K]_{co}^{st} = (k_{ij})$.

Consider now the special kernel, defined by

$$K(z, w) = \frac{q(z) - q(w)}{z - w}. \tag{5.37}$$

Proposition 5.7.1 *Given a monic polynomial q of degree n, then, with $K(z, w)$ defined by Eq. (5.37), we have.*

1. $K(z, w) = \sum_{j=1}^{n} w^{j-1} e_j(z) = \sum_{j=1}^{n} z^{j-1} e_j(w)$.

2. *For any $f \in X_d$,*

$$< K(z, \cdot), f >_w = f(z).$$

3. $$[K]_{co}^{co} = [K]_{st}^{st} = I.$$

Proof:

1. We use the fact that $z^j - w^j = (z - w)\sum_{k=1}^{j-1} z^k w^{j-k-1}$ to get

$$\begin{aligned}
q(z) - q(w) &= z^n - w^n + \sum_{j=1}^{n-1} q_j(z^j - w^j) \\
&= (z - w)\sum_{k=1}^{n-1} z^k w^{n-k-1} \\
&\quad + (z - w)\sum_{j=1}^{n} q_j \sum_{k=1}^{n} z^k w^{j-k-1} \\
&= (z - w)\sum_{j=1}^{n} w^{j-1} e_j(z)
\end{aligned}$$

and hence

$$\frac{q(z) - q(w)}{z - w} = \sum_{j=1}^{n} w^{j-1} e_j(z).$$

The equality

$$\frac{q(z) - q(w)}{z - w} = \sum_{j=1}^{n} z^{j-1} e_j(w)$$

follows by symmetry.

2. Let $f \in X_q$ and $f(z) = \sum_{j=1}^{n} f_j e_j(z)$. Then

$$
\begin{aligned}
< K(z, \cdot), f > &= < \sum_{j=1}^{n} w^{j-1} e_j(z), \sum_{k=1}^{n} f_k e_k(w) > \\
&= \sum_{j=1}^{n} \sum_{k=1}^{n} f_k e_j(z) < w^{j-1}, e_k(w) > \\
&= \sum_{j=1}^{n} \sum_{k=1}^{n} f_k e_j(z) \delta_{jk} \\
&= \sum_{k=1}^{n} f_k e_k(z) = f(z).
\end{aligned}
$$

3. Obviously, by the previous part, K acts as the identity, and hence its matrix representation is the identity. This also can be seen by direct computation. □

We say that the polynomial $K(z, w)$ is the **kernel** of the map K defined in Eq. (5.36). A kernel with the properties described in Proposition 5.7.1 is called a **reproducing kernel** for the space X_q.

Given a polynomial $M(z, w) = \sum_{i=1}^{n} \sum_{j=1}^{n} M_{ij} z^{i-1} w^{j-1}$ in the variables z, w, we have an induced quadratic form in X_q. This quadratic form is defined by

$$
\phi(f, f) = < Mf, f > = << M(z, \cdot), f >_w f >_z . \tag{5.38}
$$

Proposition 5.7.2

1. *The matrix representation of the quadratic form $\phi(f, f)$ given by Eq. (5.38) is $[M]_{co}^{st}$.*

2. *The form ϕ is positive definite if and only if $[M]_{co}^{st}$ is a positive definite matrix.*

Proof:

1. M_{ij}, the ij entry of $[M]_{co}^{st}$, is given by

$$M_{ij} \;=\; < Me_j, e_i > = << \sum_{k=1}^{n} \sum_{l=1}^{n} M_{ij} z^{k-1} w^{l-1}, e_j >_w, e_i >_z$$

$$= \; < \sum_{k=1}^{n} \sum_{l=1}^{n} M_{ij} z^{k-1} < w^{l-1}, e_j >_w, e_i >_z$$

$$= \; < \sum_{k=1}^{n} M_{kj} z^{k-1}, e_i >_z$$

$$= \; < \sum_{k=1}^{n} M_{kj} \delta_{ik} = M_{ij}.$$

2. Follows from the equality $\phi(f, f) = ([M]_{co}^{st}[f]^{co}, [f]^{co})$.   □

5.8   Exercises

1. Assume that $q(z) = z^n + q_{n-1} z^{n-1} + \cdots + q_0$ is a real or complex polynomial. Show that the solutions of the linear, homogeneous differential equation

$$y^{(n)} + q_{n-1} y^{(n-1)} + \cdots + q_0 y = 0$$

form an n-dimensional space. Use the Laplace transform $\mathcal{L}$ to show that y is a solution if and only if $\mathcal{L}(y) \in X^q$.

2. Let T be a cyclic transformation with the minimal polynomial $m(z)$ of degree n, and let $p \in F[z]$. Show that the following statements are equivalent:

   (a) The operator $p(T)$ is cyclic.

   (b) There exists a polynomial $q \in F[z]$ such that $(q \circ p)(z) = z \bmod(m)$.

   (c) The map $\Lambda_p : F[z] \longrightarrow X_m$ defined by $\Lambda_p(q) = \pi_m(q \circ p)$ is surjective.

   (d) We have $\det(\pi_{kj}) \neq 0$, where $\pi_k = \pi_m(p^k)$ and $\pi_k(z) = \sum_{j=0}^{n-1} \pi_{kj} z^j$.

3. Assume that the minimal polynomial of a cyclic operator factors into linear factors, that is, $m(z) = \prod(z - \lambda_i)^{\nu_i}$, with the λ_i distinct. Show that $p(T)$ is cyclic if and only if $\lambda_i \neq \lambda_j$ implies $p(\lambda_i) \neq p(\lambda_j)$ and $p'(\lambda_i) \neq 0$ whenever $\nu_i > 1$.

4. Show that if $\pi \in F[z]$ is irreducible and $\deg p$ is a prime number, then $p(S_\pi)$ is either scalar or a cyclic transformation.

5. Let $q(z) = z^n + q_{n-1}z^{n-1} + \cdots + q_0$, and let $C_q^\sharp$ be its companion matrix. Let $f \in X_q$, with $f(z) = f_0 + \cdots + f_{n-1}z^{n-1}$. We put

$$\mathbf{f} = \begin{pmatrix} f_0 \\ \cdot \\ \cdot \\ \cdot \\ f_{n-1} \end{pmatrix} \in F^n.$$

Show that $f(C_q) = (\mathbf{f}, C_q\mathbf{f}, \ldots, C_q^{n-1}\mathbf{f})$.

6. Given a linear transformation A in V, a vector $x \in V$, and a polynomial p, define the **Jacobson chain matrix** by

$$C_m(p, x, A) = (x, Ax, \ldots, A^{r-1}x, p(A)x, p(A)Ax, \ldots, p(A)A^{r-1}x,$$
$$\ldots, p(A)^{m-1}x, p(A)^{m-1}Ax, \ldots, p(A)^{m-1}A^{r-1}x).$$

Show the following:

(a) Let $E = C_m(p, 1, C_{p^m})$. Then E is a solution of $C_{p^m}E = EH(p^m)$.

(b) Every solution X of $C_{p^m}X = XH(p^m)$ is of the form $X = f(C_{p^m})E$ for some polynomial f of degree $< m \deg p$.

7. Let $p = z^m + p_{m-1}z^{m-1} + \cdots + p_0$ and $q(z) = z^n + q_{n-1}z^{n-1} + \cdots + q_0$. Let $H(p, q) = \begin{pmatrix} C_p^\sharp & 0 \\ N & C_q^\sharp \end{pmatrix}$, where N is the $n \times m$ matrix whose only nonzero element is $N_{1m} = 1$. Show that the general solution to the equation $C_{pq}^\sharp X = XH(pq)$ is of the form $X = f(C_{pq}^\sharp)K$, where $\deg f < \deg p + \deg q$ and

$$K = \begin{pmatrix} 1 & & & & p_0 & & & & \\ & \cdot & & & & \cdot & & & \\ & & \cdot & & & & \cdot & & \\ & & & 1 & p_{m-1} & & & \cdot & \\ & & & & 1 & & & & p_0 \\ & & & & & \cdot & & & \cdot \\ & & & & & & \cdot & & \cdot \\ & & & & & & & \cdot & \cdot \\ & & & & & & & & 1 \end{pmatrix}.$$

8. Let $D(z)$ be an $m \times m$ polynomial matrix. Show that $F^m[z]/DF^m[z]$ is a finite-dimensional vector space over F if and only if $D(z)$ is nonsingular.

9. Assume that $D(z)$ is an $m \times m$ nonsingular polynomial matrix. Define a map $\pi_D : F^m[z] \longrightarrow F^m[z]$ by

$$\pi_D f = D\pi_- D^{-1} f.$$

Show that π_D is a projection, $Ker\, \pi_D = DF^m[z]$. Let $X_D = Im\pi_D$. Define a map $S_D : X_D \longrightarrow X_D$ by

$$S_D f = \pi_D z f.$$

Let X_D have the $F[z]$-module structure induced by S_D. Show that $X_D \simeq F^m[z]/DF^m[z]$ as modules.

10. Show that $\alpha \in F$ is an eigenvalue of S_D if and only if $Ker\, D(\alpha) \neq 0$. In this case, eigenvectors are of the form $D(z)\xi/(z - \alpha)$, with $\xi \in Ker\, D(\alpha)$.

11. Show that $M \subset X_D$ is S_D-invariant if and only if $M = D_1 X_{D_2}$ for some factorization $D(z) = D_1(z)D_2(z)$, the factors being nonsingular polynomial matrices.

12. Let C be the circulant matrix

$$C = \mathrm{circ}(c_0, \ldots, c_{n-1}) = \begin{pmatrix} c_0 & c_{n-1} & \cdot & \cdot & c_1 \\ c_1 & c_0 & \cdot & \cdot & \cdot \\ \cdot & \cdot & \cdot & \cdot & \cdot \\ \cdot & \cdot & \cdot & \cdot & c_{n-1} \\ c_{n-1} & \cdot & \cdot & c_1 & c_0 \end{pmatrix}.$$

Show that $\det(C) = \prod_{i=1}^{n}(c_0 + c_1\zeta_i + \cdots + c_{n-1}\zeta_i^{n-1})$, where $1 = \zeta_1, \ldots, \zeta_n$ are the distinct n-th roots of unity, that is, the zeros of $z^n - 1$.

5.9   Notes and Remarks

Shift operators are a cornerstone of modern operator theory. Their interest lies in their universality properties. In fact, every linear transformation T acting in a finite-dimensional vector space is isomorphic to a transformation of the form S_D, with D a nonsingular polynomial matrix. A more general result has been first proved by Rota [1960] in the context of Hilbert spaces. One of the many polynomial matrices corresponding to T is $zI - T$. This has been shown in Proposition 4.5.10. It gives rise to a theory of equivalence for nonsingular polynomial matrices. Two nonsingular polynomial matrices $D, \bar{D}$ are equivalent if $S_D, S_{\bar{D}}$ are similar. It can be shown [Fuhrmann 1976] that $D, \bar{D}$ are equivalent if and only if there exist polynomial matrices $E, \bar{E}$ for which $\bar{E}D = \bar{D}E$ and certain coprimeness conditions are satisfied.

This is a generalization of Proposition 5.1.8 and Theorem 5.1.3. The analytic analog of this result is the commutant lifting theorem [Sarason 1967; Sz.-Nagy and Foias 1970], and a corresponding spectral mapping theorem [Fuhrmann 1968a,b]. We shall return to these topics in Chapter 11.

Theorem 5.1.3 highlights the importance of the Bezout equation. Since the Bezout equation can be obtained via the Euclidean algorithm, this brings up the possibility of a recursive approach to inversion algorithms for structured matrices. This will be explored further in Chapter 8. The Bezout equation reappears over the ring $\mathbf{RH}_+^\infty$ in Chapter 11.

The results and methods presented in this chapter originate in Fuhrmann [1976] and a long series of follow-up articles. In an analytic setting, Nikolskii [1985] is a very detailed study of shift operators.

6

Structure Theory of Linear Transformations

In this chapter we study the structure of linear transformations. Our aim is to understand the structure of a linear transformation in terms of its most elementary components. We do this by representing the transformation by its matrix with respect to particularly appropriate bases. The ultimate goal, which is not always achievable, is to represent a linear transformation in diagonal form.

6.1   Cyclic Transformations

We begin our study of linear transformations in a finite-dimensional linear space by studying a special subclass, namely, the class of cyclic linear transformations. Later we show that every linear transformation is isomorphic to a direct sum of cyclic transformations. Thus, the structure theory will be complete.

We begin with the following:

Definition 6.1.1 *Let U be an n-dimensional linear space over the field F. A map $A : U \longrightarrow U$ is called a* **cyclic transformation** *if there exists a vector $b \in U$ for which $\{b, Ab, \ldots, A^{n-1}b\}$ is a basis for U. Such a vector b is called a* **cyclic vector** *for A.*

Proposition 6.1.1 *Let A be a cyclic linear transformation in an n-dimensional vector space U. Then its characteristic and minimal polynomials coincide.*

Proof: Suppose that $m(z) = z^k + m_{k-1}z^{k-1} + \cdots + m_0$ is the minimal polynomial of A, and assume that $k < n$. Then $A^k = -m_0 I - \cdots - +m_{k-1}A^{k-1}$. This shows that, given an arbitrary vector $x \in U$, we have $A^k x \in L(x, Ax, \ldots, A^{k-1}x)$. By induction it follows that, for any integer p, we have $L(x, Ax, \ldots, A^p x) \subset L(x, Ax, \ldots, A^{k-1}x)$. Since $\dim L(x, Ax, \ldots, A^{k-1}x) \leq k < n$, it follows that A is not cyclic.   □

Later we prove that the coincidence of the characteristic and minimal polynomials implies cyclicity.

Given a linear transformation T in a vector space U, its **commutant** is defined as the set $\{X \in L(U)|XT = TX\}$. The next result is the characterization of the commutant of a cyclic transformation.

Proposition 6.1.2 *Let $T : U \longrightarrow U$ be a cyclic linear transformation. Then $X : U \longrightarrow U$ commutes with T if and only if $X = p(T)$ for some polynomial p.*

Proof: Let $p \in F[z]$. Then, clearly, $Tp(T) = p(T)T$.

Conversely, let X commute with T, where T is cyclic. Let $x \in U$ be a cyclic vector, that is, $x, Tx, \ldots, T^{n-1}x$ form a basis for U. Thus there exist $p_i \in F$ for which

$$Xx = \sum_{i=0}^{n-1} p_i T^i x.$$

This implies that

$$XTx = TXx = T\sum_{i=0}^{n-1} p_i T^i x = \sum_{i=0}^{n-1} p_i T^i Tx,$$

and by induction $XT^k x = \sum_{i=0}^{n-1} p_i T^i T^k x$. Since the $T^k x$ span U, this implies that $X = p(T)$, where $p(z) = \sum_{i=0}^{n-1} p_i z^i$.   □

Next we show that the class of shift transformations S_q is not that special. In fact, every cyclic linear transformation in a finite-dimensional vector space is isomorphic to a unique transformation of this class.

We now approach the study of cyclic transformations from a slightly abstract point of view.

Theorem 6.1.1 *Let V be a finite-dimensional linear space, and let $A : V \longrightarrow V$ be a cyclic transformation. Then there exists a unique monic polynomial m of degree n and a linear transformation $\phi : X_m \longrightarrow V$ for which the following diagram is commutative:*

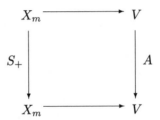

Proof: Let $b \in V$ be a cyclic vector of A. We define a map $\Phi : F[z] \longrightarrow V$ by

$$\Phi \sum_{i=0}^{k} f_i z^i = \sum_{i=0}^{k} f_i A^i b.$$

Since $\{b, Ab, \ldots, A^{n-1}b\}$ is a basis of V, the map Φ is clearly surjective. Moreover, we have

$$\begin{aligned}
\Phi(zf) &= \Phi \sum_{i=0}^{k} f_i z^{i+1} \\
&= \sum_{i=0}^{k} f_i A^{i+1} b = A \sum_{i=0}^{k} f_i A^i b = A\Phi f.
\end{aligned}$$

We now show that $Ker\,\Phi$ is an ideal in $F[z]$. Assume that $f, g \in Ker\,\Phi$; then

$$\Phi(\alpha f + \beta g) = \alpha \Phi(f) + \beta \Phi(g) = 0,$$

so $\alpha f + \beta g \in Ker\,\Phi$. Similarly, if $f \in Ker\,\Phi$, then

$$\Phi(zf) = A\Phi(f) = A0 = 0.$$

So $Ker\,\phi$ is an ideal, in fact, as is easily seen, a nontrivial ideal. Since $F[z]$ is a principal ideal domain, there exists a unique monic polynomial for which $Ker\,\Phi = mF[z]$. Let π be the natural projection of $F[z]$ onto the quotient ring $F[z]/mF[z]$, and let ϕ be the map induced by Φ. Then the following diagram is commutative:

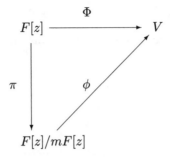

and the map ϕ is invertible. Since we have the isomorphism $X_m \simeq F[z]/mF[z]$, we can identify this diagram with the following one:

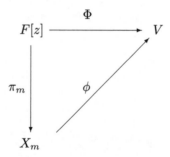

We now pass on to the proof of the commutativity, that is, we have to show that $\phi S_m = A\phi$. Let $f \in X_m$. Then $f = \pi_m f$. Therefore,

$$
\begin{aligned}
A\phi f &= A\phi\pi_m f = A\Phi f = \Phi z f \\
&= \phi\pi_m z f = \phi S_m f.
\end{aligned}
$$
$\square$

From this theorem it follows that the study of an arbitrary cyclic transformation reduces to the study of the class of model transformations of the form S_m.

In this section we exhibit canonical forms for cyclic transformations.

Theorem 6.1.2 *Let $p \in F[z]$, and let*

$$
p(z) = p_1(z)^{m_1} \cdots p_k(z)^{m_k}
$$

be the primary decomposition of p. We define

$$
\pi_i = \prod_{j \neq i} p_j^{m_j}.
$$

Then

$$
X_p = \pi_1(z)X_{p_1^{m_1}} \oplus \cdots \oplus \pi_k(z)X_{p_k^{m_k}} \tag{6.1}
$$

and

$$
S_p \simeq S_{p_1^{m_1}} \oplus \cdots \oplus S_{p_k^{m_k}}. \tag{6.2}
$$

Proof: The direct sum decomposition (6.1) was proved in Theorem 2.6.1. We define a map

$$
Z : X_{p_1^{m_1}} \oplus \cdots \oplus X_{p_k^{m_k}} \longrightarrow X_p = \pi_1(z)X_{p_1^{m_1}} \oplus \cdots \oplus \pi_k(z)X_{p_k^{m_k}}
$$

by

$$
Z(f_1, \ldots, f_k) = \sum_{i=1}^{k} \pi_i f_i.
$$

It is easily checked that the following diagram is commutative:

$$
\begin{array}{ccc}
X_{p_1^{m_1}} \oplus \cdots \oplus X_{p_k^{m_k}} & \xrightarrow{\ \ Z\ \ } & X_p \\
\Big\downarrow S_{p_1^{m_1}} \oplus \cdots \oplus S_{p_k^{m_k}} & & \Big\downarrow S_p \\
X_{p_1^{m_1}} \oplus \cdots \oplus X_{p_k^{m_k}} & \xrightarrow{\ \ Z\ \ } & X_p
\end{array}
\qquad \square
$$

Thus, by choosing appropriate bases in the spaces $X_{p_i^{m_i}}$, we get a block matrix representation for S_p. It suffices to consider the matrix representation of a typical operator S_{p^m}, where p is an irreducible polynomial. This is the content of the next proposition.

Proposition 6.1.3 *Let* $p(z) = z^r + p_{r-1}z^{r-1} + \cdots + p_0$. *Then:*

1. $$\dim X_{p^m} = mr.$$

2. *The set of polynomials*

$$
\mathcal{B}_{jo} = \\
\{1, z, \ldots, z^{r-1}, p, zp, \ldots, z^{r-1}p, \ldots, p^{m-1}, zp^{m-1}, \ldots, z^{r-1}p^{m-1}\}
$$
(6.3)

forms a basis for X_p. *We call this the* **Jordan basis** *of* X_q.

3. *The matrix representation of* S_q *with respect to this basis is of the form*

$$
[S_{p^m}]_{jo}^{jo} = \begin{pmatrix}
C_p^{\sharp} & & & & \\
N & C_p^{\sharp} & & & \\
 & N & \cdot & & \\
 & & \cdot & \cdot & \\
 & & & \cdot & \\
 & & & N & C_p^{\sharp}
\end{pmatrix},
$$
(6.4)

where $C_p^{\sharp}$ *is the companion matrix of* p *defined by Eq. (5.3), and*

$$
N = \begin{pmatrix}
0 & \cdot & \cdot & \cdot & 0 & 1 \\
\cdot & & & & & 0 \\
\cdot & & & & & \cdot \\
\cdot & & & & & \cdot \\
\cdot & & & & & \cdot \\
0 & \cdot & \cdot & \cdot & \cdot & 0
\end{pmatrix}.
$$

Proof:

1. We have
$$
\dim X_{p^m} = \deg p^m = m \deg p = mr.
$$

2. In $\mathcal{B}$ there is one polynomial of each degree and mr polynomials altogether, so necessarily they form a basis.

3. This is a simple computation. Notice however that, for $j < m - 1$,

$$
\begin{aligned}
S_{p^m} z^{r-1} p^j &= \pi_{p^m} z \cdot z^{r-1} p^j = \pi_{p^m} z^r p^j \\
&= \pi_{p^m} (z^r + p_{r-1} z^{r-1} + \cdots \\
&\quad + p_0 - p_{r-1} z^{r-1} - \cdots - p_0) p^j \\
&= \pi_{p^m} p^{j+1} - \sum_{i=0}^{r-1} p_i (z^i p^j).
\end{aligned}
$$

For $j = m - 1$ we have

$$
\begin{aligned}
S_{p^m} z^{r-1} p^{m-1} &= \pi_{p^m} z^r p^{m-1} \\
&= \pi_{p^m} (z^r + p_{r-1} z^{r-1} + \cdots \\
&\quad + p_0 - p_{r-1} z^{r-1} - \cdots - p_0) p^{m-1} \\
&= \pi_{p^m} p^m - \sum_{i=0}^{r-1} p_i (z^i p^{m-1}) \\
&= - \sum_{i=0}^{r-1} p_i (z^i p^{m-1}).
\end{aligned}
$$

$\square$

It is of interest to compute the dual basis to the Jordan basis. This leads to the Jordan form in the block upper triangular form.

Proposition 6.1.4 *Given the monic, degree r, polynomial p, let $\{e_1, \ldots, e_r\}$ be the control basis of X_p. Then:*

1. *The basis of X_{p^m} dual to the Jordan basis of Eq. (6.3) is given by*

$$
\widetilde{\mathcal{B}}_{jo} = \{p^{m-1} e_1, \ldots, p^{m-1} e_r, \ldots, e_1, \ldots, e_r\}.
$$

*Here, $e_1, \ldots, e_r$ are the elements of the control basis of X_p. We call this basis the **dual Jordan basis**.*

2. *The matrix representation of S_{q^m} with respect to the dual Jordan basis is given by*

$$
[S_{q^m}]_{\widetilde{jo}}^{\widetilde{jo}} =
\begin{pmatrix}
C_p^\flat & \tilde{N} & & & \\
 & C_p^\flat & \cdot & & \\
 & & \cdot & \cdot & \\
 & & & \cdot & \cdot \\
 & & & & \cdot & \tilde{N} \\
 & & & & & C_p^\flat
\end{pmatrix}.
\tag{6.5}
$$

3. *The change of basis transformation from the dual Jordan basis to the Jordan basis is given by*

$$R = [I]^{jo}_{\widetilde{jo}} = \begin{pmatrix} & & & K \\ & & \cdot & \\ & \cdot & & \\ K & & & \end{pmatrix},$$

where

$$K = [I]^{st}_{co} = \begin{pmatrix} p_1 & \cdot & \cdot & p_{r-1} & 1 \\ \cdot & & & \cdot & \\ \cdot & & \cdot & & \\ p_{r-1} & & \cdot & & \\ 1 & & & & \end{pmatrix}.$$

4. *The Jordan matrices in Eqs. (6.4) and (6.5) are similar. A similarity is given by R.*

Proof:

1. This is an extension of Proposition 5.6.3. We compute, with $\alpha, \beta < m$ and $i, j < r$,

$$< p^\alpha e_i, p^\beta z^{j-1} > \; = \; [p^{-m} p^\alpha \pi_+ z^{-i} p, p^\beta z^{j-1}]$$

$$= \; [\pi_+ z^{-i} p, p^{\alpha+\beta-m} z^{j-1}].$$

The last expression is clearly zero whenever $\alpha + \beta \neq m - 1$. When $\alpha + \beta = m - 1$, we have

$$[\pi_+ z^{-i} p, p^{\alpha+\beta-m} z^{j-1}] \; = \; [\pi_+ z^{-i} p, p^{-1} z^{j-1}]$$

$$= \; [z^{-i} p, p^{-1} z^{j-1}] = [z^{j-i-1}, 1] = \delta_{ij}.$$

2. Follows either by direct computation or the fact that $S_q^* = S_q$ coupled with an application of Theorem 4.4.3.

3. This results from a simple computation.

4. From $S_{p^m} I = I S_{p^m}$, we get $[S_{p^m}]^{jo}_{jo} [I]^{jo}_{\widetilde{jo}} = [I]^{jo}_{\widetilde{jo}} [S_{p^m}]^{\widetilde{jo}}_{\widetilde{jo}}$. $\qquad\qquad \square$

Note that if the polynomial p is monic of degree 1, that is, $p(z) = z - \alpha$, then with respect to the basis $\mathcal{B} = \{1, z - \alpha, \ldots, (z - \alpha)^{n-1}\}$ the matrix representation has the following form:

$$[S_{(z-\alpha)^n}]_B^B = \begin{pmatrix} \alpha & & & & & \\ 1 & \alpha & & & & \\ & 1 & \cdot & & & \\ & & \cdot & \cdot & & \\ & & & \cdot & \cdot & \\ & & & & 1 & \alpha \end{pmatrix}.$$

If our field F is algebraically closed, as is the field of complex numbers, then every irreducible polynomial is of degree 1. In case we work over the real field, the irreducible monic polynomials are either linear or of the form $(z - \alpha)^2 + \beta^2$. In this case, we make a variation on the canonical form obtained in Proposition 6.1.3. This leads to the **real Jordan form**.

Proposition 6.1.5 *Let p be the real polynomial $p(z) = (z-\alpha)^2+\beta^2$. Then:*

1. *We have*
$$\dim X_{p^n} = 2n.$$

2. *The polynomials $\mathcal{B} = \{\beta, z - \alpha, \beta p, (z - \alpha)p, \ldots, \beta p^{n-1}, (z - \alpha)p^{n-1}\}$ are a basis for X_{p^n}.*

3. *The matrix representation of S_{p^n} with respect to this basis is*

$$[S_{p^n}]_{jo}^{jo} = \begin{pmatrix} A & & & & \\ N & A & & & \\ & N & \cdot & & \\ & & \cdot & \cdot & \\ & & & \cdot & \cdot \\ & & & N & A \end{pmatrix},$$

where

$$A = \begin{pmatrix} \alpha & -\beta \\ \beta & \alpha \end{pmatrix}, \qquad N = \begin{pmatrix} 0 & \beta^{-1} \\ 0 & 0 \end{pmatrix}.$$

Proof: The proof is analogous to that of Proposition 6.1.3. □

6.2   The Invariant Factor Algorithm

The invariant factor algorithm is the tool that allows us to move from the special case of cyclic operators to the general case.

We begin by introducing an equivalence relation in the space $F[z]^{m \times n}$, that is, the space of all $m \times n$ matrices over the ring $F[z]$. We use the identification of $F[z]^{m \times n}$ with $F^{m \times n}[z]$ of matrix polynomials. Thus, given a matrix $A \in F^{n \times n}$, then $zI - A$ is a matrix polynomial of degree one, but usually we also consider it as a polynomial matrix.

Definition 6.2.1 *A matrix* $U \in F[z]^{m \times m}$ *is called* **unimodular** *if it is invertible in* $F[z]^{m \times m}$, *that is, there exists a matrix* $V \in F[z]^{m \times m}$ *for which*

$$U(z)V(z) = V(z)U(z) = I.$$

Lemma 6.2.1 $U \in F[z]^{m \times m}$ *is unimodular if and only if* $\det U(z) = \alpha \in F$ *and* $\alpha \neq 0$.

Proof: Assume that U is unimodular. Then there exist V for which $U(z)$ $V(z) = V(z)U(z) = I$. By the multiplicative rule of the determinants we have $\det U(z) \det V(z) = 1$. This implies that both $\det U$ and $\det V$ are nonzero scalars.

Conversely, if $\det U$ is a nonzero scalar, then, by using Eq. (3.5), we get that U^{-1} is a polynomial matrix. □

Definition 6.2.2 *Two matrices* $A, B \in F[z]^{m \times n}$ *are called* **equivalent** *if there exist unimodular matrices* U *and* V *such that*

$$U(z)A(z) = B(z)V(z) = I.$$

It is easily checked that this definition indeed yields an equivalence relation. The next theorem provides a canonical form for this equivalence. However, we find it convenient to prove first the following lemma:

Lemma 6.2.2 *Let* $A \in F[z]^{m \times n}$. *Then* A *is equivalent to a block matrix of the form*

$$\begin{pmatrix} a & 0 & . & . & . & 0 \\ . & & & & & \\ . & & & B & & \\ . & & & & & \\ 0 & & & & & \end{pmatrix}$$

with $a|b_{ij}$.

Proof: If A is the zero matrix, there is nothing to prove. Otherwise, interchanging rows and columns as necessary, we can bring the nonzero entry of least degree to the upper left-hand corner. We use the division rule of polynomials to write each element of the first row in the form $a_{1j} = c_{1j}a_{11} + a'_{1j}$, with $\deg a'_{1j} < \deg a_{11}$. Next we subtract the first column, multiplied by c_{1j}, from the jth column. We repeat the process with the first column. If all of the remainders are zero, we check whether a_{11} divides all other entries in the matrix. If it does, we are through. Otherwise, using column and row interchanges, we bring a lowest degree nonzero element to the upper left-hand corner and repeat the process. □

Theorem 6.2.1 (The invariant factor algorithm) *Let* $A \in F[z]^{m \times n}$. *Then:*

1. *A is equivalent to a diagonal polynomial matrix with the diagonal elements d_i monic polynomials satisfying $d_i | d_{i-1}$.*

2. *The polynomials d_i are uniquely determined and are called the* **invariant factors** *of A.*

Proof: We use the previous lemma inductively. Thus, by elementary operations, A is reducible to diagonal form with the diagonal entries satisfying $d_i | d_{i+1}$. This ordering is opposite from that in the statement of the theorem. We can get the right ordering by column and row interchanges.   □

Proposition 6.2.1 *A polynomial matrix U is unimodular if and only if it is the product of a finite number of elementary unimodular matrices.*

Proof: The product of unimodular matrices is, by the multiplicative rule for determinants, also unimodular. Conversely, assume that U is unimodular. By elementary row and column operations it can be brought to its Smith form D with the polynomials d_i on the diagonal. Since D is also unimodular, $D(z) = I$. Now let U_i and V_j be the elementary unimodular matrices representing the elementary operations in the diagonalization process. Thus, $U_k \cdots U_1 U V_1 \cdots V_l = I$, which implies $U = U_1^{-1} \cdots U_k^{-1} V_l^{-1} \cdots V_1^{-1}$. This is a representation of U as the product of elementary unimodular matrices.   □

To prove the uniqueness of the invariant factors, we introduce the determinantal divisors.

Definition 6.2.3 *Let $A \in F[z]^{m \times n}$. Then we define $D_0(A) = 1$ and $D_k(A)$ to be the g.c.d., taken to be monic, of all $k \times k$ minors of A. The $D_i(A)$ are called the* **determinantal divisors** *of A.*

Proposition 6.2.2 *Let $A, B \in F[z]^{m \times n}$. If B is obtained from A by an elementary row or column operation, then $D_k(A) = D_k(B)$.*

Proof: We consider the set of all $k \times k$ minors of A, and we consider the effect of applying elementary transformations on A on this set. Multiplication of a row or column by a nonzero scalar α can change a minor by at most a factor α. This does not change the g.c.d. Interchanging two rows or columns leaves the set of minors unchanged, except for a possible sign. The only elementary operation that has a nontrivial effect is adding the jth row, multiplied by a polynomial p to the ith row. In a minor that does not contain the ith row, or that contains elements of both the ith and jth rows, this has no effect. If a minor contains elements of the ith row but not of the jth, we expand by the ith row. We get the original minor added to the product of another minor by the polynomial p. Again this does not change the g.c.d.   □

Theorem 6.2.2 *Let $A, B \in F[z]^{m \times n}$. Then A and B are equivalent if and only if $D_k(A) = D_k(B)$.*

Proof: It is clear from the previous proposition that if a single elementary operation does not change the set of elementary divisors, then a finite sequence of elementary operation also leaves the determinantal divisors unchanged. As any unimodular matrix is the product of elementary unimodular matrices, we conclude that the equivalence of A and B implies $D_k(A) = D_k(B)$.

Assume now that $D_k(A) = D_k(B)$ for all k. Using the invariant factor algorithm, we reduce A and B to their Smith form with the invariant factors $d_1, \ldots, d_r$, ordered by $d_{i-1}|d_i$, and $e_1, \ldots, e_s$, respectively. Clearly, $D_k(A) = d_1 \cdots d_k$ and $D_k(B) = e_1 \cdots e_k$. This implies that $r = s$ and, assuming that the invariant factors are all monic, $e_i = d_i$. By transitivity of equivalence we get $A \simeq B$.   □

Corollary 6.2.1 *The invariant factors of $A \in F[z]^{m \times n}$ are uniquely determined.*

We can view the invariant factor algorithm as a far-reaching generalization of the Euclidean algorithm, for the Euclidean algorithm is the invariant factor algorithm as applied to a polynomial matrix $(p(z) \quad q(z))$.

6.3 Noncyclic Transformations

We continue now with the study of general, that is, not necessarily cyclic, transformations in finite-dimensional vector spaces. Our aim is to reduce their study to that of the cyclic ones, and we will do it using the invariant factor algorithm. However, to effectively use it, we will have to extend the modular polynomial arithmetic to the case of vector polynomials. We will not treat the case of remainders with respect to general matrix polynomials, since for our purposes it will suffice to consider two special cases.

There is a similar map we need in the following. Assume that

$$D = \begin{pmatrix} d_1 & & & \\ & \cdot & & \\ & & \cdot & \\ & & & \cdot & \\ & & & & d_n \end{pmatrix},$$

with d_i nonzero polynomials. Given a vector polynomial

$$f = \begin{pmatrix} f_1 \\ \cdot \\ \cdot \\ \cdot \\ f_n \end{pmatrix},$$

we define

$$\pi_D f = \begin{pmatrix} \pi_{d_1} f_1 \\ \cdot \\ \cdot \\ \cdot \\ \pi_{d_n} f_n \end{pmatrix},$$

where $\pi_{d_i} f_i$ are the remainders of f_i after division by d_i.

Proposition 6.3.1 *Given a nonsingular, diagonal polynomial matrix D. Then:*

1. *We have*
$$Ker\, \pi_D = DF^n[z].$$

2.
$$X_D = Im\pi_D = \left\{ \begin{pmatrix} f_1 \\ \cdot \\ \cdot \\ \cdot \\ f_n \end{pmatrix} \mid f_i \in Im\pi_{d_i} \right\}.$$

   In particular, $X_D \simeq X_{d_1} \oplus \cdots \oplus X_{d_n}$.

3. *Defining the map $S_D : X_D \longrightarrow X_D$ by $S_D f = \pi_D z f$ for $f \in X_D$, we have $S_D \simeq S_{d_1} \oplus \cdots \oplus S_{d_n}$.*

Proof:

1. Clearly, $f \in Ker\, \pi_D$ if and only if, for all i, f_i is divisible by d_i. Thus,
$$Ker\, \pi_D = \left\{ \begin{pmatrix} d_1 f_1 \\ \cdot \\ \cdot \\ \cdot \\ d_n f_n \end{pmatrix} \mid f_i \in F[z] \right\} = DF^n[z].$$

2. Follows from the definition of π_D.

3. We compute
$$S_D \begin{pmatrix} f_1 \\ \cdot \\ \cdot \\ \cdot \\ f_n \end{pmatrix} = \pi_D z \begin{pmatrix} f_1 \\ \cdot \\ \cdot \\ \cdot \\ f_n \end{pmatrix} = \begin{pmatrix} \pi_{d_1} f_1 \\ \cdot \\ \cdot \\ \cdot \\ \pi_{d_n} f_n \end{pmatrix} = \begin{pmatrix} S_{d_1} f_1 \\ \cdot \\ \cdot \\ \cdot \\ S_{d_n} f_n \end{pmatrix}.$$

$\square$

With this background material we can proceed to the following. This is the structure theorem we have been after. It shows that an arbitrary linear transformation is isomorphic to a direct sum of cyclic transformations corresponding to the invariant factors of A.

Theorem 6.3.1 *Let A be a linear transformation in F^n. Let $d_1, \ldots, d_n$ be the invariant factors of $zI - A$. Then A is isomorphic to $S_{d_1} \oplus \cdots \oplus S_{d_n}$.*

Proof: Since $d_1, \ldots, d_n$ are the invariant factors of $zI - A$, there exist unimodular polynomial matrices U and V satisfying

$$U(z)(zI - A) = D(z)V(z). \tag{6.6}$$

This equation implies that

$$U \, Ker \, \pi_{zI-A} \subset Ker \, \pi_D. \tag{6.7}$$

Since U and V are invertible as polynomial matrices, we also have

$$U(z)^{-1}D(z) = (zI - A)^{-1}V(z), \tag{6.8}$$

which implies

$$U^{-1} Ker \, \pi_D \subset Ker \, \pi_{zI-A}. \tag{6.9}$$

We define now two maps, $\Phi : F^n \longrightarrow X_D$ and $\Psi : X_D \longrightarrow F^n$, by

$$\Phi x = \pi_D U x, \qquad x \in F^n, \tag{6.10}$$

and

$$\Psi f = \pi_{zI-A} U^{-1} f, \qquad f \in X_D. \tag{6.11}$$

We now claim that Φ is invertible and its inverse is Ψ. Indeed, using Eqs. (6.10) and (6.11), we compute

$$
\begin{aligned}
\Psi \Phi x &= \pi_{zI-A} U^{-1} \pi_D U x = \pi_{zI-A} U^{-1} U x \\
&= \pi_{zI-A} x = x.
\end{aligned}
$$

Similarly, for $f \in X_D$,

$$\Phi \Psi f = \pi_D U \pi_{zI-A} U^{-1} f = \pi_D U U^{-1} f = \pi_D f = f.$$

Next we show that $S_D \Phi = \Phi A$.

Indeed, for $x \in F^n$, using Eq. (6.7) and the fact that $z Ker \, \pi_D \subset \pi_D$, we have

$$
\begin{aligned}
S_D \Phi x &= \pi_D z \cdot \pi_D U f = \pi_D z U f \\
&= \pi_D U z f = \pi_D U \pi_{zI-A} z f = \pi_D U(Af) = \Phi Af.
\end{aligned}
$$

Since Φ is an isomorphism, we have $A \simeq S_D$. But $S_D \simeq S_{d_1} \oplus \cdots \oplus S_{d_n}$, and the result follows. $\qquad \square$

Corollary 6.3.1 *Let A be a linear transformation in an n-dimensional vector space V. Let $d_1, \ldots, d_n$ be the invariant factors of A. Then A is similar to the diagonal block matrix composed of the companion matrices corresponding to the invariant factors, that is,*

$$A \simeq \begin{pmatrix} C_{d_1}^\sharp & & & \\ & \cdot & & \\ & & \cdot & \\ & & & \cdot \\ & & & & C_{d_n}^\sharp \end{pmatrix}. \tag{6.12}$$

Proof: By Theorem 6.3.1, A is isomorphic to $S_{d_1} \oplus \cdots \oplus S_{d_n}$ acting in $X_{d_1} \oplus \cdots \oplus X_{d_n}$. Choosing the standard basis in each X_{d_i} and taking the union of these basis elements, we have

$$[S_{d_1} \oplus \cdots \oplus S_{d_n}]_{st}^{st} = \text{diag}\,([S_{d_1}]_{st}^{st}, \ldots, [S_{d_n}]_{st}^{st}).$$

We conclude by recalling that $[S_{d_i}]_{st}^{st} = C_{d_i}^\sharp$.   □

Theorem 6.3.2 *Let A and B be linear transformations in n-dimensional vector spaces U and V, respectively. Then the following statements are equivalent:*

1. A and B are similar.

2. The polynomial matrices $zI - A$ and $zI - B$ are equivalent.

3. The invariant factors of A and B coincide.

Proof:

1. Assume that A and B are similar. Let X be an invertible matrix for which $XA = BX$. This implies that $X(zI - A) = (zI - B)X$ and therefore the equivalence of $zI - A$ and $zI - B$.

2. The equivalence of the polynomial matrices $zI - A$ and $zI - B$ shows that their invariant factors coincide.

3. Assuming that the invariant factors coincide, both A and B are similar to $S_{d_1} \oplus \cdots \oplus S_{d_n}$, and the claim follows from the transitivity of similarity.   □

Corollary 6.3.2 *Let A be a linear transformation in an n-dimensional vector space U over F. Let $d_1, \ldots, d_n$ be the invariant factors of A, ordered so that $d_i | d_{i-1}$. Then:*

1. The minimal polynomial of A, m_A, satisfies

$$m_A(z) = d_1(z).$$

2. d_A, the characteristic polynomial of A, satisfies

$$d_A(z) = \prod_{i=1}^{n} d_i(z). \tag{6.13}$$

Proof:

1. Since $A \simeq S_{d_1} \oplus \cdots \oplus S_{d_n}$, for an arbitrary polynomial p, we have $p(A) \simeq p(S_{d_1}) \oplus \cdots \oplus p(S_{d_n})$. So $p(A) = 0$ if and only if $p(S_{d_i}) = 0$ for all i or, equivalently, $d_i|p$. In particular, $d_1|p$. But d_1 is the minimal polynomial of S_{d_1}, so it is also the minimal polynomial of $S_{d_1} \oplus \cdots \oplus S_{d_n}$ and, by similarity, also of A.

2. Using Proposition 6.3.1 and the matrix representation (6.12), Eq. (6.13) follows. □

Next we give a characterization of cyclicity in terms of the characteristic and minimal polynomials.

Proposition 6.3.2 *A transformation T in U is cyclic if and only if its characteristic and minimal polynomials coincide.*

Proof: It is obvious that if T is cyclic, its characteristic and minimal polynomials have to coincide.

Conversely, if the characteristic and minimal polynomials coincide, then the only nontrivial invariant factor d is equal to the characteristic polynomial. By Theorem 6.3.1 we have $T \simeq S_d$, and, as S_d is cyclic, so is T.
□

6.4 Diagonalization

The maximal simplification one can hope to obtain for the structure of an arbitrary linear transformation in a finite-dimensional vector space is diagonalizability. Unhappily, that is too much to ask. The following is a complete characterization of diagonalizability:

Theorem 6.4.1 *Let T be a linear transformation in a finite-dimensional vector space U. Then T is diagonalizable if and only if m_T, the minimal polynomial of T, splits into distinct linear factors.*

Proof: This follows from Proposition 5.1.7 taken in conjunction with Theorem 6.3.1. However, we find it of interest to give a direct proof.

That the existence of a diagonal representation implies that the minimal polynomial of T splits into distinct linear factors is obvious.

Assume now that the minimal polynomial splits into distinct linear factors. So $m_T(z) = \Pi_{i=1}^{k}(z - \lambda_i)$, where the λ_i are distinct. Let π_i be the corresponding Lagrange interpolation polynomials. Clearly, $(z - \lambda_i)\pi_i(z) = m_T(z)$. Since, for $i \neq j$, $z - \lambda_i | \pi_j(z)$, we get

$$m_T | \pi_i \pi_j. \tag{6.14}$$

Similarly, the polynomial $\pi_i^2 - \pi_i$ vanishes at all the points $\lambda_i, i = 1, \ldots, k$. So,

$$m_T | \pi_i^2 - \pi_i. \tag{6.15}$$

The two division relations now imply that

$$\pi_i(T)\pi_j(T) = 0, \qquad i \neq j, \tag{6.16}$$

and

$$\pi_i(T)^2 = \pi_i(T). \tag{6.17}$$

So the $\pi(T)$ are projections on independent subspaces. Set $U_i = \mathrm{Im}\,\pi_i(T)$. Then, since the equality $1 = \sum_{i=1}^{k} \pi_i(z)$ implies $I = \sum_{i=1}^{k} \pi_i(T)$, we get $U = U_1 \oplus \cdots \oplus U_k$. Moreover, as $T\pi_i(T) = \pi_i(T)T$, the subspaces U_i are T-invariant. From the equality $(z - \lambda_i)\pi_i(z) = m_T(z)$ we get $(T - \lambda_i I)\pi_i(T) = 0$, or $U_i \subset Ker(\lambda_i I - T)$. This can be restated as $T|U_i = \lambda_i I_{U_i}$. Choosing bases in the subspaces U_i and taking their union leads to a basis of U made up of eigenvectors. With respect to this basis, T has a diagonal matrix representation. $\qquad \square$

The following is the basic structure theorem that yields a spectral decomposition for a linear transformation. Again, the result follows from our previous considerations and can be raed off from the Jordan form. Still we find it of interest to also give another proof.

Theorem 6.4.2 *Let T be a linear transformation in a finite-dimensional vector space U. Let m_T be the minimal polynomial of T and let $m_T = p_1^{\nu_1} \cdots p_k^{\nu_k}$ be the primary decomposition of m_T. Set $U_i = Ker\, p_i(T)^{\nu_i}, i = 1, \ldots, k$, and define polynomials by $\pi_i(z) = \Pi_{j \neq i} p_i^{\nu_i}(z)$. Then:*

1. *There exists a unique representation*

$$1 = \sum_{j=1}^{k} \pi_j(z)a_j(z), \tag{6.18}$$

   with $a_j \in X_{p_j^{\nu_j}}$.

2. *We have*

$$m_T | \pi_i \pi_j, \quad for \ i \neq j, \tag{6.19}$$

   and

$$m_T | \pi_i^2 a_i^2 - \pi_i a_i. \tag{6.20}$$

3. *The operators $P_i = \pi_i(T)a_i(T)$, $i = 1, \ldots, k$, are projection operators satisfying*

$$\left\{ \begin{array}{rcl} P_i P_j &=& \delta_{ij} P_j \\[2mm] I &=& \displaystyle\sum_{i=1}^{k} P_i \\[2mm] T P_i &=& P_i T P_i \\[2mm] \mathrm{Im} P_i &=& \mathrm{Ker}\, p_i^{\nu_i}(T). \end{array} \right.$$

4. *Define $U_i = \mathrm{Im} P_i$. Then $U = U_1 \oplus \cdots \oplus U_k$.*

5. *The subspaces U_i are T-invariant.*

6. *Let $T_i = T|U_i$. Then the minimal polynomial of T_i is $p_i^{\nu_i}$.*

Proof:

1. Clearly, the π_i are relatively prime and $p_i^{\nu_i} \pi_i = m_T$. By the Chinese remainder theorem, there exists a unique representation $1 = \sum_{j=1}^{k} \pi_j(z)a_j(z)$, with $a_j \in X_{p_j^{\nu_j}}$. The a_j can be computed as follows. Since for $i \neq j$ we have $\pi_{p_i^{\nu_i}} \pi_j f = 0$, we have

$$1 = \pi_{p_i^{\nu_i}} \sum_{j=1}^{k} \pi_j a_j = \pi_{p_i^{\nu_i}} \pi_i a_i = \pi_i(S_{p_i^{\nu_i}})a_i,$$

and hence $a_i = \pi_i(S_{p_i^{\nu_i}})^{-1} 1$. That the operator $\pi_i(S_{p_i^{\nu_i}}) : X_{p_i^{\nu_i}} \longrightarrow X_{p_i^{\nu_i}}$ is invertible follows, by Theorem 5.1.3, from the coprimeness of π_i and $p_i^{\nu_i}$.

2. To see Eq. (6.19) note that, if $j \neq i$, $p_i^{\nu_i} | \pi_j$, and hence m_T is a factor of $\pi_i \pi_j$. Naturally, we also get $m_T | \pi_i a_i \pi_j a_j$.

   To see Eq. (6.20), it suffices to show that, for all j, $p_i^{\nu_i} | \pi_j^2 a_j^2 - \pi_j a_j$ or, equivalently, that $\pi_j^2 a_j^2$ and $\pi_j a_j$ have the same remainder after division by $p_i^{\nu_i}$. For $j \neq i$ this is trivial, as $p_i^{\nu_i} | \pi_j$. So we assume that $j = i$ and compute

$$\pi_{p_i^{\nu_i}}(\pi_i^2 a_i^2) = \pi_{p_i^{\nu_i}}(\pi_i a_i)\pi_{p_i^{\nu_i}} \pi_i a_i = 1,$$

for $\pi_{p_i^{\nu_i}} \pi_i a_i = \pi_i(S_{p_i^{\nu_i}})a_i = 1$.

3. Properties (6.19) and (6.20) now have the following consequences. For $i \neq j$, we have $\pi_i(T)a_i(T)\pi_j(T)a_j(T) = 0$ and $(\pi_i(T)a_i(T))^2 = \pi_i(T)a_i(T)$. This implies that $P_i = \pi_i(T)a_i(T)$ are commuting projections on independent subspaces.

4. Now set $U_i = \mathrm{Im} P_i = \mathrm{Im} \pi_i(T)a_i(T)$. Equation (6.18) implies that $I = \sum_{i=1}^{k} P_i$, and hence we get the direct sum decomposition $U = U_1 \oplus \cdots \oplus U_k$.

5. Since P_i commutes with T, the subspaces U_i actually reduce T.

6. Let $T_i = T|U_i$. Since $p_i^{\nu_i}\pi_i = m_T$, it follows that, for all $x \in U$,

$$p_i^{\nu_i}(T)P_ix = p_i^{\nu_i}(T)\pi_i(T)a_i(T)x = 0,$$

that is, the minimal polynomial of T_i divides $p_i^{\nu_i}$. To see the converse, let q be any polynomial such that $q(T_i) = 0$. Then $q(T)\pi_i(T) = 0$. It follows that $q\pi_i$ is divisible by the minimal polynomial of T. But that means that $p_i^{\nu_i}\pi_i|q\pi_i$ or $p_i^{\nu_i}|q$. We conclude that the minimal polynomial of T_i is $p_i^{\nu_i}$.   □

6.5 Exercises

1. Let $T(z)$ be an $n \times n$ nonsingular polynomial matrix. Show that there exists a unimodular matrix P such that PT has the upper triangular form

$$\begin{pmatrix} t_{11} & . & . & . & t_{1n} \\ 0 & . & . & . & . \\ . & . & . & . & . \\ . & . & . & . & . \\ 0 & . & . & 0 & t_{nn} \end{pmatrix},$$

with $t_{ii} \neq 0$ and $\deg(t_{ji}) < \deg(t_{ii})$.

2. Let T be a linear operator on F^2. Prove that any nonzero vector which is not an eigenvector of T is a cyclic vector for T. Prove that T is either cyclic or a multiple of the identity.

3. Let T be a diagonalizable operator in an n-dimensional vector space. Show that if T has a cyclic vector, then T has n distinct eigenvalues. If T has n distinct eigenvalues, and if $x_1, \ldots, x_n$ is a basis of eigenvectors, show that $x = x_1 + \cdots + x_n$ is a cyclic vector for T.

4. Let T be a linear transformation in the finite-dimensional vector space V. Prove that $\mathrm{Im}T$ has a complementary invariant subspace if and only if $\mathrm{Im}T$ and $\mathrm{Ker}\,T$ are independent subspaces. Show that if $\mathrm{Im}T$ and $\mathrm{Ker}\,T$ are independent subspaces, then $\mathrm{Ker}\,T$ is the unique complementary invariant subspace for $\mathrm{Im}T$.

5. How many possible Jordan forms are there for a 6×6 complex matrix with characteristic polynomial $(z+2)^4(z-1)^2$?

6. Given a linear transformation A in an n-dimensional space V, let d and m be its characteristic and minimal polynomials, respectively. Show that d divides m^n.

7. Let $p(z) = p_0 + p_1 z + \cdots + p_{k-1} z^{k-1}$. Solve the equation

$$\pi_{z^k} q p = 1.$$

8. Let $A = \begin{pmatrix} a & c \\ b & d \end{pmatrix}$ be a complex matrix. Apply the Gram–Schmidt process to the columns of A. Show that this results in a basis for $\mathbf{C}^2$ if and only if $\det A \neq 0$.

9. Let A be an $n \times n$ matrix over the field F. We say that A is **diagonalizable** if there exists a matrix representation that is diagonal. Show that *diagonalizable* and *cyclic* are independent concepts, that is, show the existence of matrices that are $DC, \overline{D}C, D\overline{C}, \overline{D}\,\overline{C}$. (Here $\overline{D}C$ means nondiagonalizable and cyclic.)

10. Let A be an $n \times n$ Jordan block matrix, that is,

$$A = \begin{pmatrix} \lambda & 1 & & & \\ & \cdot & \cdot & & \\ & & \cdot & \cdot & \\ & & & \cdot & 1 \\ & & & & \lambda \end{pmatrix}$$

and

$$N = \begin{pmatrix} 0 & 1 & & & \\ & \cdot & \cdot & & \\ & & \cdot & \cdot & \\ & & & \cdot & 1 \\ & & & & 0 \end{pmatrix},$$

that is, $A = \lambda I + N$. Let p be a polynomial of degree r. Show that

$$p(A) = \sum_{k=0}^{n-1} \frac{1}{k!} p^{(k)}(\lambda) N^k.$$

11. Show that if A is diagonalizable and $p \in F[z]$, then $p(A)$ is diagonalizable.

12. Let $p \in F[z]$ be of degree n.

   (a) Show that the dimension of the smallest invariant subspace of X_p that contains f is $n - r$ where $r = \deg(f \wedge p)$.

   (b) Conclude that f is a cyclic vector of S_p if and only if f and p are coprime.

13. Let $A : V \longrightarrow V$ and let $A_1 : V_1 \longrightarrow V_1$ be linear transformations with minimal polynomials m and m_1, respectively.

(a) Show that if there exists a surjective (onto) map $Z : V \longrightarrow V_1$ satisfying $ZA = A_1 Z$, then $m_1 | m$. (Is the converse true? Under what conditions?)

(b) Show that if there exists an injective (1-1) map $Z : V_1 \longrightarrow V$ satisfying $ZA_1 = AZ$, then $m_1 | m$. (Is the converse true? Under what conditions?)

(c) Show that the same is true replacing the minimal polynomials by the characteristic polynomials.

14. Let $A : V \longrightarrow V$ and $M \subset V$ be an invariant subspace of A. Show that the minimal and characteristic polynomials of $A|M$ divide the minimal and characteristic polynomials of A, respectively.

15. Let $A : V \longrightarrow V$ and $M \subset V$ be an invariant subspace of A. Let $d_1, \ldots, d_n$ be the invariant factors of A and $e_1, \ldots, e_m$ be the invariant factors of $A|M$.

(a) Show that $e_i | d_i$ for $i = 1, \ldots, m$.

(b) Show that, given arbitrary vectors $v_1, \ldots, v_k$ in V, we have

$$\dim \operatorname{span} \{A^i v_j | j = 1, \ldots, k, \ \ i = 0, \ldots, n - 1\} \leq \sum_{j=1}^{k} \deg d_j.$$

16. Let $A : V \longrightarrow V$ and assume that

$$V = \operatorname{span} \{A^i v_j | j = 1, \ldots, k, \ \ i = 0, \ldots, n - 1\}.$$

Show that the number of nontrivial invariant factors of A is $\leq k$.

17. Let $A(z)$ and $B(z)$ be polynomial matrices and let $C(z) = A(z)B(z)$. Let a_i, b_i, c_i be the invariant factors of A, B, C, respectively. Show that $a_i | c_i$ and $b_i | c_i$.

18. Let A and B be cyclic linear transformations. Show that if $ZA = BZ$ and Z has rank r, then the degree of the greatest common divisor of the characteristic polynomials of A and B is at least r.

19. Show that if the minimal polynomials of A_1 and A_2 are coprime, then the only solution to $ZA_1 = A_2 Z$ is the zero solution.

20. If there exists a rank r solution to $ZA_1 = A_2 Z$, what can you say about the characteristic polynomials of A_1 and A_2?

21. Let V be an n-dimensional linear space. Assume that $T : V \longrightarrow V$ is diagonalizable. Show that:

(a) If T is cyclic, then T has n distinct eigenvalues.

(b) If T has n distinct eigenvalues, then T is cyclic.

(c) If $b_1, \ldots, b_n$ are the eigenvectors corresponding to the distinct eigenvalues, then $b = b_1 + \cdots + b_n$ is a cyclic vector for T.

22. Show that if T^2 is cyclic, then T is cyclic. Is the converse true?

23. Let V be an n-dimensional linear space over the field F. Show that every nonzero vector in V is a cyclic vector of T if and only if the characteristic polynomial of T is irreducible over F.

24. Show that if m is the minimal polynomial of a linear transformation $A : V \longrightarrow V$, then there exists a vector $x \in V$ such that

$$\{p \in F[z] \mid p(A)x = 0\} = mF[z].$$

25. Let $A : V \longrightarrow V$ be a linear transformation. Show that:

   (a) For each x there exists a smallest A-invariant subspace M_x containing x.

   (b) x is a cyclic vector for $A|M_x$.

26. Show that if A_1, A_2 are cyclic maps with cyclic vectors b_1, b_2, respectively, and if the minimal polynomial of A_1 and A_2 are coprime, then $b_1 \oplus b_2$ is a cyclic vector for $A_1 \oplus A_2$.

27. Let A be a cyclic transformation in V and let $M \subset V$ be an invariant subspace of T. Show that M has a complementary invariant subspace if and only if the characteristic polynomials of $T|_M$ and $T_{V/M}$ are coprime.

28. Show that, if $T^2 - T = 0$, then T is similar to a matrix of the form $\operatorname{diag}(1, \ldots, 1, 0, \ldots, 0)$.

29. The following series of exercises provides an alternative approach to the structure theory of a linear transformation, which avoids the use of determinants. It is based on Axler [1995], although we do not assume the field to be algebraically closed.

   Let T be a linear transformation in a finite-dimensional vector space X over the field F. A nonzero polynomial $p \in F[z]$ is called a **prime** of T if p is monic, irreducible, and there exists a nonzero vector $x \in X$ for which $p(T)x = 0$. Such a vector x is called a p-null vector. Clearly, these notions generalize eigenvalues and eigenvectors.

   (a) Show that every linear transformation T in a finite-dimensional vector space X has at least one prime.

   (b) Let p be a prime of T and let x be a p-null vector. If q is any polynomial for which $q(T)x = 0$, then $p|q$. In particular, if $\deg q < \deg p$ and $q(T)x = 0$, then necessarily $q = 0$.

(c) Show that nonzero p_i-null vectors corresponding to distinct primes are linearly independent.

(d) Let p be a prime of T with $\deg p = \nu$. Show that $\{x | p(T)^k x = 0, \text{ for some } k = 1, 2, \ldots\} = \{x | p(T)^{[n/\nu]} x = 0\}$. We call elements of $Ker\, p(T)^{[n/\nu]}$ generalized p-null vectors.

(e) Show that X is spanned by generalized p-null vectors.

(f) Show that generalized p-null vectors corresponding to distinct primes are linearly independent.

(g) Let $p_1, \ldots, p_m$ be the distinct primes of T, and let $X_i = Ker\, p_i(T)^{[n/\nu_i]}$. Show that

  i. X has the direct sum representation $X = X_1 \oplus \cdots \oplus X_m$.

  ii. X_i are T-invariant subspaces.

  iii. The operators $p_i(T)|_{X_i}$ are nilpotent.

  iv. The operator $T|_{X_i}$ has only one prime, namely, p_i.

30. Given a nilpotent operator N, show that it is similar to a matrix of the form

$$\begin{pmatrix} N_1 & & & \\ & \cdot & & \\ & & \cdot & \\ & & & N_k \end{pmatrix},$$

with N_i of the form

$$\begin{pmatrix} 0 & 1 & & \\ & \cdot & \cdot & \\ & & \cdot & 1 \\ & & & 0 \end{pmatrix}.$$

31. Let T be a linear operator in an n-dimensional complex vector space X. Let $\lambda_1, \ldots, \lambda_m$ be the distinct eigenvalues of T, and let $m(z) = \prod_{i=1}^{m} (z - \lambda_i)^{\nu_i}$ be its minimal polynomial. Let $H_i(z)$ be the solution to the Hermite interpolation problem $H_i^{(k)}(\lambda_j) = \delta_{0\,k} \delta_{ij}$, $i, j = 1, \ldots, m$, $k = 0, \ldots, \nu_j - 1$. Define $P_k = H_k(T)$. Show that for any polynomial p, we have

$$p(T) = \sum_{k=1}^{m} \sum_{j=0}^{\nu_k - 1} \frac{p^{(j)}(\lambda_k)}{j!} (T - \lambda_k I)^j P_k.$$

32. Let T be a linear operator in an n-dimensional vector space X. Let d_i be the invariant factors of T and let $\delta_i = \deg d_i$. Show that the commutant of T has dimension equal to $\sum_{i=1}^{n} (2i - 1)\delta_i$. Conclude that the commutant of T is $L(X)$ if and only if T is scalar, that is, $T = \alpha I$ for some $\alpha \in F$. Also show that the dimension of the commutant of T is equal to n if and only if T is cyclic.

6.6   Notes and Remarks

The study of a linear transformation on a vector space via the study of the polynomial module structure induced by it on that space appears already in Van der Waerden [1931]. Although it is very natural, it did not become the standard approach in the literature, most notably in books aimed at a broader mathematical audience, probably due to the perception that the study of modules is too abstract.

Our approach, based on polynomial and rational models and shift operators, is a concretization of that approach. It was originated in Fuhrmann [1976], which in turn was based on ideas stemming from the theory of operators on Hilbert space. In this connection the reader is advised to consult Sz.-Nagy and Foias [1970] and Fuhrmann [1981a]. The passage from the cyclic case to the noncyclic case uses a special case of general polynomial model techniques. This is indicated in exercises 5.7 to 5.10.

Section 5.2 on circulant matrices is based on results taken from Davis [1979].

7

Inner Product Spaces

In this chapter we focus on the study of linear spaces and linear transformations that relate to notions of distance, angle, and orthogonality. We restrict ourselves throughout to the case of the real field $\mathbf{R}$ or the complex field $\mathbf{C}$.

7.1   Geometry of Inner Product Spaces

Definition 7.1.1 *Let U be a vector space over $\mathbf{C}$ (or $\mathbf{R}$). An **inner product** on U is a function $(\cdot, \cdot) : U \times U \longrightarrow U$ that satisfies the following conditions:*

1. *$(x, x) \geq 0$ and $(x, x) = 0$ if and only if $x = 0$.*

2. *For $\alpha_1, \alpha_2 \in \mathbf{C}$ and $x_1, x_2, y \in U$, we have*

$$(\alpha_1 x_1 + \alpha_2 x_2, y) = \alpha_1(x_1, y) + \alpha_2(x_2, y).$$

3. *We have*
$$(x, y) = \overline{(y, x)}.$$

We note that these axioms imply the antilinearity of the inner product in the right variable, that is,

$$(x, \alpha_1 y_1 + \alpha_2 y_2) = \overline{\alpha_1}(x, y_1) + \overline{\alpha_2}(x, y_2).$$

A function $\phi(x, y)$ that is linear in x and antilinear in y and satisfies $\phi(x, y) = \overline{\phi(y, x)}$ is called a **Hermitian form**. Thus, the inner product is a Hermitian form in U.

We define the **norm** of a vector $x \in U$ by

$$\|x\| = (x, x)^{\frac{1}{2}},$$

where we take the nonnegative square root. A vector x is called normalized, or a unit vector, if $\|x\| = 1$. The existence of an inner product allows us to introduce and utilize the all-important notion of orthogonality.

We say that $x, y \in U$ are **orthogonal**, and write $x \perp y$, if $(x, y) = 0$. Given a set S, we write $x \perp S$ if $(x, s) = 0$ for all $s \in S$. We define the set $S^\perp$ by

$$S^\perp = \{x \mid (x, s) = 0, \text{ for all } s \in S\} = \bigcap_{s \in S} \{x \mid (x, s) = 0\}.$$

A set of vectors $\{x_i\}$ is called **orthogonal** if $(x_i, x_j) = 0$ whenever $i \neq j$. A set of vectors $\{x_i\}$ is called **orthonormal** if $(x_i, x_j) = \delta_{ij}$.

Theorem 7.1.1 (The Pythagorean theorem) *Let $\{x_i\}_{i=1}^k$ be an orthogonal set of vectors in U. Then*

$$\|\sum_{i=1}^{k} x_i\|^2 = \sum_{i=1}^{k} \|x_i\|^2.$$

Proof:

$$\|\sum_{i=1}^{k} x_i\|^2 = (\sum_{i=1}^{k} x_i, \sum_{j=1}^{k} x_j) = \sum_{i=1}^{k}\sum_{j=1}^{k} (x_i, x_j) = \sum_{i=1}^{k} \|x_i\|^2. \qquad \square$$

Theorem 7.1.2 (The Schwarz inequality) *For all $x, y \in U$ we have*

$$|(x, y)| \leq \|x\| \cdot \|y\|. \tag{7.1}$$

Proof: If $y = 0$, the equality holds trivially. So assume that $y \neq 0$. We set $e = y/\|y\|$. We note that $x = (x, e)e + (x - (x, e)e)$ and $(x - (x, e)e) \perp (x, e)e$. Applying the Pythagorean theorem, we get

$$\|x\|^2 = \|(x, e)e\|^2 + \|x - (x, e)e\|^2 \geq \|(x, e)e\|^2 = |(x, e)|^2,$$

or $|(x, e)| \leq \|x\|^2$. Substituting $y/\|y\|$ for e, inequality (7.1) follows. $\qquad \square$

Theorem 7.1.3 (The triangle inequality) *For all $x, y \in U$ we have*

$$\|x + y\| \leq \|x\| + \|y\|. \tag{7.2}$$

Proof: We compute

$$\begin{aligned}
\|x+y\|^2 &= (x+y, x+y) = (x,x) + (x,y) + (y,x) + (y,y) \\
&= \|x\|^2 + 2\mathrm{Re}(x,y) + \|y\|^2 \\
&\leq \|x\|^2 + 2|(x,y)| + \|y\|^2 \\
&\leq \|x\|^2 + 2\|x\|\|y\| + \|y\|^2 = (\|x\| + \|y\|)^2.
\end{aligned}$$ □

Next, we prove two important identities. The proof of both is computational and we omit it.

Proposition 7.1.1 *For all vectors* $x, y \in U$ *we have the following:*

1. **The polarization identity:**

$$(x,y) = \tfrac{1}{4}\{(\|x+y\|)^2 - (\|x-y\|)^2 + i(\|x+iy\|)^2 - i(\|x-iy\|)^2\}. \quad (7.3)$$

2. **The parallelogram identity:**

$$(\|x+y\|)^2 + (\|x-y\|)^2 = 2\left((\|x\|)^2 + (\|y\|)^2\right). \quad (7.4)$$

Proposition 7.1.2 (The Bessel inequality) *Let* $\{e_i\}_{i=1}^k$ *be an orthonormal set of vectors in* U. *Then*

$$\|x\|^2 \geq \|\sum_{i=1}^k |(x, e_i)|^2.$$

Proof: Let x be an arbitrary vector in U. Obviously, we can write $x = \sum_{i=1}^k (x, e_i)e_i + \left(x - \sum_{i=1}^k (x, e_i)e_i\right)$. Since the vector $x - \sum_{i=1}^k (x, e_i)e_i$ is orthogonal to all e_j, $j = 1, \ldots, k$, we can use the Pythagorean theorem to obtain

$$\begin{aligned}
\|x\|^2 &= \|x - \sum_{i=1}^k (x, e_i)e_i\|^2 + \|\sum_{i=1}^k (x, e_i)e_i\|^2 \\
&\geq \|\sum_{i=1}^k (x, e_i)e_i\|^2 = \|\sum_{i=1}^k |(x, e_i)|^2.
\end{aligned}$$ □

Proposition 7.1.3 *An orthogonal set of nonzero vectors* $\{x_i\}_{i=1}^k$ *is linearly independent.*

Proof: Assume that $\sum_{i=1}^k c_i x_i = 0$. Taking the inner product with x_j, we get

$$0 = \sum_{i=1}^k c_i(x_i, x_j) = c_j \|x_j\|^2.$$

This implies that $c_j = 0$ for all j. □

Since orthogonality implies linear independence, we can search for bases consisting of orthogonal, or even better orthonormal, vectors. Thus, an **orthogonal basis** for U is an orthogonal set that is also a basis for U. Similarly, an **orthonormal basis** for U is an orthonormal set that is also a basis for U.

The next theorem, generally known as the **Gram–Schmidt orthonormalization process**, gives a constructive method for computing orthonormal bases.

Theorem 7.1.4 (Gram–Schmidt orthonormalization) *Let* $\{x_1, \ldots, x_k\}$ *be linearly independent vectors in* U. *Then there exists an orthonormal set* $\{e_1, \ldots, e_k\}$ *such that*

$$L(e_1, \ldots, e_j) = L(x_1, \ldots, x_j), \qquad 1 \le j \le k. \qquad (7.5)$$

Proof: We prove this by induction. For $j = 1$, we set $e_1 = x_1/\|x_1\|$. Assume that we constructed $e_1, \ldots, e_{j-1}$ as required. We consider next $x'_j = x_j - \sum_{i=1}^{j-1}(x_j, e_i)e_i$. Clearly, $x'_j \ne 0$, for otherwise $x_j \in L(e_1, \ldots, e_{j-1}) = L(x_1, \ldots, x_{j-1})$, contrary to the assumption of linear independence of the x_i. We now define

$$e_j = \frac{x_j}{\|x_j\|} = \frac{x_j - \sum_{i=1}^{j-1}(x_j, e_i)e_i}{\|x_j - \sum_{i=1}^{j-1}(x_j, e_i)e_i\|}.$$

Obviously, $e_j \in L(e_1, \ldots, e_{j-1}, x_j) = L(x_1, \ldots, x_j)$. That is, for $i = 1, \ldots, j-1$, $e_i \in L(e_1, \ldots, e_{j-1}, x_j)$ is obvious. So $L(e_1, \ldots, e_j) \subset L(x_1, \ldots, x_j)$.

To get the inverse inclusion, we observe that $L(x_1, \ldots, x_{j-1}) \subset L(e_1, \ldots, e_j)$ follows from the induction hypothesis. Finally, $x_j = \sum_{i=1}^{j-1}(x_j, e_i)e_i + \|x_j - \sum_{i=1}^{j-1}(x_j, e_i)e_i\|e_j$ implies that $x_j \in L(e_1, \ldots, e_j)$. So $L(x_1, \ldots, x_j) \subset L(e_1, \ldots, e_j)$, and these two inclusions imply Eq. (7.5). $\qquad \square$

Corollary 7.1.1 *Every finite-dimensional inner product space has an orthonormal basis.*

Corollary 7.1.2 *Every orthonormal set in a finite-dimensional inner product space* U *can be extended to an orthonormal basis.*

Proof: Let $\{e_1, \ldots, e_m\}$ be an orthonormal set, and thus necessarily linearly independent. Hence, by Theorem 2.4.2, it can be extended to a basis $\{e_1, \ldots, e_m, x_{m+1}, \ldots, x_n\}$ for U. Applying the Gram–Schmidt orthonormalization process, we get an orthonormal basis $\{e_1, \ldots, e_n\}$. The first m vectors remain unchanged. $\qquad \square$

Orthonormal bases are particularly convenient for vector expansions or, alternatively, for computing coordinates.

Proposition 7.1.4 *Let $\{e_1, \ldots, e_n\}$ be an orthonormal basis for a finite-dimensional inner product space U. Then every vector $x \in U$ has a unique representation in the form*

$$x = \sum_{i=1}^{n} (x, e_i) e_i.$$

We consider next an important approximation problem. We define the distance of a vector $x \in U$ to a subspace M by

$$\delta(x, M) = \inf\{\|x - m\| \mid m \in M\}.$$

A best approximant for x in M is a vector $m_0 \in M$ that satisfies $\|x - m_0\| = \delta(x, M)$.

Theorem 7.1.5 *Let U be a finite-dimensional inner product space and M a subspace. Let $x \in U$. Then:*

1. *A best approximant for x in M exists.*

2. *A best approximant for x in M is unique.*

3. *A vector $m_0 \in M$ is a best approximant for x in M if and only if $x - m_0$ is orthogonal to M.*

Proof:

1. Let $\{e_1, \ldots, e_m\}$ be an orthonormal basis for M. Extend it to an orthonormal basis $\{e_1, \ldots, e_n\}$ for U. We claim that $m_0 = \sum_{i=1}^{m} (x, e_i) e_i$ is a best approximant. Clearly, $m_0 \in M$. Any other vector $m \in M$ has a unique representation of the form $m = \sum_{i=1}^{m} c_i e_i$. Therefore,

$$x - \sum_{i=1}^{m} c_i e_i = x - \sum_{i=1}^{m} (x, e_i) e_i + \sum_{i=1}^{m} [(x, e_i) - c_i] e_i.$$

Now, the vector $x - \sum_{i=1}^{m} (x, e_i) e_i$ is orthogonal to $L(e_1, \ldots, e_m)$. Hence, applying the Pythagorean theorem, we get

$$\left\| x - \sum_{i=1}^{m} c_i e_i \right\|^2 = \left\| x - \sum_{i=1}^{m} (x, e_i) e_i \right\|^2 + \left\| \sum_{i=1}^{m} [(x, e_i) - c_i] e_i \right\|^2$$

$$\geq \left\| x - \sum_{i=1}^{m} (x, e_i) e_i \right\|^2.$$

2. Let $\sum_{i=1}^{m} c_i e_i$ be another best approximant. By the previous computation, we must have $\left\| \sum_{i=1}^{m} [(x, e_i) - c_i] e_i \right\|^2 = \sum_{i=1}^{m} |(x, e_i) - c_i|^2$, and this happens if and only if $c_i = (x, e_i)$ for all i. Thus the two approximants coincide.

3. Assume that $m_0 \in M$ is the best approximant. We saw that $m = \sum_{i=1}^{m} (x, e_i) e_i$, and hence $x - \sum_{i=1}^{m} (x, e_i) e_i$ is orthogonal to $M = L(e_1, \ldots, e_m)$.

Conversely, assume that $m_0 \in M$ and $x - m_0 \perp M$. Then, for any vector $m \in M$, we have $x - m = (x - m_0) + (m - m_0)$. Since $m - m_0 \in M$, using the Pythagorean theorem, we have

$$\|x - m\|^2 = \|x - m_0\|^2 + \|m - m_0\|^2 \geq \|x - m_0\|^2.$$

Hence, m_0 is the best approximant. $\square$

In the special case of inner product spaces, we reserve the notation $U = M \oplus N$ for the case of orthogonal direct sums, that is, $U = M + N$ and $M \perp N$. We note that $M \perp N$ implies $M \cap N = \{0\}$.

Theorem 7.1.6 *Let U be a finite-dimensional inner product space and M a subspace. Then we have*

$$U = M \oplus M^{\perp}. \tag{7.6}$$

Proof: The subspaces M and $M^{\perp}$ are orthogonal. It suffices to show that they span U. Thus, let $x \in U$, and let m be its best approximant in M. We can write $x = m + (x - m)$. By Theorem 7.1.5, we have $x - m \in M^{\perp}$. $\square$

Inner product spaces are a particularly convenient setting for the development of duality theory. The key to this is the identification of the dual space to an inner product space with the space itself. This is done by identifying a linear functional with an inner product with a vector.

Theorem 7.1.7 *Let U be a finite-dimensional inner product space. Then f is a linear functional if and only if there exists a vector $\xi_f \in U$ such that*

$$f(x) = (x, \xi_f). \tag{7.7}$$

Proof: If f is defined by Eq. (7.7), then obviously it is a linear functional.

To prove the converse, we note first that if F is the zero functional, then we just choose $\xi_f = 0$. Thus, without loss of generality, we can assume that f is a nonzero functional. In this case, $M = \text{Ker } f$ is a subspace of codimension 1. This implies that $\dim M^{\perp} = 1$. Let us choose an arbitrary nonzero vector $y \in M^{\perp}$ and set $\xi_f = \alpha y$. Since $(y, \xi_f) = (y, \alpha y) = \overline{\alpha} \|y\|^2$, by choosing $\alpha = \overline{f(y)} / \|y\|^2$, we get $f(y) = (y, \xi_f)$ and the same holds, by linearity, for all vectors in $M^{\perp}$. For $x \in M$, it is clear that $f(x) = (x, \xi_f) = 0$. Thus, Eq. (7.7) follows. $\square$

7.2 Operators in Inner Product Spaces

In this section we study some important classes of operators in inner product spaces. We begin by studying the adjoint transformation. Due to the availability of inner products and the representation of linear functionals, the definition is slightly different than that given for general transformations.

The Adjoint Transformation

Theorem 7.2.1 *Let U, V be inner product spaces and let $T : U \longrightarrow V$ be a linear transformation. Then there exists a unique linear transformation $T^* : V \longrightarrow U$ such that*

$$(Tx, y) = (x, T^*y) \tag{7.8}$$

for all $x \in U$ and $y \in V$.

Proof: Fix a vector $y \in V$. Then $f(x) = (Tx, y)$ defines a linear functional on U. Thus, there exists a unique vector $\xi \in U$ such that $(Tx, y) = (x, \xi)$. We define T^* by $T^*y = \xi$. Therefore, T^* is a map from V to U. It remains to show that T^* is a linear map. To this end we compute

$$
\begin{aligned}
(x, T^*(\alpha_1 y_1 + \alpha_2 y_2)) &= (Tx, \alpha_1 y_1 + \alpha_2 y_2) \\
&= \overline{\alpha_1}(Tx, y_1) + \overline{\alpha_2}(Tx, y_2) \\
&= \overline{\alpha_1}(x, T^*y_1) + \overline{\alpha_2}(x, T^*y_2) \\
&= (x, \alpha_1 T^*y_1) + (x, \alpha_2 T^*y_2) \\
&= (x, \alpha_1 T^*y_1 + \alpha_2 T^*y_2).
\end{aligned}
$$

By the uniqueness of the representing vector, we get

$$T^*(\alpha_1 y_1 + \alpha_2 y_2) = \alpha_1 T^*y_1 + \alpha_2 T^*y_2. \qquad \square$$

We call the transformation T^* the **adjoint** or, if emphasis is needed, the **Hermitian adjoint** of T. Given a complex matrix $A = (a_{ij})$, its adjoint $A^* = (a_{ij}^*)$ is defined by $a_{ij}^* = \overline{a_{ji}}$ or, alternatively, by $A^* = \overline{A}$.

Proposition 7.2.1 *Let $T : U \longrightarrow V$ be a linear transformation. Then*

$$(\operatorname{Im} T)^{\perp} = \operatorname{Ker} T^*.$$

Proof: We have for $x \in U, y \in V$, $(Tx, y) = (x, T^*y)$. So $y \perp \operatorname{Im} T$ if and only if $T^*y \perp V$, that is, if and only if $y \in \operatorname{Ker} T^*$. $\qquad \square$

Corollary 7.2.1 *Let T be a linear transformation in an inner product space U. Then we have the following direct sum decompositions:*

$$
\begin{cases}
U = \operatorname{Im} T \oplus \operatorname{Ker} T^* \\
\\
U = \operatorname{Im} T^* \oplus \operatorname{Ker} T.
\end{cases}
$$

Moreover, $\operatorname{rank} T = \operatorname{rank} T^*$ and $\dim \operatorname{Ker} T = \dim \operatorname{Ker} T^*$.

Proof: Follows from Theorem 7.1.6 and Proposition 7.2.1. $\square$

Proposition 7.2.2 *Let T be a linear transformation in U. Then M is a T-invariant subspace of U if and only if $M^\perp$ is a T^*-invariant subspace.*

Proof: Follows from Eq. (7.8). $\square$

We proceed to study the matrix representation of a linear transformation T with respect to orthonormal bases. These are particularly easy to compute.

Proposition 7.2.3 *Let $T : U \longrightarrow V$ be a linear transformation, and let $\mathcal{B} = \{e_1, \ldots, e_n\}$ and $\mathcal{B}_1 = \{f_1, \ldots, f_m\}$ be orthonormal bases for U and V, respectively. Then:*

1. *For the matrix representation $[T]_{\mathcal{B}}^{\mathcal{B}_1} = (t_{ij})$, we have*

$$t_{ij} = (Te_j, f_i).$$

2. *We have for the adjoint T^**

$$[T^*]_{\mathcal{B}_1}^{\mathcal{B}} = ([T]_{\mathcal{B}}^{\mathcal{B}_1})^*.$$

Proof:

1. The matrix representation of T with respect to the given bases is defined by $Te_j = \sum_{k=1}^{m} t_{ki} f_k$. We compute

$$
\begin{aligned}
(Te_j, f_i) &= \left(\sum_{k=1}^{m} t_{kj} f_k, f_i \right) = \sum_{k=1}^{m} t_{kj} (f_k, f_i) \\
&= \sum_{k=1}^{m} t_{kj} \delta_{ik} = t_{ij}.
\end{aligned}
$$

2. Let $[T^*]_{\mathcal{B}_1}^{\mathcal{B}} = (t_{ij}^*)$. Then,

$$t_{ij}^* = (T^* f_j, e_i) = (f_j, Te_i) = \overline{(Te_i, f_j)} = \overline{t_{ji}}.   \square$$

Next we study the properties of the map from $T \mapsto T^*$ as a function from $L(U, V)$ into $L(V, U)$.

Proposition 7.2.4 *Let U, V be inner product spaces. Then the adjoint map has the following properties:*

1. $(T + S)^* = T^* + S^*$.

2. $(\alpha T)^* = \overline{\alpha} T^*$.

3. $(ST)^* = T^* S^*$.

4. $(T^*)^* = T$.

5. For the identity map in U, I_U, we have $I_U^* = I_U$.

6. If $T : U \longrightarrow V$ is invertible, then $(T^{-1})^* = (T^*)^{-1}$.

Proof:

1. Let $x \in U$ and $y \in V$. We compute

$$
\begin{aligned}
(x, (T + S)^* y) &= ((T + S)x, y) = (Tx + Sx, y) = (Tx, y) + (Sx, y) \\
&= (x, T^* y) + (x, S^* y) = (x, T^* y + S^* y) \\
&= (x, (T^* + S^*)y).
\end{aligned}
$$

By uniqueness we get $(T + S)^* = T^* + S^*$.

2. Computing

$$
\begin{aligned}
(x, (\alpha T)^* y) &= ((\alpha T)x, y) = (\alpha T x, y) \\
&= \alpha(Tx, y) = \alpha(x, T^* y) = (x, \overline{\alpha} T^* y),
\end{aligned}
$$

we get, by using uniqueness, that $(\alpha T)^* = \overline{\alpha} T^*$.

3. We compute

$$
(x, (ST)^* y) = (STx, y) = (Tx, S^* y) = (x, T^* S^* y)
$$

or $(ST)^* = T^* S^*$.

4.
$$
(Tx, y) = (x, T^* y) = \overline{(T^* y, x)} = \overline{(y, T^{**} x)} = (T^{**} x, y),
$$
and this implies $T^{**} = T$.

5. Let I_U be the identity map in U. Then

$$
(I_U x, y) = (x, y) = (x, I_U y),
$$

so $I_U^* = I_U$.

6. Assume that $T : U \longrightarrow V$ is invertible. Then $T^{-1}T = I_U$ and $TT^{-1} = I_V$. The first equality implies that $I + U = I_U^* = T^*(T^{-1})^*$, that is, $(T^{-1})^*$ is a right inverse of T^*. The second equality implies that it is a left inverse, and $(T^{-1})^* = (T^*)^{-1}$ follows. $\qquad\square$

The availability of the norm function on an inner product space allows us to consider a numerical measure of the size of a linear transformation.

Definition 7.2.1 Let U_1, U_2 be two inner product spaces. Given a linear transformation $T : U_1 \longrightarrow U_2$, we define its **norm** by

$$\|T\| = \sup_{\|x\| \leq 1} \|Tx\|.$$

Clearly, for an arbitrary linear transformation T, the norm $\|T\|$ is finite, for it is defined as the supremum of a continuous function on the unit ball $\{x \in U_1 | \|x\| \leq 1\}$, which is a compact set. It follows that the supremum is actually attained and thus there exists a not necessarily unique unit vector x for which $\|Tx\| = \|T\|$.

The following proposition, whose standard proof we omit, sums up the basic properties of the operator norm:

Proposition 7.2.5 *We have the following properties of the operator norm:*

1. $\|T + S\| \leq \|T\| + \|S\|$.

2. $\|\alpha T\| = |\alpha| \cdot \|T\|$.

3. $\|TS\| \leq \|T\| \cdot \|S\|$.

4. $\|T^*\| = \|T\|$.

5. $\|I\| = 1$.

6. $\|T\| = \sup_{\|x\|=1} \|Tx\| = \sup_{\|x\|,\|y\|\leq 1} |(Tx,y)| = \sup_{\|x\|,\|y\|=1} |(Tx,y)|$.

7.3   Unitary Operators

The extra structure that inner product spaces have over linear spaces, namely, the existence of inner products, leads to a new definition, that of an isomorphism of two inner product spaces.

Definition 7.3.1 *Let V and W be two inner product spaces over the same field. A linear transformation $T : V \longrightarrow W$ is an* **isometry** *if it preserves inner products, that is, if $(Tx, Ty) = (x, y)$, for all $x, y \in V$. An* **isomorphism of inner product spaces** *is an isomorphism that is isometric. We also call it an* **isometric isomorphism** *or, alternatively, a* **unitary isomorphism**.

Proposition 7.3.1 *Given a linear transformation $T : V \longrightarrow W$, the following properties are equivalent:*

1. *T preserves inner products.*

2. *For all $x \in V$ we have $\|Tx\| = \|x\|$.*

3. *We have $T^*T = I$.*

Proof: Assume that T preserves inner products, that is, $(Tx, Ty) = (x, y)$, for all $x, y \in V$. Choosing $y = x$, we get $\|Tx\|^2 = (Tx, Tx) = (x, x) = \|x\|^2$.

If $\|Tx\| = \|x\|$, then, using the polarization identity (7.3), we get

$$
\begin{aligned}
(Tx, Ty) &= \tfrac{1}{4}\{(\|Tx + Ty\|)^2 - (\|Tx - Ty\|)^2 \\
&\quad + i(\|Tx + iTy\|)^2 - i(\|Tx - iTy\|)^2\} \\
&= \tfrac{1}{4}\{(\|x + y\|)^2 - (\|x - y\|)^2 + i(\|x + iy\|)^2 - i(\|x - iy\|)^2\} \\
&= (x, y).
\end{aligned}
$$

Next observe that if T preserves inner products, then $((I - T^*T)x, y) = 0$, which implies that $T^*T = I$. Finally, if the equality $T^*T = I$ holds, then clearly T preserves inner products. $\qquad\square$

Corollary 7.3.1 *An isometry* $T : V \longrightarrow W$ *maps orthonormal sets into orthonormal sets.*

The next result is the counterpart, for inner product spaces, of Theorem 4.1.11.

Corollary 7.3.2 *Two inner product spaces are isomorphic if and only if they have the same dimension.*

Proof: Assume that a unitary isomorphism $U : V \longrightarrow W$ exists. Since it maps bases into bases, the dimensions of V and W coincide.

Conversely, suppose that V and W have the same dimensions. We choose orthonormal bases $\{e_1, \ldots, e_n\}$ and $\{f_1, \ldots, f_n\}$, in V and W, respectively. We define a linear map $U : V \longrightarrow W$ by $Ue_i = f_i$. It is immediate that the operator U so defined is a unitary isomorphism. $\qquad\square$

Proposition 7.3.2 *Let* $U : V \longrightarrow W$ *be a linear transformation. Then* U *is unitary if and only if* $U^* = U^{-1}$.

Proof: If $U^* = U^{-1}$, we get $U^*U = UU^* = I$, that is, U is an isometric isomorphism.

Conversely, if U is unitary, we have $U^*U = I$. Since U is an isomorphism and the inverse transformation, if it exists, is unique, we obtain $U^* = U^{-1}$. $\qquad\square$

Next we consider matrix representations. We say that a complex matrix is unitary if $A^*A = I$. Here A^* is defined by $a_{ij}^* = \overline{a_{ji}}$. We proceed to consider matrix representations for unitary maps. We say that a complex $n \times n$ matrix A is unitary if $A^*A = I$, where A^* is the Hermitian adjoint of A. We expect that the matrix representation of a unitary matrix, when taken with respect to orthonormal bases, should reflect the unitarity property. This is indeed the case.

Proposition 7.3.3 *Let $U : V \longrightarrow W$ be a linear transformation. Let $\mathcal{B} = \{e_1, \ldots, e_n\}$ and $\mathcal{B}_1 = \{f_1, \ldots, f_n\}$ be orthonormal bases in V and W, respectively. Then U is unitary if and only if its matrix representation $[T]_{\mathcal{B}}^{\mathcal{B}_1}$ is a unitary matrix.*

Proof: Taking matrix representations with respect to the given bases, and using Proposition 7.2.3, we obtain

$$([U]_{\mathcal{B}}^{\mathcal{B}_1})^*[U]_{\mathcal{B}}^{\mathcal{B}_1} = [U^*]_{\mathcal{B}_1}^{\mathcal{B}}[U]_{\mathcal{B}}^{\mathcal{B}_1} = [U^*U]_{\mathcal{B}}^{\mathcal{B}}.$$

This clearly shows that $U^*U = I$ if and only if $[U]_{\mathcal{B}}^{\mathcal{B}_1}$ is a unitary matrix.
$\square$

The notion of similarity can be strengthened in the context of inner product spaces.

Definition 7.3.2 *Let V and W be two inner product spaces over the same field, and let $T : V \longrightarrow V$ and $S : W \longrightarrow W$ be linear transformations. We say that T and S are **unitarily equivalent** if there exists a unitary map $U : V \longrightarrow W$ for which $UT = SU$ or, equivalently, for which the following diagram is commutative:*

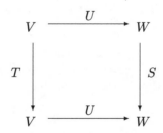

The unitary transformations play, in the algebra of linear transformations on an inner product space, the same role that complex numbers of absolute value 1 play in the complex number field $\mathbf{C}$. This is reflected in the eigenvalues of unitary transformations as well as in their structure.

Proposition 7.3.4 *Let U be a unitary operator in an inner product space V. Then:*

1. *All eigenvalues of U have absolute value 1.*

2. *If $Ux = \lambda x$, then $U^*x = \bar{\lambda}x$.*

3. *Eigenvectors corresponding to different eigenvectors are orthogonal.*

4. *We have*

$$\mathrm{Ker}\, (U - \lambda_1 I)^{\nu_1} \cdots (U - \lambda_k I)^{\nu_k} = \mathrm{Ker}\, (U - \lambda_1 I) \cdots (U - \lambda_k I).$$

5. *The minimal polynomial m_U of U splits into distinct linear factors.*

Proof:

1. Let λ be an eigenvalue of U and x a corresponding eigenvector. Then $Ux = \lambda x$ implies that $(Ux, x) = (\lambda x, x) = \lambda(x, x)$. Now, using the unitarity of U, we have

$$\|x\|^2 = \|Ux\|^2 = \|\lambda x\|^2 = |\lambda|^2 \|x\|^2.$$

This implies that $|\lambda| = 1$.

2. Assume that $Ux = \lambda x$. Using the fact that $U^*U = UU^* = I$, we compute

$$
\begin{aligned}
\|U^*x - \overline{\lambda}x\|^2 &= (U^*x - \overline{\lambda}x, U^*x - \overline{\lambda}x) \\
&= (U^*x, U^*x) - \overline{\lambda}(x, x) - \lambda(x, Ux) + |\lambda|^2 \|x\|^2 \\
&= (UU^*x, x) - \overline{\lambda}(Ux, x) - \lambda(x, Ux) + \|x\|^2 \\
&= \|x\|^2 - \overline{\lambda}(\lambda x, x) - \lambda(x, \lambda x) + \|x\|^2 \\
&= \|x\|^2 - 2|\lambda|^2 \|x\|^2 + \|x\|^2 = 0.
\end{aligned}
$$

3. Let λ, μ be distinct eigenvalues and x, y the corresponding eigenvectors. Then

$$\lambda(x, y) = (Ux, y) = (x, U^*y) = (x, \overline{\mu}y) = \mu(x, y).$$

Hence, $(\lambda - \mu)(x, y) = 0$. Since $\lambda \neq \mu$, we conclude that $(x, y) = 0$ or $x \perp y$.

4. Since U is unitary, U and U^* commute. Therefore, $U - \lambda I$ and $U^* - \overline{\lambda}I$ also commute. Assume now that $(U - \lambda I)^\nu x = 0$. Without loss of generality we may assume that $(U - \lambda I)^{2^n} x = 0$. This implies that

$$0 = (U^* - \overline{\lambda}I)^{2^n}(U - \lambda I)^{2^n} x = [(U^* - \overline{\lambda}I)^{2^{n-1}}(U - \lambda I)^{2^{n-1}}]^2 x.$$

So

$$
\begin{aligned}
0 &= ([(U^* - \overline{\lambda}I)^{2^{n-1}}(U - \lambda I)^{2^{n-1}}]^2 x, x) \\
&= \|(U^* - \overline{\lambda}I)^{2^{n-1}}(U - \lambda I)^{2^{n-1}} x\|^2,
\end{aligned}
$$

or $(U^* - \overline{\lambda}I)^{2^{n-1}}(U - \lambda I)^{2^{n-1}} x = 0$. This in turn implies that $\|(U - \lambda I)^{2^{n-1}} x\| = 0$, and hence $(U - \lambda I)^{2^{n-1}} x = 0$. By repeating the argument, we conclude that $(U - \lambda I)x = 0$.

Next, we use the fact that all factors $U - \lambda_i I$ commute. Assuming that $(U - \lambda_1 I)^{\nu_1} \cdots (U - \lambda_k I)^{\nu_k} x = 0$, we conclude that

$$
\begin{aligned}
0 &= (U - \lambda_1 I)(U - \lambda_2 I)^{\nu_2} \cdots (U - \lambda_k I)^{\nu_k} x \\
&= (U - \lambda_2 I)^{\nu_2}(U - \lambda_1 I)(U - \lambda_3 I)^{\nu_3} \cdots (U - \lambda_k I)^{\nu_k} x.
\end{aligned}
$$

This implies that $(U - \lambda_1 I)(U - \lambda_2 I)(U - \lambda_3 I)^{\nu_3} \cdots (U - \lambda_k I)^{\nu_k} x = 0$. Proceeding inductively, we get $(U - \lambda_1 I) \cdots (U - \lambda_k I)x = 0$.

5. Follows from part 4. $\qquad\square$

7.4   Self-Adjoint Operators

Probably the most important class of operators in an inner product space is that of self-adjoint operators. An operator T in an inner product space U is called **self-adjoint** or **Hermitian** if $T^* = T$. Similarly, a complex matrix A is called self-adjoint or Hermitian if it satisfies $A^* = A$ or, equivalently, $a_{ij} = \overline{a_{ji}}$.

The next proposition shows that self-adjoint operators in an inner product space play a role in $L(U)$ similar to that of the real numbers in the complex field.

Proposition 7.4.1 *Every linear transformation T in an inner product space can be written in the form*

$$T = T_1 + iT_2$$

with T_1, T_2 self-adjoint. Such a representation is unique.

Proof: Write

$$T = \frac{1}{2}(T + T^*) + i\frac{1}{2i}(T - T^*),$$

and note that $T_1 = \frac{1}{2}(T + T^*)$ and $T_2 = \frac{1}{2i}(T - T^*)$ are both self-adjoint.

Conversely, if $T = T_1 + iT_2$ with T_1, T_2 self-adjoint, then $T^* = T_1 - iT_2$. By elimination we get that T_1, T_2 necessarily have the form given previously. □

In preparation for the spectral theorem, we prove a few properties of self-adjoint operators that are of intrinsic interest.

Proposition 7.4.2 *Let T be a self-adjoint operator in an inner product space U. Then:*

1. *All eigenvalues of T are real.*

2. *Eigenvectors corresponding to different eigenvectors are orthogonal.*

3. *We have*

$$\text{Ker}\,(T - \lambda_1 I)^{\nu_1} \cdots (T - \lambda_k I)^{\nu_k} = \text{Ker}\,(T - \lambda_1 I) \cdots (T - \lambda_k I).$$

4. *The minimal polynomial m_T of T splits into distinct linear factors.*

Proof:

1. Let λ be an eigenvalue of T and x a corresponding eigenvector. Then $Tx = \lambda x$ implies that $(Tx, x) = (\lambda x, x) = \lambda(x, x)$. Now, using the self-adjointness of T, we have $\overline{(Tx, x)} = (x, Tx) = (Tx, x)$, that is, (Tx, x) is real, and so is $\lambda = (Tx, x)/(x, x)$.

2. Let λ, μ be distinct eigenvalues of T and x, y corresponding eigenvectors. Using the fact that eigenvalues of T are real, we get $(Tx, y) = (\lambda x, y) = \lambda(x, y)$ and

$$(Tx, y) = (x, Ty) = (x, \mu y) = \overline{\mu}(x, y) = \mu(x, y).$$

By subtraction we get $(\lambda - \mu)(x, y) = 0$ and, as $\lambda \neq \mu$, we conclude that $(x, y) = 0$ or $x \perp y$.

3. We prove this by induction on the number of factors. So, let $k = 1$ and assume that $(T - \lambda_1 I)^{\nu_1} x = 0$. Then, by multiplying this equation by $(T - \lambda_1 I)^{\mu}$, we may assume without loss of generality that $(T - \lambda_1 I)^{2^n} x = 0$. Then we get

$$
\begin{aligned}
0 &= ((T - \lambda_1 I)^{2^n} x, x) = ((T - \lambda_1 I)^{2^{n-1}} x, (T - \lambda_1 I)^{2^{n-1}} x) \\
&= \|(T - \lambda_1 I)^{2^{n-1}} x\|,
\end{aligned}
$$

which implies that $(T - \lambda_1 I)^{2^{n-1}} x = 0$. Repeating this argument, we finally obtain $(T - \lambda_1 I) x = 0$. Assume now that the statement holds for up to $k - 1$ factors and that $\Pi_{i=1}^{k}(T - \lambda_i I)^{\nu_i} x = 0$. Therefore, $(T - \lambda_k I)^{\nu_k} \Pi_{i=1}^{k-1}(T - \lambda_i I)^{\nu_i} x = 0$. By the argument for $k = 1$ we conclude that

$$0 = (T - \lambda_i I) \prod_{i=1}^{k-1}(T - \lambda_i I)^{\nu_i} x = \prod_{i=1}^{k-1}(T - \lambda_i I)^{\nu_i}[(T - \lambda_k I)x].$$

By the induction hypothesis we get

$$0 = \prod_{i=1}^{k-1}(T - \lambda_i I)[(T - \lambda_i I)x] = \prod_{i=1}^{k}(T - \lambda_i I)x,$$

or, since x is arbitrary, $\Pi_{i=1}^{k}(T - \lambda_i I) = 0$.

4. Let $\lambda_1, \ldots, \lambda_k$ be the distinct eigenvalues of T. Then $\Pi_{i=1}^{k}(z - \lambda_i)$ divides $m_T(z)$. On the other hand, by Part 3, $(T - \lambda_1 I) \cdots (T - \lambda_k I) = 0$. Thus, $m_T(z) = \Pi_{i=1}^{k}(z - \lambda_i)$. $\square$

Theorem 7.4.1 *Let T be self-adjoint. Then there exists an orthonormal basis consisting of eigenvectors of T.*

Proof: Let m_T be the minimal polynomial of T. Then $m_T(z) = \Pi_{i=1}^{k}(z - \lambda_i)$, with the λ_i distinct. Let $\pi_i(z)$ be the corresponding Lagrange interpolation polynomials. We observe that the $\pi_i(T)$ are orthogonal projections. That they are projections has been proved in Theorem 6.4.1. That they are orthogonal projections follows from their self-adjointness, noting that, given a real polynomial p and a self-adjoint operator T, necessarily $p(T)$ is self-adjoint. This implies that $U = \text{Ker}(\lambda_1 I - T) \oplus \cdots \oplus \text{Ker}(\lambda_k I - T)$

is an orthogonal direct sum decomposition. Choosing an orthonormal basis in each subspace $\text{Ker}\,(\lambda_i I - T)$, we get an orthonormal basis made of eigenvectors. □

As a result we obtain the spectral theorem.

Theorem 7.4.2 *Let T be self-adjoint. Then it has a unique representation of the form*

$$T = \sum_{i=1}^{s} \lambda_i P_i,$$

where the $\lambda_i \in \mathbf{R}$ are distinct and P_i are orthogonal projections satisfying

$$P_i P_j = \delta_{ij} P_j$$
$$\sum_{i=1}^{s} P_i = I.$$

Proof: Let λ_i be the distinct eigenvalues of T. We define the P_i to be the orthogonal projections on $\text{Ker}\,(\lambda_i I - T)$.

Conversely, if such a representation exists, it follows that, necessarily, the λ_i are the eigenvalues of T and $\text{Im}\,P_i = \text{Ker}\,(\lambda_i I - T)$. □

In the special case of self-adjoint operators, the computation of the norm has a further characterization.

Proposition 7.4.3 *Let T be a self-adjoint operator in a finite-dimensional inner product space U. Then:*

1. *We have $\|T\| = \sup_{\|x\| \le 1} |(Tx, x)| = \sup_{\|x\|=1} |(Tx, x)|$.*

2. *Let $\lambda_1, \ldots, \lambda_n$ be the eigenvalues of T ordered so that $|\lambda_i| \ge |\lambda_{i+1}|$. Then $\|T\| = |\lambda_1|$, that is, the norm equals the modulus of the largest eigenvalue.*

Proof:

1. Since, for x satisfying $\|x\| \le 1$, we have

$$|(Tx, x)| \le \|Tx\| \cdot \|x\| \le \|T\| \cdot \|x\|^2 \le \|T\|,$$

it follows that $\sup_{\|x\|=1} |(Tx, x)| \le \sup_{\|x\|\le 1} |(Tx, x)| \le \|T\|$.

To prove the converse inequality, we argue as follows. We use the following version of the polarization identity:

$$4\text{Re}(Tx, y) = (T(x + y), x + y) - (T(x - y), x - y),$$

which implies that

$$|(\mathrm{Re}(Tx, y)| \le \tfrac{1}{4} \sup_{\|z\| \le 1} |(Tz, z)| [\|x + y\|^2 + \|x - y\|^2]$$

$$|(\mathrm{Re}(Tx, y)| \le \tfrac{1}{4} \sup_{\|z\| \le 1} |(Tz, z)| [2\|x\|^2 + 2\|y\|^2]$$

$$\le \sup_{\|z\| \le 1} |(Tz, z)|.$$

Choosing $y = Tx/\|Tx\|$, it follows that $\|Tx\| \le \sup_{\|z\| \le 1} |(Tz, z)|$. The equality $\sup_{\|z\| \le 1} |(Tz, z)| = \sup_{\|z\| = 1} |(Tz, z)|$ is obvious.

2. Let λ_1 be the eigenvalue of T of largest modulus and x_1 a corresponding normalized eigenvector. Then

$$
\begin{aligned}
|\lambda_1| &= |(\lambda_1(x_1, x_1)| = |(\lambda_1 x_1, x_1)| \\
&= |(Tx_1, x_1)| \le \|T\| \cdot \|x_1\|^2 = \|T\|.
\end{aligned}
$$

So the absolute values of all eigenvalues of T are bounded by $\|T\|$. Assume now that x is a vector for which $\|Tx\| = \|T\|$. We will show that it is an eigenvector of T corresponding to either $\|T\|$ or $-\|T\|$. The equality $\sup |(Tx, x)| = \|T\|$ implies that either $\sup(Tx, x) = \|T\|$ or $\inf(Tx, x) = -\|T\|$. We assume that the first is satisfied. Thus, there exists a unit vector x for which $(Tx, x) = \|T\|$. We compute

$$
\begin{aligned}
0 &\le \|Tx - \|T\|x\|^2 = \|Tx\|^2 - 2\|T\|(Tx, x) + \|T\|^2\|x\|^2 \\
&\le \|T\|^2\|x\|^2 - 2\|T\|^2 + \|T\|^2\|x\|^2 = 0.
\end{aligned}
$$

So necessarily $Tx = \|T\| \cdot \|x\|$ or $\|T\| = \lambda_1$. If $\inf(Tx, x) = -\|T\|$, the same argument can be used. $\qquad\square$

The Minimax Principle

For self-adjoint operators we have a very nice characterization of eigenvalues, known as the **minimax principle**.

Theorem 7.4.3 *Let T be a self-adjoint operator on an n-dimensional inner product space U. Let $\lambda_1 \ge \cdots \ge \lambda_n$ be the eigenvalues of T. Then*

$$\lambda_k = \min_{\dim M = n-k+1} \max_{x \in M} \{(Tx, x)| \; \|x\| = 1\}.$$

Proof: Now let M be an arbitrary subspace of U of dimension $n - k + 1$. Let $M_k = \{e_1, \ldots, e_k\}$; then, since $\dim M_k + \dim N_k = k + (n - k + 1) = n + 1$, their intersection $M_k \cap N_k$ is nontrivial and contains a unit vector x. For this vector we have

$$(Tx, x) = \sum_{i=1}^{k} \lambda_i |(x, e_i)|^2 \ge \lambda_k \sum_{i=1}^{k} |(x, e_i)|^2 = \lambda_k.$$

This shows that $\min_{\dim M = n-k+1} \max_{x \in M} \{(Tx, x)| \; \|x\| = 1\} \ge \lambda_k$.

To complete the proof we need to exhibit at least one subspace on which the reverse inequality holds. Let $\{e_1, \ldots, e_n\}$ be an orthonormal basis of U consisting of eigenvectors of T. Let $M_k = L(e_k, \ldots, e_n)$. Then

$$\max_{x \in M}\{(Tx, x)| \; \|x\| = 1\} \; = \; \sum_{i=k}^{n} \lambda_i |(x, e_i)|^2$$

$$\leq \; \lambda_k \sum_{i=k}^{n} |(x, e_i)|^2 \leq \lambda_k \|x\|^2 = \lambda_k. \quad \square$$

The Cayley Transform

If one compares the properties of unitary and self-adjoint operators, the similarity between them is remarkable. As we pointed out, these two classes generalize the sets of complex mumbers of absolute value 1 and the set of real numbers, respectively.

Now the fractional linear transformation $w = (z - i)/(z + i)$ maps the upper half-plane onto the unit disk and in particular the real line onto the unit circle. Naturally, one wonders whether this can be extended to a map of self-adjoint operators onto unitary operators. The next theorem focuses on this map.

Theorem 7.4.4 *Let A be a self-adjoint operator in a finite-dimensional inner product space V. Then:*

1. *For each vector $x \in V$, we have*

$$\|(A + iI)x\|^2 = \|(A - iI)x\|^2 = \|Ax\|^2 + \|x\|^2.$$

2. *The operators $A + iI, A - iI$ are both injective and hence invertible.*

3. *The operator U defined by*

$$U = (A - iI)(A + iI)^{-1} \tag{7.9}$$

is a unitary operator for which 1 is not an eigenvalue. U is called the **Cayley transform** *of A.*

4. *Given a unitary map U in a finite-dimensional inner product space, such that 1 is not an eigenvalue, then the operator A defined by*

$$A = i(I + U)(I - U)^{-1} \tag{7.10}$$

is a self-adjoint operator. This map is called the **inverse Cayley transform**.

Proof:

1. We compute

$$\begin{aligned} \|(A+iI)x\|^2 &= \|Ax\|^2 + \|x\|^2 + (ix, Ax) + (Ax, ix) \\ &= \|Ax\|^2 + \|x\|^2 = \|(A-iI)x\|^2. \end{aligned}$$

2. The injectivity of both $A + iI$ and $A - iI$ is an immediate consequence of the previous equality. This shows the invertibility of both operators.

3. Let $x \in V$ be arbitrary. Then there exists a unique vector z such that $x = (A+iI)x$ or $z = (A+iI)^{-1}x$. Therefore,

$$\|(A-iI)(A+iI)^{-1}x\| = \|(A-iI)z\| = \|(A-iI)z\| = \|x\|.$$

This shows that U defined by Eq. (7.9) is unitary.

To see that 1 is not an eigenvalue of U, assume that $Ux = x$. Thus, $(A-iI)(A+iI)^{-1}x = x = (A+iI)(A+iI)^{-1}x$, and hence, $2i(A+iI)^{-1}x = 0$. This implies that $x = 0$.

4. Assume that U is a unitary map such that 1 is not an eigenvalue. Then $I - U$ is invertible. Defining A by Eq. (7.10), and using the fact that $U^* = U^{-1}$, we compute

$$\begin{aligned} A^* &= -i(I+U^*)(I-U^*)^{-1} = -i(I+U^{-1})UU^{-1}(I-U^{-1})^{-1} \\ &= -i(U+I)(U-I)^{-1} = i(I+U)(I-U)^{-1} = A. \end{aligned}$$

So A is self-adjoint. $\qquad\qquad\square$

Normal Operators

The analysis of the classes of unitary and self-adjoint operators in inner product spaces showed remarkable similarities. In particular, both classes admitted the existence of orthonormal bases made up of eigenvectors. One wonders if the set of all operators having this property can be characterized. This can in fact be done, and the corresponding class is that of normal operators, which we proceed to introduce.

Definition 7.4.1 *Let T be a linear transformation in a finite-dimensional inner product space. We say T is* **normal** *if it satisfies*

$$T^*T = TT^*.$$

Proposition 7.4.4 *The operator T is normal if and only if, for every $x \in V$, we have*

$$\|Tx\| = \|T^*x\|. \tag{7.11}$$

Proof: We compute, assuming T normal,

$$\begin{aligned}
\|Tx\|^2 &= (Tx, Tx) = (T^*Tx, x) \\
&= (TT^*x, x) = (T^*x, T^*x) = \|T^*x\|^2,
\end{aligned}$$

which implies Eq. (7.11).

Conversely, assuming the equality (7.11), we have $(TT^*x, x) = (T^*Tx, x)$ and hence, by Proposition 7.4.3, we get $TT^* = T^*T$, that is, T is normal. $\quad\square$

Corollary 7.4.1 *Let T be a normal operator. Let α be an eigenvalue of T. Then $Tx = \alpha x$ implies $T^*x = \bar{\alpha}x$.*

Proof: If T is normal, so is $T - \alpha I$, and we apply Proposition 7.4.4. $\quad\square$

Proposition 7.4.5 *Let T be a normal operator in a complex inner product space. If, for some complex number λ, we have $(T - \lambda I)^\nu x = 0$, then $(T - \lambda I)x = 0$.*

Proof: The proof is exactly as in the case of unitary operators. $\quad\square$

Lemma 7.4.1 *Let T be a normal operator and assume that λ, μ are distinct eigenvalues of T. If x, y are corresponding eigenvectors, then $x \perp y$.*

Proof: We compute

$$\lambda(x, y) = (\lambda x, y) = (Tx, y) = (x, T^*y) = (x, \bar{\mu}y) = \mu(x, y).$$

Hence, $(\lambda - \mu)(x, y) = 0$, and, since $\lambda \neq \mu$, we conclude that $(x, y) = 0$. Thus, x, y are orthogonal. $\quad\square$

The following is known as the spectral theorem for normal operators:

Theorem 7.4.5 *Let T be a normal operator in a finite-dimensional complex inner product space V. Then there exists an orthonormal basis of V consisting of eigenvectors of T.*

Proof: Let $m_T(z) = \Pi_{i=1}^k (z - \lambda_i)^{\nu_i}$ be the primary decomposition of the minimal polynomial of T. We set $M_i = \text{Ker}\,(\lambda_i I - T)^{\nu_i}$. By Proposition 7.4.5 we have $M_i = \text{Ker}\,(\lambda_i I - T)$, that is, it is the eigenspace corresponding to the eigenvalue λ_i. By Lemma 7.4.1, the subspaces M_i are mutually orthogonal. We now apply Theorem 6.4.2 to conclude that $V = M_1 \oplus \cdots \oplus M_k$, where this is now an orthogonal direct sum decomposition. Choosing an orthonormal basis in each subspace M_i and taking their union, we get an orthonormal basis for V made of eigenvectors. $\quad\square$

Theorem 7.4.6 *Let T be a normal operator in a finite-dimensional inner product space V. Then:*

    1. *A subspace $M \subset V$ is invariant under T if and only if it is invariant under T^*.*

2. *Let M be an invariant subspace for T. Then the direct sum decomposition $V = M \oplus M^{\perp}$ reduces T.*

3. *T reduced to an invariant subspace M, that is, $T|_M$, is a normal operator.*

Proof:

1. Let M be invariant under T. Since $T|_M$ has at least one eigenvalue and a corresponding eigenvector, say $Tx_1 = \lambda_1 x_1$, then also $T^* x_1 = \overline{\lambda_1} x_1$. Let M_1 be the subspace spanned by x_1. We consider $M \ominus M_1 = M \cap M_1^{\perp}$. This is also invariant under T. Proceeding by induction, we conclude that M is spanned by eigenvectors of T and hence, by Proposition 7.4.5, by eigenvectors of T^*. This shows that M is T^* invariant. The converse follows by symmetry.

2. Since M is invariant under both T and T^*, so is $M^{\perp}$.

3. Let $\{e_1, \ldots, e_m\}$ be an orthonormal basis of M consisting of eigenvectors. Thus, $Te_i = \lambda_i e_i$ and $T^* e_i = \overline{\lambda_i} e_i$. Let $x = \sum_{i=1}^{m} \alpha_i e_i \in M$. Then

$$
\begin{aligned}
T^* T x &= T^* T \sum_{i=1}^{m} \alpha_i e_i = T^* \sum_{i=1}^{m} \alpha_i \lambda_i e_i \\
&= T^* \sum_{i=1}^{m} \alpha_i \lambda_i e_i \\
&= \sum_{i=1}^{m} \alpha_i |\lambda_i|^2 e_i = T \sum_{i=1}^{m} \alpha_i \overline{\lambda_i} e_i \\
&= T T^* \sum_{i=1}^{m} \alpha_i e_i = T T^* x.
\end{aligned}
$$

This shows that $(T|_M)^* = T^*|_M$. $\qquad\square$

Theorem 7.4.7 *The operator T is normal if and only if $T^* = p(T)$ for some polynomial p.*

Proof: If $T^* = p(T)$, then obviously T and T^* commute, that is, T is normal.

Conversely, if T is normal, there exists an orthonormal basis $\{e_1, \ldots, e_n\}$ consisting of eigenvectors corresponding to the eigenvalues $\lambda_1, \ldots, \lambda_n$. Let p be any polynomial that interpolates the values $\overline{\lambda_i}$ at the points λ_i. Then

$$
p(T)e_i = p(\lambda_i)e_i = \overline{\lambda_i} e_i = T^* e_i.
$$

This shows that $T^* = p(T)$. $\qquad\square$

Positive Operators

We consider next a subclass of self-adjoint operators, that of positive operators.

Definition 7.4.2 *An operator T on an inner product space V is called* **nonnegative** *if, for all $x \in V$, we have $(Tx, x) \geq 0$. T is called* **positive** *if it is nonnegative and $(Tx, x) = 0$ for $x = 0$ only.*

Similarly, a complex Hermitian matrix A is **nonnegative** *if, for all $x \in \mathbf{C}^n$, we have $(Ax, x) \geq 0$.*

Given vectors $x_1, \ldots, x_n$ in an inner product space V, we define the corresponding **Gram matrix**, or **Gramian**, $G = (g_{ij})$ by $g_{ij} = (x_i, x_j)$.

The next results connect Gramians with positivity.

Theorem 7.4.8 *Let V be an n-dimensional inner product space. A necessary and sufficient condition for a $k \times k$, with $k \leq n$, Hermitian matrix G to be nonnegative definite is that it is the Gram matrix of k vectors in V.*

Proof: Assume that $G = (g_{ij})$, with $g_{ij} = (x_i, x_j)$. Let $\mathcal{B} = \{e_1, \ldots, e_n\}$ be an orthonormal basis in V. Thus, we have the expansions $x_i = \sum_{j=1}^{n} (x_i, e_j) e_j$. Define a linear transformation $T : V \longrightarrow V$ by

$$Te_i = \begin{cases} x_i & 1 \leq i \leq k \\ 0 & k < i \leq n. \end{cases}$$

We compute

$$
\begin{aligned}
([T^*T]_{\mathcal{B}}^{\mathcal{B}})_{ij} &= (T^*Te_j, e_i) = (Te_j, Te_i) \\
&= \left(\sum_{k=1}^{n} (x_j, e_k)e_k, \sum_{l=1}^{n} (x_i, e_l)e_l \right) \\
&= \sum_{k=1}^{n} \sum_{l=1}^{n} (x_j, e_k)\overline{(x_i, e_l)}(e_k, e_l) \\
&= \sum_{k=1}^{n} \sum_{l=1}^{n} (x_j, e_k)\overline{(x_i, e_l)}\delta_{kl} \\
&= \sum_{k=1}^{n} (x_j, e_k)\overline{(x_i, e_k)} = (x_j, x_i).
\end{aligned}
$$

Proof: Let

$$\begin{pmatrix} \xi_1 \\ \cdot \\ \cdot \\ \cdot \\ \xi_n \end{pmatrix} \in \mathbf{C}^n.$$

Then

$$
\begin{aligned}
(G\xi, \xi) &= \sum_{i=1}^{n} \sum_{j=1}^{n} g_{ij}\xi_j\overline{\xi}_i = \sum_{i=1}^{n} \sum_{j=1}^{n} (x_i, x_j)\xi_j\overline{\xi}_i \\
&= \left(\sum_{i=1}^{n} \overline{\xi}_i x_i, \sum_{j=1}^{n} \overline{\xi}_i x_j \right) \geq 0,
\end{aligned}
$$

which shows that G is nonnegative. $\square$

Proposition 7.4.6

1. *A Hermitian operator T in an inner product space is nonnegative if and only if all of its eigenvalues are nonnegative.*

2. *A Hermitian operator T in an inner product space is positive if and only if all its eigenvalues are positive.*

Proof:

1. Assume that T is Hermitian and nonnegative. Let λ be an arbitrary eigenvalue of T and x a corresponding eigenvector. Then

$$\lambda\|x\|^2 = \lambda(x,x) = (\lambda x, x) = (Tx, x) \geq 0,$$

so $\lambda \geq 0$.

Conversely, assume that T is Hermitian and all of its eigenvalues are nonnegative. Let $\{e_1, \ldots, e_n\}$ be an orthonormal basis consisting of eigenvectors corresponding to the eigenvalues $\lambda_i \geq 0$. An arbitrary vector x has the expansion $x = \sum_{i=1}^n (x, e_i)e_i$, and hence

$$
\begin{aligned}
(Tx, x) &= \left(T\sum_{i=1}^n (x, e_i)e_i, \sum_{j=1}^n (x, e_j)e_j \right) \\
&= \sum_{i=1}^n \sum_{j=1}^n (x, e_i)\overline{(x, e_j)}(Te_i, e_j) \\
&= \sum_{i=1}^n \sum_{j=1}^n (x, e_i)\overline{(x, e_j)}(\lambda_i e_i, e_j) \\
&= \sum_{i=1}^n \lambda_i |(x, e_i)|^2 \geq 0.
\end{aligned}
$$

2. The proof is the same except for the inequalities being strict.   □

An easy way of producing nonnegative operators is to consider operators of the form T or TT^*. The next result shows that this is the only way.

Proposition 7.4.7 $T \in L(U)$ *is nonnegative if and only if $T = A^*A$ for some operator A in U.*

Proof: That A^*A is nonnegative is immediate.

Conversely, assume that T is nonnegative. By Theorem 7.4.1, there exists an orthonormal basis $\mathcal{B} = \{e_1, \ldots, e_n\}$ in U made of eigenvectors corresponding to the eigenvalues $\lambda, \ldots, \lambda_n$. Since T is nonnegative, all of the λ_i are nonnegative and have nonnegative square roots $\lambda_i^{1/2}$. Define a linear transformation S by letting $Se_i = \lambda_i^{1/2}e_i$. Clearly, $S^2 e_i = \lambda_i e_i = Te_i$. The operator S so defined is self-adjoint, for, given $x, y \in U$, we have

$$
\begin{aligned}
(Sx, y) &= \left(S\sum_{i=1}^{n}(x, e_i)e_i, \sum_{j=1}^{n}(y, e_j)e_j \right) \\
&= \left(\sum_{i=1}^{n}\lambda_i^{\frac{1}{2}}(x, e_i)e_i, \sum_{j=1}^{n}(y, e_j)e_j \right) \\
&= \left(\sum_{i=1}^{n}(x, e_i)e_i, \sum_{j=1}^{n}\lambda_i^{\frac{1}{2}}(y, e_j)e_j \right) \\
&= \left(\sum_{i=1}^{n}(x, e_i)e_i, S\sum_{j=1}^{n}(y, e_j)e_j \right) \\
&= (x, Sy).
\end{aligned}
$$

So $T = S^2 = S^*S$. $\square$

Effectively, in the proof of the previous theorem, we have constructed a nonnegative square roof. We formalize this.

Theorem 7.4.9 *A nonnegative operator T has a unique nonnegative square root.*

Proof: The existence of a nonnegative square root has been proved in Proposition 7.4.7.

To prove uniqueness, let $\lambda_1 \geq \cdots \geq \lambda_n$ be the eigenvalues of T and $\mathcal{B} = \{e_1, \ldots, e_n\}$ an orthonormal basis in U made of eigenvectors. Let A be an arbitrary nonnegative square root of T, that is, $T = A^2$, since

$$
0 = (\lambda_i I - A^2)e_i = (\lambda_i^{1/2}I + A)(\lambda_i^{1/2}I - A)e_i.
$$

Now, in case $\lambda_i > 0$, the operators $\lambda_i^{1/2}I + A$ are invertible, and hence $Ae_i = \lambda_i^{1/2}e_i$. In case $\lambda_i = 0$, we compute

$$
\|Ae_i\|^2 = (Ae_i, Ae_i) = (A^2e_i, e_i) = (Te_i, e_i) = 0.
$$

So $Ae_i = 0$. Thus, a nonnegative square root is completely determined by T and hence uniqueness. $\square$

The Polar Decomposition

In analogy with the polar representation of a complex number in the form $z = re^{i\theta}$, we have a representation of an arbitrary linear transformation in an inner product space as the product of a unitary operator and a nonnegative one.

Theorem 7.4.10 *Let T be a linear transformation in an inner product space U. Then there exist nonnegative operators P_1, P_2 and unitary operators V_1, V_2 such that*

$$T = P_1 V_1 = V_2 P_2.$$

If T is nonsingular, then the V_i and P_i are uniquely determined.

Proof: If $T = P_2 V_2$, then $T^* T = P_2 V_2^* V_2 P_2 = P_2^2$. Hence, for the polar decomposition to hold, it is necessary that $P_2 = (T^* T)^{1/2}$. Therefore, given T, we let $P_2 = (T^* T)^{1/2}$ be the unique nonnegative square root of $T^* T$. We have the orthogonal direct sum decompositions $U = \operatorname{Im} P_2 \oplus \operatorname{Ker} P_2$ and $U = \operatorname{Im} T \oplus \operatorname{Ker} T^*$. Moreover, for every $x \in U$, we have

$$\|P_2 x\|^2 = ((T^* T)^{1/2} x, T^* T)^{1/2} x) = (T^* T x, x) = (T x, T x) = \|T x\|^2, \tag{7.12}$$

which shows that $\operatorname{Ker} P_2 = \operatorname{Ker} T$ and hence

$$\dim \operatorname{Ker} P_2 = \dim \operatorname{Ker} T = \dim \operatorname{Ker} T^*. \tag{7.13}$$

By Eq. (7.12), there exists an isometric map V_2 from $\operatorname{Im} P_2$ onto $\operatorname{Im} T$, given by $V_2 P_2 X = T x$. By Eq. (7.13), we can extend V_2, choosing orthonormal bases in $\operatorname{Ker} P_2$ and $\operatorname{Ker} T^*$, to a unitary map in U. Thus, $T = V_2 P_2$ follows.

It is clear from the construction of V_2 that it is uniquely determined on $\operatorname{Im} P_2$; therefore, V_2 is unique if and only if $\operatorname{Ker} P_2 = \{0\}$. However, $\operatorname{Ker} P_2 = \operatorname{Ker} T$, so uniqueness is equivalent to the invertibility of T.

The polar decomposition $T = P_1 V_1$ is proved analogously, or it can be obtained by duality considerations, starting from T^*.   □

Partial Isometries

Definition 7.4.3 *Let U_1, U_2 be inner product spaces. An operator $U :$ $U_1 \longrightarrow U_2$ is called a **partial isometry** if there exists a subspace $M \subset U_1$ such that*

$$\|U x\| = \begin{cases} \|x\| & x \in M \\ 0 & x \perp M. \end{cases}$$

*The space M is called the **initial space** of U and the image of U the **final space**.*

In the following proposition we collect the basic facts on partial isometries:

Proposition 7.4.8 *Let U_1, U_2 be inner product spaces and $U : U_1 \longrightarrow U_2$ a linear transformation. Then:*

1. *U is a partial isometry if and only if $U^* U$ is a projection onto the initial space of U.*

2. U is a partial isometry with initial space M if and only if U^* is a partial isometry with initial space UM.

3. UU^* is the orthogonal projection on the final space of U.

Proof:

1. Assume that U is a partial isometry with initial space M. Let P be the orthogonal projection of U_1 on M. Then, for $x \in M$, we have

$$(U^*Ux, x) = \|Ux\|^2 = \|x\|^2 = (x, x) = (Px, x).$$

On the other hand, if $x \perp M$, then

$$(U^*Ux, x) = \|Ux\|^2 = 0 = (Px, x).$$

So $((U^*U - P)x, x) = 0$ for all x, and hence, $U^*U = P$.

Conversely, suppose that $P = U^*U$ is a necessarily orthogonal projection in U_1. Let $M = \{x | U^*Ux = x\}$. Then, for any $x \in U_1$,

$$\|Ux\|^2 = (U^*Ux, x) = (Px, x) = \|Px\|^2.$$

This shows that U is a partial isometry with initial space M.

2. Assume that U is a partial isometry with initial space M. Note that we have $U_2 = \text{Im}\, U \oplus \text{Ker}\, U^* = UM \oplus \text{Ker}\, U^*$. Obviously, $U^* | \text{Ker}\, U^* = 0$, whereas, if $x \in M$, then $U^*Ux = Px = x$. So $\|U^*Ux\| = \|x\| = \|Ux\|$, that is, U^* is a partial isometry with initial space UM. By the previous part, it follows that UU^* is the orthogonal projection on UM, the final space of U.

3. From part 1, it follows that UU^* is the orthogonal projection on the initial space of U^*, which is UM, the final space of U.   □

7.5   Singular Vectors and Singular Values

In this section we study the basic properties of singular vectors and singular values. We use this to give a characterization of singular vectors as approximation numbers. These results will be applied in Chapter 11 to Hankel norm approximation problems.

Starting from a polar decomposition of a linear transformation, and noting that the norm of a unitary operator, and, for that matter, also isometries and nontrivial partial isometries, U is 1, it seems plausible that the operators $(T^*T)^{1/2}$ and $(TT^*)^{1/2}$ provide a measure on the size of T.

Proposition 7.5.1 *Let T be a linear transformation and let $(T^*T)^{1/2}$ and $(TT^*)^{1/2}$ be the nonnegative square roots of T^*T and TT^*, respectively. Then:*

1. $(T^*T)^{1/2}$ *and* $(TT^*)^{1/2}$ *have the same eigenvalues, including multiplicities.*

2. *Let* $\mu_1 \geq \cdots \geq \mu_n \geq 0$ *be the singular values of* $(T^*T)^{1/2}$. *Let* $\{\phi_1, \ldots, \phi_n\}$ *be an orthonormal basis of* U, *for which* $\{\phi_1, \ldots, \phi_r\}$ *is an orthonormal basis of* $\mathrm{Im}\,(T^*T)^{1/2}$, *consisting of eigenvectors of* $(T^*T)^{1/2}$ *that correspond to the singular values* $\mu_1 \geq \cdots \geq \mu_r \geq 0$, *and* $\{\phi_{r+1}, \ldots, \phi_n\}$ *is an orthonormal basis of* $\mathrm{Ker}\,T^*$. *Define, for* $i = 1, \ldots, r$, $\psi_i = (T\phi_i)/\mu_i$ *and take* $\{\psi_{r+1}, \ldots, \psi_n\}$ *as any orthonormal basis of* $\mathrm{Ker}\,T$. *Then* ψ_i *are eigenvectors of* $(TT^*)^{1/2}$ *corresponding to the eigenvalues* μ_i.

3. *If we define* $V\phi_i = \psi_i$, *then* V *is unitary and* $T = V(T^*T)^{1/2}$.

Proof: We know that $\dim \mathrm{Ker}\,T = \dim \mathrm{Ker}\,T^*$. Define a linear transformation $V : U \longrightarrow U$ by $V\phi_i = \psi_i$. Clearly, V is a unitary transformation. We compute

$$
\begin{aligned}
TT^* &= TT^* \frac{T\phi_i}{\mu_i} = \frac{T}{\mu_i}(T^*T\phi_i) = \frac{T}{\mu_i}(\mu_i^2 \phi_i) \\
&= \mu_i^2 \frac{T\phi_i}{\mu_i} = \mu_i^2 \psi_i,
\end{aligned}
$$

and hence also $(TT^*)^{1/2}\psi_i = \mu_i \psi_i$, that is, the μ_i are the eigenvalues of $(TT^*)^{1/2}$ and the multiplicities coincide with those eigenvalues of $(T^*T)^{1/2}$.

We now compute

$$
T\phi_i = \mu_i \psi_i = \mu_i V\phi_i = V\mu_i \phi_i = V(T^*T)^{1/2}\phi_i,
$$

that is, $T = V(T^*T)^{1/2}$, and we have the polar decomposition of T. $\square$

The introduction of the eigenvectors of $(T^*T)^{1/2}$ and $(TT^*)^{1/2}$ was done in a nonsymmetric way. We note however that in the proof we showed that $T\phi_i = \mu_i \psi_i$ and

$$
T^*\psi_i = T^* \frac{T\phi_i}{\mu_i} = \frac{\mu_i^2 \phi_i}{\mu_i} = \mu_i \phi_i.
$$

Moreover, suppose that we have a scalar μ and a pair of vectors ϕ, ψ satisfying

$$
\begin{cases}
T\phi &= \mu\psi \\
T^*\psi &= \mu\phi.
\end{cases}
$$

Then it follows that $T^*T\phi = \mu T^*\psi = \mu^2 \phi$ and $TT^*\psi = \mu T\psi = \mu^2 \psi$, that is, ϕ, ψ are a pair of eigenvectors of T and T^* corresponding to the same eigenvalue μ.

Any such nonzero scalar μ is necessarily real, and furthermore, we have $\|\phi\| = \|\psi\|$. This is seen from the following computation:

$$\mu(\psi, \psi) = (T\phi, \psi) = (\phi, T^*\psi) = (\phi, \mu\psi) = \overline{\mu}(\phi, \phi).$$

Then it follows that $T^*T\phi = \mu T^*\psi = \mu^2\phi$ and, without loss of generality, by multiplication by -1, we may assume that $\mu > 0$.

This leads us to the following definition:

Definition 7.5.1 *Let $T : U \longrightarrow U_1$ be a linear transformation. A pair of vectors $\{\phi, \psi\}$, with $\phi \in U$ and $\psi \in U_1$, is called a **Schmidt pair** of T corresponding to the **singular value** μ if*

$$\begin{cases} T\phi &= \mu\psi \\ T^*\psi &= \mu\phi \end{cases} \tag{7.14}$$

holds.

We also refer to ϕ, ψ as the **singular vectors** of T and T^*, respectively.

The minimax principle leads to the following characterization of singular values:

Theorem 7.5.1 *Let U_1, U_2 be inner product spaces, and let $T : U_1 \longrightarrow U_2$ be a linear transformation. Let $\mu_1 \geq \cdots \geq \mu_n \geq 0$ be its singular values. Then*

$$\mu_k = \min_{\operatorname{codim} M = k-1} \max_{x \in M} \frac{\|Tx\|}{\|x\|}.$$

Proof: The μ_i^2 are the eigenvalues of T^*T. Applying the minimax principle, we have

$$\mu_k^2 = \min_{\operatorname{codim} M = k-1} \max_{x \in M} \frac{(T^*Tx, x)}{(x, x)}$$

$$= \min_{\operatorname{codim} M = k-1} \max_{x \in M} \frac{\|Tx\|^2}{\|x\|^2},$$

which is equivalent to the statement of the theorem.   □

With the use of the polar decomposition and the spectral representation of self-adjoint operators, we get a convenient representation of arbitrary operators.

Theorem 7.5.2 *Let $T : U \longrightarrow U_1$ be a linear transformation, let $\mu_1 \geq \cdots \geq \mu_n \geq 0$ be its singular values, and let $\{\phi_i, \psi_i\}$ be the corresponding orthonormal sets of Schmidt pairs. Then, for every vector $x \in U, y \in U_1$, we have*

$$Tx = \sum_{i=1}^{n} \mu_i(x, \phi_i)\psi_i, \tag{7.15}$$

and

$$T^*y = \sum_{i=1}^{n} \mu_i(y, \psi_i)\phi_i. \tag{7.16}$$

Proof: Let $\mu_1 \geq \cdots \geq \mu_r > 0$ be the nonzero singular values and $\{\phi_i, \psi_i\}$ be the corresponding Schmidt pairs.

We have seen that, with V defined by $V\phi_i = \psi_i$, and of course $V^*\psi_i = \phi_i$, we have $T = V(T^*T)^{1/2}$. Now, for $x \in U$, we have

$$(T^*T)^{1/2}x = \sum_{i=1}^{n} (x, \phi_i)\phi_i,$$

and hence

$$Tx = V(T^*T)^{1/2}x = V\sum_{i=1}^{n}(x, \phi_i)\phi_i = \sum_{i=1}^{n}\mu_i(x, \phi_i)\psi_i.$$

The other representation for T^* can be derived analogously or, alternatively, can be obtained by computing the adjoint of T using Eq. (7.15). $\qquad\square$

The availability of the previously obtained representations for T and T^* leads directly to an extremely important characterization of singular values in terms of approximation properties. The question we address is the following. Given a linear transformation of rank m between two inner product spaces, how well can it be approximated, in the operator norm, by transformations of lower rank? We have the following characterization:

Theorem 7.5.3 *Let $T : U_1 \longrightarrow U_2$ be a linear transformation of rank m. Let $\mu_1 \geq \cdots \geq \mu_m$ be its nonzero singular values. Then*

$$\mu_k = \inf\{\|T - T'\| \mid \operatorname{rank} T' \leq k - 1\}.$$

Proof: By Theorem 7.5.2, we may assume that $Tx = \sum_{i=1}^{n}\mu_i(x, \phi_i)\psi_i$. We define a linear transformation T' by $T'x = \sum_{i=1}^{k-1}\mu_i(x, \phi_i)\psi_i$. Obviously, $\operatorname{rank}(T') = k - 1$ and $(T - T')x = \sum_{i=k}^{n}\mu_i(x, \phi_i)\psi_i$, and therefore we conclude that $\|T - T'\| = \mu_k$.

Conversely, let us assume that $\operatorname{rank}(T') = k - 1$, that is, $\operatorname{codim} \operatorname{Ker} T' = k - 1$. By Theorem 7.5.1, we have

$$\begin{aligned} \mu_k &\leq \max_{x \in \operatorname{Ker}T'} \frac{\|Tx\|}{\|x\|} \\ &= \max_{x \in \operatorname{Ker}T'} \frac{\|(T - T')x\|}{\|x\|} \leq \|T - T'\|. \qquad\square \end{aligned}$$

7.6   Unitary Embeddings

The existence of nonnegative square roots is a central tool in modern operator theory. In this section we discuss some related embedding theorems.

The basic question we address is, given a linear transformation A in an inner product space U, when can it be embedded in a 2×2 block unitary operator matrix of the form $V = \begin{pmatrix} A & B \\ C & D \end{pmatrix}$?

This block matrix initially is to be considered an operator defined on the inner product space $U \oplus U$ by $\begin{pmatrix} A & B \\ C & D \end{pmatrix} \begin{pmatrix} x \\ y \end{pmatrix} = \begin{pmatrix} Ax+By \\ Cx+Dy \end{pmatrix}$. It is clear that if V given before is unitary, then, obseving that P defined by $P \begin{pmatrix} x \\ y \end{pmatrix} = \begin{pmatrix} x \\ 0 \end{pmatrix}$ is an orthogonal projection, it follows that $A = PV|_{U \oplus \{0\}}$, and we have

$$\|A\| = \|PV|_{U \oplus \{0\}}\| \le \|V\| = 1,$$

that is, A is necessarily a contraction. The following theorem focuses on the converse:

Theorem 7.6.1 *A linear transformation A can be embedded in a 2×2 block unitary matrix if and only if it is a contraction.*

Proof: That embeddability in a unitary matrix implies that A is contractive has been shown before.

Thus, assume that A is contractive, that is, $\|Ax\| \le \|x\|$ for all $x \in U$, since

$$\begin{aligned} \|x\|^2 - \|Ax\|^2 &= (x, x) - (Ax, Ax) = (x, x) - (A^*Ax, x) \\ &= ((I - A^*A)x, x) = \|(I - A^*A)^{\frac{1}{2}} x\|^2. \end{aligned}$$

Here we used the fact that the nonnegative operator $I - A^*A$ has a unique nonnegative square root. This implies that, setting $C = (I - A^*A)^{1/2}$, the transformation given by the block matrix $\begin{pmatrix} A \\ C \end{pmatrix}$ is isometric, that is, it satisfies

$$(\; A^* \quad C^* \;) \begin{pmatrix} A \\ C \end{pmatrix} = I.$$

Similarly, using the requirement that

$$(\; A \quad B \;) \begin{pmatrix} A^* \\ B^* \end{pmatrix} = I,$$

we set $B = (I - AA^*)^{1/2}$. Computing now

$$\begin{pmatrix} I & 0 \\ 0 & I \end{pmatrix} =$$

$$\begin{pmatrix} A & (I - AA^*)^{1/2} \\ (I - A^*A)^{1/2} & D \end{pmatrix} \begin{pmatrix} A^* & (I - A^*A)^{1/2} \\ (I - AA^*)^{1/2} & D^* \end{pmatrix},$$

we have $(I - A^*A)^{1/2}A^* + D(I - AA^*)^{1/2} = 0$. Using the equality $(I - A^*A)^{1/2}A^* = A^*(I - AA^*)^{1/2}A^*$, we get $(A^* + D)(I - AA^*)^{1/2} = 0$. This indicates that we should choose $D = -A^*$. It remains to check that

$$V = \begin{pmatrix} A & (I - AA^*)^{1/2} \\ (I - A^*A)^{1/2} & -A^* \end{pmatrix}$$

is a required embedding. This embedding is not unique. In fact, if K and L are unitary operators in U, then

$$\begin{pmatrix} A & (I - AA^*)^{1/2}L \\ K(I - A^*A)^{1/2} & -KA^*L \end{pmatrix}$$

parametrize all such embeddings. $\qquad\qquad\qquad\qquad\qquad\square$

The embedding presented in the previous theorem is possibly redundant, as it requires a space of twice the dimension of the original space U. For example, if A is unitary to begin with, then no extension is necessary. Thus the dimension of the minimal space where a unitary extension is possible seems to be related to the distance of the contraction A from unitary operators. This generally is measured by the ranks of the operators $(I - A^*A)^{1/2}$ and $(I - AA^*)^{1/2}$. The next theorem handles this situation.

However, before embarking on that route, we give an important dilation result. This is an extension of a result originally obtained in Halmos [1950]. That construction played a central role, via the Schäffer matrix [Sz.-Nagy and Foias 1970] in the theory of unitary dilations.

Theorem 7.6.2 *Given a contraction A in $\mathbf{C}^n$. Then there exists a unitary matrix $\begin{pmatrix} A & B \\ C & D \end{pmatrix}$, with B injective and C surjective, if and only if this matrix has the form*

$$\begin{pmatrix} A & (I - AA^*)^{1/2}V \\ U(I - A^*A)^{1/2} & -UA^*V \end{pmatrix}, \qquad (7.17)$$

where $U : \mathbf{C}^n \longrightarrow \mathbf{C}^p$ is a partial isometry with initial space $\mathrm{Im}\,(I - A^*A)^{1/2}$ and where $V : \mathbf{C}^p \longrightarrow \mathbf{C}^n$ is an isometry with final space $\mathrm{Im}\,(I - AA^*)^{1/2}$.

Proof: Let $P_{\mathrm{Im}\,(I-AA^*)^{1/2}}$ and $P_{\mathrm{Im}\,(I-A^*A)^{1/2}}$ be the orthogonal projections on the appropriate spaces. Assume that U, V are partial isometries as described in the theorem. Then we have

$$\begin{array}{ll} UU^* = I, & U^*U = P_{\mathrm{Im}\,(I-A^*A)^{1/2}} \\ V^*V = I, & VV^* = P_{\mathrm{Im}\,(I-AA^*)^{1/2}}. \end{array} \qquad (7.18)$$

These identities imply the following:

$$\begin{array}{l} VV^*(I - AA^*)^{1/2} = (I - AA^*)^{1/2} \\ (I - AA^*)^{1/2}VV^* = (I - AA^*)^{1/2}. \end{array} \qquad (7.19)$$

We now compute the product

$$\begin{pmatrix} A & (I - AA^*)^{1/2}V \\ U(I - A^*A)^{1/2} & -UA^*V \end{pmatrix}$$

$$\times \begin{pmatrix} A^* & (I - A^*A)^{1/2}U^* \\ V^*(I - AA^*)^{1/2} & -V^*AU^* \end{pmatrix}.$$

We use the identities in Eq. (7.19). So

$$AA^* + (I - AA^*)^{1/2}VV^*(I - AA^*)^{1/2}$$
$$= AA^* + (I - AA^*)^{1/2}(I - AA^*)^{1/2} = I.$$

Next,

$$A(I - A^*A)^{1/2}U^* - (I - AA^*)^{1/2}VV^*AU^*$$
$$= A(I - A^*A)^{1/2}U^* - (I - AA^*)^{1/2}AU^* = 0.$$

Similarly,

$$U(I - A^*A)^{1/2}A^* - UA^*VV^*(I - AA^*)^{1/2}$$
$$= U[(I - A^*A)^{1/2}A^* - A^*(I - AA^*)^{1/2}] = 0.$$

Finally,

$$U(I - A^*A)^{1/2}(I - A^*A)^{1/2}U^* + UA^*VV^*AU^*$$
$$= U(I - A^*A)U^* + UA^*AU^* = UU^* = I.$$

This shows that the matrix in Eq. (7.17) is indeed unitary.

Conversely, suppose that we are given a contraction A. If C satisfies $A^*A + C^*C = I$, then $C^*C = I - A^*A$. This in turn implies that for every vector x we have

$$\|Cx\|^2 = \|(I - A^*A)^{1/2}x\|^2.$$

Therefore, there exists a partial isometry U defined on $\operatorname{Im}(I - A^*A)^{1/2}$ for which

$$C = U(I - A^*A)^{1/2}. \tag{7.20}$$

Obviously, $\operatorname{Im} C \subset \operatorname{Im} U$. Since C is surjective, this forces U to be surjective. Thus, $UU^* = I$ and $U^*U = P_{\operatorname{Im}(I-A^*A)^{1/2}}$. In a similar fashion, we start from the equality $AA^* + BB^* = I$ and see that, for every x,

$$\|B^*x\|^2 = \|(I - AA^*)^{1/2}x\|^2.$$

So we conclude that there exists a partial isometry V^* with initial space $\operatorname{Im}(I - AA^*)^{1/2}$ for which

$$B^* = V^*(I - AA^*)^{1/2}$$

or

$$B = (I - AA^*)^{1/2}V. \tag{7.21}$$

As B is injective, so is V. So we get $V^*V = I$ and $VV^* = P_{\operatorname{Im}(I-AA^*)^{1/2}}$.

Next we use the equality $DB^* + CA^* = 0$ to get

$$DV^*(I - AA^*)^{1/2} + U(I - A^*A)^{1/2}A^* = (DV^* + UA^*)(I - AA^*)^{1/2} = 0.$$

So $DV^* + UA^*|_{\text{Im}\,(I-AA^*)^{1/2}} = 0$. Now, we have $\mathbf{C}^n = \text{Im}\,(I - AA^*)^{1/2} \oplus \text{Ker}\,(I - AA^*)^{1/2}$. Moreover, the identity $A(I - A^*A)^{1/2} = (I - AA^*)^{1/2}A$ implies that $A\text{Im}\,(I - A^*A)^{1/2} \subset \text{Im}\,(I - AA^*)^{1/2}$ as well as $A\text{Ker}\,(I - A^*A)^{1/2} \subset \text{Ker}\,(I - AA^*)^{1/2}$. Similarly, $A^*\text{Ker}\,(I - AA^*)^{1/2} \subset \text{Ker}\,(I - A^*A)^{1/2}$.

So now let $x \in \text{Ker}\,(I - AA^*)^{1/2}$. Then $A^*x \in \text{Ker}\,(I - A^*A)^{1/2}$. As U is a partial isometry with initial space $\text{Im}\,(I - A^*A)^{1/2}$, necessarily $UA^*x = 0$. On the other hand, V^* is a partial isometry with initial space $(I - AA^*)^{1/2}$, and hence $V^*|_{\text{Ker}\,(I-AA^*)^{1/2}} = 0$. So $DV^* + UA^*|_{\text{Ker}\,(I-AA^*)^{1/2}} = 0$, and hence we can conclude that $DV^* + UA^* = 0$. Using the fact that $V^*V = I$, this implies that

$$D = -UA^*V. \tag{7.22}$$

It is easy to check now, as in the sufficiency part that with B, C, D satisfying Eqs. (7.20), (7.21), and (7.22), respectively, that all other equations are satisfied. $\qquad\square$

7.7 Exercises

1. Show that the rational function $q(z) = \sum_{k=-n}^{n} q_k z^k$ is nonnegative on the unit circle if and only if $q(z) = p(z)p^\sharp(z)$ for some polynomial p. Here, $p^\sharp(z) = \overline{p(\bar{z}^{-1})}$. This theorem, the simplest example of spectral factorization, is due to Fejer [1915].

2. Show that a polynomial $q(z)$ is nonnegative on the imaginary axis if and only if, for some polynomial p, we have $q(z) = p(z)\overline{p(-\bar{z})}$.

3. Given vectors $x_1, \ldots, x_k$ in an inner product space U. The Gram determinant is defined by $G(x_1, \ldots, x_k) = \det((x_i, x_j))$. Show that $x_1, \ldots, x_k$ are linearly independent if and only if $\det G(x_1, \ldots, x_k) \neq 0$.

4. Let X be an inner product space of functions. Let $f_0, f_1, \ldots$ be a sequence of linearly independent functions in X, and let $\phi_0, \phi_1, \ldots$ be the sequence of functions obtained from it by way of the Gram–Schmidt orthonormalization procedure.

   (a) Show that

   $$\phi_n(x) = \begin{vmatrix} (f_0, f_0) & \cdots & (f_n, f_0) \\ \vdots & & \vdots \\ (f_0, f_{n-1}) & \cdots & (f_n, f_{n-1}) \\ f_0(x) & \cdots & f_n(x) \end{vmatrix}.$$

(b) Let

$$\phi_n(x) = \sum_{i=0}^{n} c_{n,i} f_i(x),$$

and define

$$\begin{cases} D_{-1} = 1 \\ \\ D_n = \begin{vmatrix} (f_0, f_0) & \cdot & \cdot & \cdot & (f_n, f_0) \\ \cdot & \cdot & \cdot & \cdot & \cdot \\ \cdot & \cdot & \cdot & \cdot & \cdot \\ \cdot & \cdot & \cdot & \cdot & \cdot \\ \cdot & \cdot & \cdot & \cdot & \cdot \\ (f_0, f_n) & \cdot & \cdot & \cdot & (f_n, f_n) \end{vmatrix}. \end{cases}$$

Show that

$$c_{n,n} = \left(\frac{D_{n-1}}{D_n}\right)^{1/2}.$$

5. In the space of real polynomials $\mathbf{R}[x]$, an inner product is defined by

$$(p, q) = \int_{-1}^{1} p(x)q(x)dx.$$

Apply the Gram–Schmidt orthogonalization procedure to the polynomials $1, x, x^2, \ldots$. Show that, up to multiplicative constants, the resulting polynomials coincide with the Legendre polynomials

$$P_n(x) = \frac{1}{2^n n!} \frac{d^n (x^2 - 1)^n}{dx^n}.$$

6. Show that the rank of a skew-symmetric matrix in a real inner product space is even.

7. Let K be a skew-Hermitian operator, that is, $K^* = -K$. Show that $U = (I + K)^{-1}(I - K)$ is unitary. Show that every unitary operator is of the form $U = \lambda(I + K)^{-1}(I - K)$, where λ is a complex number of absolute value 1.

8. Show that if A and B are normal and $\operatorname{Im} A \perp \operatorname{Im} B$, then $A + B$ is normal.

9. Show that A is normal if and only if, for some unitary U, we have $A^* = AU$.

10. Let N be a normal operator and let $AN = NA$. Show that $AN^* = N^*A$.

11. Given a normal operator N and any positive integer k. Show that there exists an operator A satisfying $A^k = N$.

12. Show that a circulant matrix is normal.

13. Let T be a real orthogonal matrix, that is, it satisfies $\tilde{T}T = I$. Show that T is similar to a block diagonal matrix $\operatorname{diag}(I, R_1, \ldots, R_k)$, where
$$R_i = \left(\begin{array}{cc} \cos\theta_i & \sin\theta_i \\ -\sin\theta_i & \cos\theta_i \end{array} \right).$$

14. Show that a linear operator A in an inner product space, having eigenvalues $\lambda_1, \ldots, \lambda_n$, is normal if and only if trace $(A^*A) = \sum_{i=1}^{n} |\lambda_i|^2$.

15. Given contractive operators in finite-dimensional inner product spaces U_1 and U_2, respectively. Show that there exists an operator $X : U_2 \longrightarrow U_1$ for which $\left(\begin{smallmatrix} A & X \\ 0 & B \end{smallmatrix} \right)$ is contractive if and only if $X = (I - AA^*)^{1/2}C(I - B^*B)^{1/2}$ and $C : U_2 \longrightarrow U_1$ a contraction.

16. Define the Jacobi tridiagonal matrices by
$$J_k = \left(\begin{array}{cccccc} a_1 & -b_1 & \cdot & \cdot & & 0 \\ -c_1 & a_2 & \cdot & & & \cdot \\ \cdot & \cdot & \cdot & \cdot & & \cdot \\ \cdot & & \cdot & \cdot & \cdot & -b_{k-1} \\ 0 & & \cdot & \cdot & -c_{k-1} & a_k \end{array} \right),$$
assuming that $b_i, c_i > 0$. Let $D_k(\lambda) = \det(\lambda I - J_k)$.

(a) Show that the following recurrence relation is satisfied:
$$D_k(\lambda) = (\lambda - a_k)D_{k-1}(\lambda) - b_{k-1}c_{k-1}D_{k-2}(\lambda).$$

(b) Show that J_k is similar to a self-adjoint transformation.

17. Let $\sigma_1, \ldots, \sigma_n$ be the singular values of A. Show that the eigenvalues of $\left(\begin{smallmatrix} 0 & A \\ A^* & 0 \end{smallmatrix} \right)$ are $\sigma_1, \ldots, \sigma_n, -\sigma_1, \ldots, -\sigma_n$.

7.8 Notes and Remarks

Most of the results in this chapter go over in one way or another to the context of Hilbert spaces, where in fact most of them were proved initially. Young [1988] is a good modern source for the basics of Hilbert space theory.

The spectral theorem in Hilbert space was one of the major early achievements of functional analysis and operator theory. For an extensive source on these topics and the history of the subject see Dunford and Schwartz [1958, 1963].

8
Quadratic Forms

In this chapter we discuss both the general theory of quadratic forms, the notion of congruence and its invariants, and the applications of the theory to the analysis of special forms. We focus on quadratic forms induced by rational functions, most notably the Hankel and Bezout forms, because of their connection to system-theoretic problems like stability and signature-symmetric realizations. These forms use as their data different representations of rational functions, power series, and coprime factorizations, respectively. But we also will discuss the partial fraction representation in relation to the computation of the Cauchy index of a rational function, the proof of the Hermite–Hurwitz theorem, and the continued fraction representation as a tool in the computation of signatures of Hankel matrices as well as in the problem of Hankel matrix inversion. Thus, different representations of rational functions, that is, different encodings of the information carried by a rational function, provide efficient starting points for different methods. The results obtained for rational functions are applied to root location problems for polynomials in the next chapter.

8.1 Preliminaries

Given a linear transformation T in an inner product space U, we define a field-valued function ϕ on $U \times U$ by

$$\phi(x, y) = (Tx, y). \tag{8.1}$$

Clearly, ϕ is linear in the variable x and antilinear in the variable y. Such a function will be called a **sesquilinear form** or just a **form**. If the field is the field $\mathbf{R}$ of real numbers, then a form is actually linear in both variables, that is, it is a **bilinear form**.

It might seem that the forms defined by Eq. (8.1) are rather special. This is not the case, as is seen from the following:

Theorem 8.1.1 *Let U be a finite-dimensional inner product space. The ϕ is a form on U if and only if there exists a linear transformation T in U such that Eq. (8.1) holds. The operator T is uniquely determined by ϕ.*

Proof: We utilize Theorem 7.1.7 on the representation of linear functionals on inner product spaces. Thus, if we fix a vector $y \in U$, the function $\phi(x, y)$ is a linear functional and hence is given by an inner product with a vector $\eta_{T,y}$, that is, $\phi(x, y) = (x, \eta_{T,y})$. It is clear that $\eta_{T,y}$ depends linearly on Y. Therefore, there exists a linear transformation S in U for which $\eta_{T,y} = Sy$. We complete the proof by defining $T = S^*$.

To see uniqueness, assume that T_1, T_2 represent the same form, that is, $(T_1 x, y) = (T_2 x, y)$. This implies that $((T_1 - T_2)x, y) = 0$ for all $x, y \in U$. Choosing $y = (T_1 - T_2)x$, we get $\|(T_1 - T_2)x\| = 0$ for all $x \in U$. This shows that $T_2 = T_1$. $\qquad\square$

Suppose that we choose a basis $\mathcal{B} = \{f_1, \ldots, f_n\}$ in U. Then arbitrary vectors in U can be written as $x = \sum_{j=1}^{n} \xi_j f_j$ and $x = \sum_{i=1}^{n} \eta_i f_i$. In terms of the coordinates, we can compute the form by

$$
\begin{aligned}
\phi(x, y) &= (Tx, y) = \left(T \sum_{j=1}^{n} \xi_j f_j, \sum_{i=1}^{n} \eta_i f_i \right) \\
&= \sum_{j=1}^{n} \sum_{i=1}^{n} \xi_j \overline{\eta_i} (T f_j, f_i) \\
&= \sum_{j=1}^{n} \sum_{i=1}^{n} \phi_{ij} \xi_j \overline{\eta_i},
\end{aligned}
$$

where we define $\phi_{ij} = (T f_j, f_i)$. We call the matrix (ϕ_{ij}) the **matrix representation of the form** ϕ in the basis $\mathcal{B}$, and we denote this matrix by $[\phi]_{\mathcal{B}}^{\mathcal{B}}$. Using the standard inner product in $\mathbf{C}^n$, we can write

$$
\phi(x, y) = ([\phi]_{\mathcal{B}}^{\mathcal{B}} [x]^{\mathcal{B}}, [x]^{\mathcal{B}}).
$$

Next we consider how the matrix of a quadratic form changes with a change of basis in U. Let $\mathcal{B}' = \{g_1, \ldots, g_n\}$ be another basis in U. Then $[x]^{\mathcal{B}} = [I]_{\mathcal{B}}^{\mathcal{B}'} [x]^{\mathcal{B}'}$, and therefore,

$$
\begin{aligned}
\phi(x, y) &= ([\phi]_{\mathcal{B}}^{\mathcal{B}} [x]^{\mathcal{B}}, [y]^{\mathcal{B}}) = ([\phi]_{\mathcal{B}}^{\mathcal{B}} [I]_{\mathcal{B}}^{\mathcal{B}'} [x]^{\mathcal{B}'}, [I]_{\mathcal{B}}^{\mathcal{B}'} [y]^{\mathcal{B}'}) \\
&= (([I]_{\mathcal{B}}^{\mathcal{B}'})^* [\phi]_{\mathcal{B}}^{\mathcal{B}} [I]_{\mathcal{B}}^{\mathcal{B}'} [x]^{\mathcal{B}'}, [y]^{\mathcal{B}'}) = ([\phi]_{\mathcal{B}'}^{\mathcal{B}'} [x]^{\mathcal{B}'}, [y]^{\mathcal{B}'}),
\end{aligned}
$$

which shows that

$$[\phi]_{\mathcal{B}'}^{\mathcal{B}'} = ([I]_{\mathcal{B}}^{\mathcal{B}'})^* [\phi]_{\mathcal{B}}^{\mathcal{B}} [I]_{\mathcal{B}}^{\mathcal{B}'}.$$

The next result clarifies the connection between the matrix representation of a form and the matrix representation of the corresponding linear transformation.

Proposition 8.1.1 *Let U be an inner product space and ϕ a Hermitian form on U. Let T be the uniquely defined linear transformation in U for which $\phi(x,y) = (Tx,y)$, and let $\mathcal{B} = \{f_1, \ldots, f_n\}$ for U. Then $[\phi]_{\mathcal{B}}^{\mathcal{B}} = [T]_{\mathcal{B}}^{\mathcal{B}^*}$.*

Proof: Let $\mathcal{B}^* = \{g_1, \ldots, g_n\}$ be the basis dual to $\mathcal{B}$. Then $[T]_{\mathcal{B}}^{\mathcal{B}^*} = (t_{ij})$, where the t_{ij} are defined through $Tf_j = \sum_{k=1}^n t_{kj} g_k$. Computing

$$(Tf_j, f_i) = \left(\sum_{k=1}^n t_{kj} g_k, f_i \right) = \sum_{k=1}^n t_{kj} \delta_{ik} = t_{ij},$$

the result follows.   □

We say that a complex, square matrix A_1 is **congruent** to a matrix A if there exists a nonsingular matrix R such that $A_1 = R^* A R$. In case we deal with the real field, R^*, the Hermitian adjoint is replaced by the transpose $\tilde{R}$.

Proposition 8.1.2 *In the ring of square, real, or complex matrices, congruence is an equivalence relation.*

A **quadratic form** in an inner product space U is a function of the form

$$\hat{\phi}(x) = \phi(x, x),$$

where $\phi(x, y)$ is a symmetric bilinear form on U. We say that a form on U is a **Hermitian form** if we have the additional property

$$\phi(x, y) = \overline{\phi(y, x)}.$$

Proposition 8.1.3 *A form ϕ on an inner product space U is Hermitian if and only if there exists a Hermitian operator T in U for which $\phi(x, y) = (Tx, y)$.*

Proof: Assume that T is Hermitian, that is, $T^* = T$. Then

$$\phi(x, y) = (Tx, y) = (x, Ty) = \overline{(Ty, x)} = \overline{\phi(y, x)}.$$

Conversely, assume that $\phi(x, y) = \overline{\phi(y, x)}$. Let T be the linear transformation for which $\phi(x, y) = (Tx, y)$. We compute

$$(Tx, y) = \phi(x, y) = \overline{\phi(y, x)} = \overline{(Ty, x)} = (x, Ty),$$

and this for all $x, y \in U$. This shows that T is Hermitian.   □

We are interested in the possibility of reducing Hermitian forms, using congruence transformations, to a simple form.

Theorem 8.1.2 *Let ϕ be a Hermitian form on a finite-dimensional inner product space U. Then there exists an orthonormal basis $\mathcal{B} = \{e_1, \ldots, e_n\}$ in U for which $[\phi]_{\mathcal{B}}^{\mathcal{B}}$ is diagonal with real entries.*

Proof: Let $T : U \longrightarrow U$ be the Hermitian operator representing the form ϕ. By Theorem 7.4.1, there exists an orthonormal basis $\mathcal{B} = \{e_1, \ldots, e_n\}$ in U consisting of eigenvectors of T corresponding to the real eigenvalues $\lambda_1, \ldots, \lambda_n$. With respect to this basis, the matrix of the form is given by

$$\phi_{ij} = (Te_j, e_i) = \lambda_j(e_j, e_i) = \lambda_j \delta_{ij}. \qquad \square$$

We can utilize the existence of square roots in the complex field to get a further reduction.

Corollary 8.1.1 *Let ϕ be a Hermitian form on the complex inner product space U. Then there exists an orthogonal basis $\mathcal{B} = \{f_1, \ldots, f_n\}$ for which the matrix $[\phi]_{\mathcal{B}}^{\mathcal{B}}$ is diagonal with*

$$\phi_{ii} = \begin{cases} 1 & 0 \leq i \leq r \\ 0 & r < i \leq n. \end{cases}$$

Just as positive operators are a subset of self-adjoint operators, so positive forms are a subset of Hermitian forms. We say that a form ϕ on an inner product space U is nonnegative if it is Hermitian and $\phi(x, x) \geq 0$ for all $x \in U$. We say that a form ϕ on an inner product space U is positive if it is Hermitian and $\phi(x, x) > 0$ for all $0 \neq x \in U$.

Proposition 8.1.4 *A form ϕ on an inner product space U is positive if and only if there exists a positive operator P for which $\phi(x, y) = (Px, y)$.*

Of course, to check positivity of a form it suffices to compute all of the eigenvalues and check them for positivity. However, computing eigenvalues is not, in general, an algebraic process. So our next goal is to search for a computable criterion for positivity.

8.2 Sylvester's Law of Inertia

If we deal with real forms, then, due to the different definition of congruence, we cannot apply the complex result directly.

The proof of Theorem 8.1.2 is particularly simple. However, it uses the spectral theorem, and that theorem is constructive only if we have access to the eigenvalues of the appropriate self-adjoint operator. This is generally not the case. Thus, we would like to reprove the diagonalization by the congruence result again. It is satisfying that this can be done algebraically, and the next theorem, known as the **Lagrange reduction method**, accomplishes this.

Theorem 8.2.1 *Let ϕ be a quadratic form on a real inner product space U. Then there exists a basis $\mathcal{B} = \{f_1, \ldots, f_n\}$ of U for which $[\phi]_\mathcal{B}^\mathcal{B}$ is diagonal.*

Proof: Let $\phi(x, y)$ be a quadratic form on U. We prove the theorem by induction on the dimension n of U. If $n = 1$, then obviously $[\phi]_\mathcal{B}^\mathcal{B}$, being a 1×1 matrix, is diagonal.

Assume that we proved the theorem for spaces of dimension $\leq n - 1$. Now let ϕ be a quadratic form in an n-dimensional space U. Assume that $\mathcal{B} = \{f_1, \ldots, f_n\}$ is an arbitrary basis of U.

We consider two cases. In the first, we assume that not all of the diagonal elements of $[\phi]_\mathcal{B}^\mathcal{B}$ are zero, that is, there exists an index i for which $\phi(f_i, f_i) \neq 0$. The second case is the complementary one where all of the $\phi(f_i, f_i)$ are zero.

Case 1: Without loss of generality, by using symmetric transpositions of rows and columns, we assume that $\phi(f_i, f_i) \neq 0$. We define a new basis $\mathcal{B}' = \{f_1', \ldots, f_n'\}$, where

$$
f_i' = \begin{cases} f_1 & i = 1 \\ f_i - \dfrac{\phi(f_1, f_i)}{\phi(f_1, f_1)} f_1 & 1 < i \leq n. \end{cases}
$$

Clearly, $\mathcal{B}'$ is indeed a basis for U and

$$
\phi(f_1', f_i') = \begin{cases} \phi(f_1, f_1) & i = 1 \\ 0 & 1 < i \leq n. \end{cases}
$$

Thus, with the subspace $M = L(f_2', \ldots, f_n')$, we have

$$
[\phi]_{\mathcal{B}'}^{\mathcal{B}'} = \begin{pmatrix} \phi(f_1, f_1) & 0 \\ 0 & [\phi|M]_{\mathcal{B}_0'}^{\mathcal{B}_0'} \end{pmatrix}.
$$

Here, $\phi|M$ is the restriction of ϕ to M, and $\mathcal{B}_0' = \{f_2', \ldots, f_n'\}$ is a basis for M. The proof, in this case, is completed by using the induction hypothesis.

Case 2: Assume now that $\phi(f_i, f_i) = 0$ for all i. If $\phi(f_i, f_j) = 0$ for all i, j, then ϕ is the zero form and hence $[\phi]_\mathcal{B}^\mathcal{B}$ is the zero matrix, which is certainly diagonal. So we assume that there exists a pair of distinct indices i, j for which $\phi(f_i, f_j) \neq 0$. Without loss of generality, we assume that $i = 1, j = 2$. To simplify notation, we write $a = \phi(f_1, f_2)$. First, we note that, defining $f_1' = (f_1 + f_2)/2$ and $f_2' = (f_1 - f_2)/2$, then

$$
\begin{aligned}
\phi(f_1', f_1') &= \phi(f_1 + f_2, f_1 + f_2) = \frac{a}{2} \\
\phi(f_1', f_2') &= \phi(f_1 + f_2, f_1 - f_2) = 0 \\
\phi(f_2', f_2') &= \phi(f_1 - f_2, f_1 - f_2) = -\frac{a}{2}.
\end{aligned}
$$

We choose the other vectors in the new basis to be of the form $f_i' = f_i - \alpha_i f_1 - \beta_i f_2$, with the requirement that, for $i = 3, \ldots, n$, we have $\phi(f_1', f_i') = \phi(f_2', f_i') = 0$. Therefore, we get the pair of equations

$$
\begin{aligned}
0 &= \phi(f_1 + f_2, f_i - \alpha_i f_1 - \beta_i f_2) \\
&= \phi(f_1, f_i) + \phi(f_2, f_i) - \alpha_i \phi(f_2, f_1) - \beta_i \phi(f_2, f_1) \\
0 &= \phi(f_1 - f_2, f_i - \alpha_i f_1 - \beta_i f_2) \\
&= \phi(f_1, f_i) + \phi(f_2, f_i) + \alpha_i \phi(f_2, f_1) - \beta_i \phi(f_2, f_1).
\end{aligned}
$$

This system has the solution $\alpha_i = [\phi(f_2, f_i)]/a$ and $\beta_i = [\phi(f_1, f_i)]/a$. We now define $\mathcal{B}' = \{f_1', \ldots, f_n'\}$, where

$$
\begin{aligned}
f_1' &= \frac{f_1 + f_2}{2} \\
f_2' &= \frac{f_1 - f_2}{2} \qquad\qquad\qquad\qquad (8.2) \\
f_i' &= f_i - \frac{\phi(f_2, f_i)}{a} f_1 - \frac{\phi(f_1, f_i)}{a} f_2.
\end{aligned}
$$

It is easy to check that $\mathcal{B}'$ is a basis for U. With respect to this basis, we have the block diagonal matrix

$$
[\phi]_{\mathcal{B}'}^{\mathcal{B}'} = \begin{pmatrix} \dfrac{a}{2} & 0 & \\ 0 & -\dfrac{a}{2} & \\ & & [\phi|M]_{\mathcal{B}_0'}^{\mathcal{B}_0'} \end{pmatrix}.
$$

Here, $\mathcal{B}_0' = \{f_3', \ldots, f_n'\}$ and $M = L(f_3', \ldots, f_n')$. We complete the proof by applying the induction hypothesis to $\phi|M$. □

Corollary 8.2.1 *Let ϕ be a quadratic form defined on a real inner product space U. Let $\mathcal{B} = \{f_1, \ldots, f_n\}$ be an arbitrary basis of U, and let $\mathcal{B}' = \{g_1, \ldots, g_n\}$ be a basis of U that diagonalizes ϕ. Let $\phi_i = \phi(g_i, g_i)$. If*

$$
[x]^{\mathcal{B}} = \begin{pmatrix} \xi_1 \\ \cdot \\ \cdot \\ \cdot \\ \xi_n \end{pmatrix}, \qquad [x]^{\mathcal{B}'} = \begin{pmatrix} \eta_1 \\ \cdot \\ \cdot \\ \cdot \\ \eta_n \end{pmatrix}
$$

and $[I]_{\mathcal{B}}^{\mathcal{B}'} = (a_{ij})$, then

$$
\phi(x, x) = \sum_{i=1}^{n} \phi_i \Big(\sum_{j=1}^{n} a_{ij} \xi_j\Big)^2. \qquad (8.3)
$$

Proof: We have

$$
\begin{aligned}
\phi(x, x) &= ([\phi]_{\mathcal{B}}^{\mathcal{B}} [x]^{\mathcal{B}}, [x]^{\mathcal{B}}) = ([\phi]_{\mathcal{B}'}^{\mathcal{B}'} [x]^{\mathcal{B}'}, [x]^{\mathcal{B}'}) \\
&= ([\phi]_{\mathcal{B}'}^{\mathcal{B}'} [I]_{\mathcal{B}}^{\mathcal{B}'} [x]^{\mathcal{B}'}, [[I]_{\mathcal{B}}^{\mathcal{B}'} x]^{\mathcal{B}'}).
\end{aligned}
$$

Also, $[x]^{B'} = [I]_B^{B'}[x]^B$ implies that $\eta_i = \sum_{j=1}^n a_{ij}\xi_j$. So

$$\phi(x,x) = ([\phi]_{B'}^{B'}[x]^{B'}, [x]^{B'}) = \sum_{i=1}^n \phi_i \eta_i^2 = \sum_{i=1}^n \phi_i \left(\sum_{j=1}^n a_{ij}\xi_j\right)^2. \qquad \square$$

We say that Eq. (8.3) is a representation of the quadratic form as the **sum of squares**. Such a representation is far from unique. In different representations the diagonal elements as well as the linear forms may differ. We have however the following result, known as Sylvester's law of inertia, which characterizes the congruence invariant.

Theorem 8.2.2 *Let U be a real inner product space and $\phi(x,y)$ a symmetric bilinear form on U. Then there exists a basis $B = \{f_1, \ldots, f_n\}$ for U such that the matrix $[\phi]_B^B$ is diagonal, with $\phi(f_i, f_j) = \epsilon_i \delta_{ij}$, with*

$$\epsilon_i = \begin{cases} 1 & 0 \leq i \leq k \\ -1 & k < i \leq r \\ 0 & r < i \leq n. \end{cases}$$

Moreover, the numbers k and r are uniquely determined by ϕ.

Proof: As the form ϕ is symmetric, there exists a self-adjoint operator T in U for which $\phi(x,y) = (Tx, y)$. In this case, given any basis B of U, the matrix $[\phi]_B^B$ is real symmetric. Let $B = \{e_1, \ldots, e_n\}$ be an orthonormal basis consisting of eigenvectors of T corresponding to the real eigenvalues $\lambda_1, \ldots, \lambda_n$. Without loss of generality, we may assume that $\lambda_i > 0$ for $1 \leq i \leq k$, $\lambda_i < 0$ for $k < i \leq r$, and $\lambda_i = 0$ for $r < i \leq n$. We set $\mu_i = |\lambda_i|^{1/2}$ and $\epsilon_i = \text{sign}\,\lambda_i$. Hence, we have, for all i, $\lambda_i = \epsilon_i \mu_i^2$. We consider now the basis $B_1 = \{\mu_1^{-1}e_1, \ldots, \mu_n^{-1}e_n\}$. Clearly, with respect to this basis, the matrix $[\phi]_{B_1}^{B_1}$ has the required form.

To prove the uniqueness of k and r, we note first that r is equal to the rank of $[\phi]_B^B$. So we have to prove uniqueness for k only. Suppose that we have another basis $B' = \{g_1, \ldots, g_n\}$ such that $\phi(g_j, g_i) = \epsilon_i'\delta_{ij}$, where

$$\epsilon_i' = \begin{cases} 1 & 0 \leq i \leq k' \\ -1 & k' < i \leq r \\ 0 & r < i \leq n. \end{cases}$$

Without loss of generality, we assume that $k < k'$. Given $x \in U$, we have $x = \sum_{i=1}^n \xi_i f_i = \sum_{i=1}^n \eta_i g_i$, that is, the ξ_i and η_i are the coordinates of x with respect to B and B', respectively. Computing the quadratic form we get

$$\phi(x,x) = \sum_{i=1}^n \epsilon_i \xi_i^2 = \sum_{i=1}^n \epsilon_i' \eta_i^2,$$

or

$$\xi_1^2 + \cdots + \xi_k^2 - \xi_{k+1}^2 - \cdots - \xi_r^2 = \eta_1^2 + \cdots + \eta_{k'}^2 - \eta_{k'+1}^2 - \cdots - \eta_r^2.$$

This can be rewritten as

$$\xi_1^2 + \cdots + \xi_k^2 + \eta_{k'+1}^2 + \cdots + \eta_r^2 = \eta_1^2 + \cdots + \eta_{k'}^2 + \xi_{k+1}^2 + \cdots + \xi_r^2. \quad (8.4)$$

Now we consider the set of equations $\xi_1 = \cdots = \xi_k = \eta_{k'+1} = \cdots = \eta_r = 0$. We have here $r - k' + k = r - (k' - k) < r$ equations. Thus, there exists a nontrivial vector x for which not all of the coordinates $\xi_{k+1} = \cdots = \xi_r$ are zero. However, by Eq. (8.4), we get $\xi_{k+1} = \cdots = \xi_r = 0$, which is a contradiction. Thus, we must have $k' \geq k$, and by symmetry we get $k' = k$. $\quad\square$

Definition 8.2.1 *The numbers r and $s = 2k - r$ determined by Sylvester's law of inertia are called the* **rank** *and* **signature**, *respectively, of the quadratic form ϕ. We denote the signature of a quadratic form by $\sigma(\phi)$. Similarly, the signature of a symmetric (Hermitian) matrix A is denoted by $\sigma(A)$.*

Corollary 8.2.2

1. *Two quadratic forms on a real inner product space U are congruent if and only if they have the same rank and signature.*

2. *Two $n \times n$ symmetric matrices are congruent if and only if they have the same rank and signature.*

Theorem 8.2.3 *Let A be an $m \times m$ real symmetric matrix, X an $m \times n$ matrix of full row rank, and B the $n \times n$ symmetric matrix defined by*

$$B = \tilde{X}AX. \quad (8.5)$$

Then the rank and signature of A and B coincide.

Proof: The statement concerning ranks is trivial. From Eq. (8.5) we obtain the following two inclusions: $\operatorname{Ker} B \supset \operatorname{Ker} X$ and $\operatorname{Im} B \subset \operatorname{Im} \tilde{X}$. Let us choose a basis for $\mathbf{R}^n$ that is compatible with the direct sum decomposition

$$\mathbf{R}^n = \operatorname{Ker} X \oplus \operatorname{Im} \tilde{X}.$$

In that basis, $X = (0 \quad X_1)$ with X_1 invertible and

$$B = \begin{pmatrix} 0 & 0 \\ 0 & B_1 \end{pmatrix} = \begin{pmatrix} 0 \\ \tilde{X}_1 \end{pmatrix} A \begin{pmatrix} 0 & X_1 \end{pmatrix},$$

or $B_1 = \tilde{X}_1 A X_1$. Thus, $\sigma(B) = \sigma(B_1) = \sigma(A)$, which proves the theorem. $\quad\square$

Theorem 8.2.4

1. *Let $v_1, \ldots v_k \in \mathbf{C}^n$, $\lambda_1, \ldots, \lambda_k \in \mathbf{R}$, and let A be the Hermitian matrix $\sum_{i=1}^k \lambda_i v_i v_i^*$. Then A is Hermitian congruent to $\Lambda = \operatorname{diag}(\lambda_1, \ldots, \lambda_k, 0, \ldots, 0)$.*

2. *If A_1 and A_2 are Hermitian and*

$$\text{rank}\,(A_1 + A_2) = \text{rank}\,(A_1) + \text{rank}\,(A_2),$$

then

$$\sigma(A_1 + A_2) = \sigma(A_1) + \sigma(A_2).$$

Proof:

1. Extend $\{v_1, \ldots, v_k\}$ to a basis $\{v_1, \ldots, v_n\}$ of $\mathbf{C}^n$. Let V be the matrix whose columns are $v_1, \ldots, v_n$. Then $V \Lambda V^* = A$.

2. Let $\lambda_1, \ldots, \lambda_k$ be the nonzero eigenvalues of A_1, with corresponding eigenvectors $v_1, \ldots v_k$, and let $\lambda_{k+1}, \ldots, \lambda_{k+l}$ be the nonzero eigenvalues of A_2, with corresponding eigenvectors $v_{k+1}, \ldots, v_{k+l}$. Thus $A_1 = \sum_{i=1}^{k} \lambda_i v_i v_i^*$ and $A_2 = \sum_{i=k+1}^{k+l} \lambda_i v_i v_i^*$. Since, by our assumption, $\text{Im}\,A_1 \cap \text{Im}\,A_2 = \{0\}$, the vectors $v_1, \ldots v_{k+l}$ are linearly independent. Hence, by part 1, $A_1 + A_2$ is Hermitian congruent to $\text{diag}\,(\lambda_1, \ldots, \lambda_{k+l}, 0, \ldots, 0)$. $\square$

The positive definiteness of a quadratic form can be determined from any of its matrix representations.

Theorem 8.2.5 *Let ϕ be a quadratic form on an n-dimensional linear space U, and let $[\phi]_{\mathcal{B}}^{\mathcal{B}} = (\phi_{ij})$ be its matrix representation with respect to the basis $\mathcal{B}$. Then ϕ is positive definite if and only if the determinants of all of the principal minors are positive, that is,*

$$\begin{vmatrix} \phi_{11} & \cdot & \cdot & \cdot & \phi_{1i} \\ \cdot & \cdot & \cdot & \cdot & \cdot \\ \cdot & \cdot & \cdot & \cdot & \cdot \\ \cdot & \cdot & \cdot & \cdot & \cdot \\ \phi_{i1} & \cdot & \cdot & \cdot & \phi_{ii} \end{vmatrix} > 0,$$

for $i = 1, \ldots, n$.

Proof: If ϕ is positive, then its restriction to any subspace M is also positive. This implies that $\det(\phi|_M) > 0$, for it is equal to the product of all eigenvalues.

Conversely, assume that all of the principal minors have positive determinants. Let (ϕ_{ij}) be the matrix of the qudratic form. Let R be the matrix that reduces ϕ to upper triangular form via the Gauss elimination procedure. If we do not normalize the diagonal elements, then R is lower triangular with all elements on the diagonal equal to 1. Thus, $R\phi\tilde{R}$ is diagonal with the determinants of the principal minors on the digonal. As $R\phi\tilde{R}$ is positive by assumption, so is ϕ. $\square$

In the sequel, we focus our interest on some important quadratic forms induced by rational functions, namely, Hankel and Bezout forms.

8.3   Hankel Operators and Forms

We begin by studying Hankel operators. Although our interest is mainly in Hankel operators with rational symbols, we need to enlarge the setting in order to give Kronecker's characterization of rationality in terms of infinite Hankel matrices.

Let $g(z) = \sum_{j=-\infty}^{n} g_j z^{-j} \in F((z^{-1}))$. Let π_+ and π_- be the projections, defined in Eqs. (1.17) and (1.18), respectively, which correspond to the direct sum representation $F((z^{-1})) = F[z] \oplus z^{-1}F[[z^{-1}]]$. We define the **Hankel operator** $H_g : F[z] \longrightarrow z^{-1}F[[z^{-1}]]$ by

$$H_g f = \pi_- g f \qquad \text{for} \qquad f \in F[z].$$

g is called the **symbol** of the Hankel operator. Clearly, H_g is a well-defined linear transformation.

We recall that $F[z]$ and $z^{-1}F[[z^{-1}]]$ also carry $F[z]$-module structures. In $F[z]$, this is simply given in terms of polynomial multiplication. However, $z^{-1}F[[z^{-1}]]$ is not an $F[z]$-submodule of $F((z^{-1}))$; therefore, we define

$$p \cdot h = \pi_- p h \qquad h \in z^{-1}F[[z^{-1}]].$$

For the polynomial z we denote the corresponding operators by $S_+ : F[z] \longrightarrow F[z]$ and $S_- : z^{-1}F[[z^{-1}]] \longrightarrow z^{-1}F[[z^{-1}]]$.

Proposition 8.3.1 *Let $g \in F((z^{-1})$. Then:*

1. *$H_g : F[z] \longrightarrow z^{-1}F[[z^{-1}]]$ is an $F[z]$-module homomorphism.*

2. *Any $F[z]$-module homomorphism from $F[z]$ to $z^{-1}F[[z^{-1}]]$ is a Hankel operator.*

3. *Let $g_1, g_2 \in F((z^{-1}))$. Then $H_{g_1} = H_{g_2}$ if and only if $g_1 - g_2 \in F[z]$.*

Proof:

1. It suffices to show that

$$H_g S_+ = S_- H_g. \tag{8.6}$$

   We do this by computing

   $$H_g S_+ f = \pi_- g z f = \pi_- z(gf) = \pi_- z \pi_-(gf) = S_- H_g f.$$

   Here we used the fact that $S_+ \mathrm{Ker}\, \pi_- \subset \mathrm{Ker}\, \pi_-$.

2. Let $H : F[z] \longrightarrow z^{-1}F[[z^{-1}]]$ be an $F[z]$-module homomorphism. We set $g = H1 \in z^{-1}F[[z^{-1}]]$. Then, as H is a homomorphism, for any polynomial f we have

   $$Hf = H(f \cdot 1) = f \cdot H1 = \pi_- fg = \pi_g f = H_g f.$$

   Thus, we have identified H with a Hankel operator.

3. As $H_{g_1} - H_{g_2} = H_{g_1 - g_2}$, it suffices to show that $H_g = 0$ if and only if $g \in F[z]$. Assume therefore that $H_g = 0$. Then $\pi_- g1 = 0$, which means that $g \in F[z]$. The converse is obvious. $\qquad\square$

We call Eq. (8.6) the **functional equation of Hankel operators**.

Corollary 8.3.1 *Let* $g \in F((z^{-1}))$. *Then* H_g *as a map from* $F[z]$ *to* $z^{-1}F[[z^{-1}]]$ *is not invertible.*

Proof: Were H_g invertible, it would follow from Eq. (8.6) that S_+ and S_- are similar transformations. This is impossible, since for S_- each $\alpha \in F$ is an eigenvalue, whereas S_+ has no eigenvalues at all. $\qquad\square$

Corollary 8.3.2 *Let* H_g *be a Hankel operator. Then* $\operatorname{Ker} H_g$ *is a submodule of* $F[z]$ *and* $\operatorname{Im} H_g$ *a submodule of* $z^{-1}F[[z^{-1}]]$.

Proof: Follows from Eq. (8.6). $\qquad\square$

Now, submodules of $F[z]$ are simply ideals and hence of the form $qF[z]$. We also have a characterization of finite-dimensional submodules of $F_-(z) \subset z^{-1}F[[z^{-1}]]$. This leads to Kronecker's theorem.

Theorem 8.3.1 (Kronecker)

1. *A Hankel operator* H_g *has finite rank if and only if* g *is a rational function.*

2. *If* $g = p/q$ *with* p, q *coprime, then*

$$\operatorname{Ker} H_{p/q} = qF[z],$$

and

$$\operatorname{Im} H_{p/q} = X^q.$$

Proof: Assume that g is rational. So $g = p/q$, and we may assume that p and q are coprime. Clearly, $qF[z] \subset \operatorname{Ker} H_g$ and, because of coprimeness, we actually have the equality $qF[z] = \operatorname{Ker} H_g$. Now we have $F[z] = X_q \oplus qF[z]$ and hence $\operatorname{Im} H_g = \{\pi_- q^{-1} pf | f \in X_q\}$. We compute

$$\pi_- q^{-1} pf = q^{-1}(q\pi_- q^{-1} pf) = q^{-1}\pi_- qpf \in X^q.$$

Using coprimeness, and Theorem 5.1.3, we have $\{\pi_q pf | f \in X_q\} = X_q$ and therefore $\operatorname{Im} H_g = X^q$, which shows that H_g has finite rank.

Conversely, assume that H_g has finite rank. So $M = \operatorname{Im} H_g$ is a finite-dimensional subspace of $z^{-1}F[[z^{-1}]]$. Let $A = S_-|M$. Let q be the minimal polynomial of A. Therefore, for every $f \in F[z]$ we have $0 = \pi_- q\pi_- gf = \pi_- qgf$. Choosing $f = 1$, we conclude that $qg = p$ is a polynomial and hence $g = p/q$ is rational. $\qquad\square$

In terms of expansions in powers of z, the Hankel operator has a infinite matrix representation, namely,

$$H = \begin{pmatrix} g_1 & g_2 & g_3 & \cdot & \cdot \\ g_2 & g_3 & g_4 & \cdot & \cdot \\ g_3 & g_4 & g_5 & \cdot & \cdot \\ \cdot & \cdot & \cdot & \cdot & \cdot \\ \cdot & \cdot & \cdot & \cdot & \cdot \end{pmatrix}.$$

Any infinite matrix (g_{ij}) with $g_{ij} = g_{i+j-1}$ is called a Hankel matrix. In terms of Hankel matrices, Kronecker's theorem can be restated as follows:

Theorem 8.3.2 (Kronecker) *An infinite Hankel matrix H has finite rank if and only if its symbol $g(z) = \sum_{j=1}^{\infty} g_j z^{-j}$ is a rational function.*

Definition 8.3.1 *Given a proper rational function g, we define its* **McMillan degree**, $\delta(g)$, *by*

$$\delta(g) = \operatorname{rank} H_g.$$

Proposition 8.3.2 *Let $g = p/q$, with p, q coprime, be a proper rational function. Then*

$$\delta(g) = \deg q.$$

Proof: By Theorem 8.3.1 we have $\operatorname{Im} H_{\frac{p}{q}} = X^q$, and so

$$\delta(g) = \operatorname{rank} H_g = \dim X^q = \deg q. \qquad \square$$

Proposition 8.3.3 *Let $g_i = p_i/q_i$ be rational functions, with p_i, q_i coprime. Then:*

1. *We have*
$$\operatorname{Im} H_{g_1+g_2} \subset \operatorname{Im} H_{g_1} + \operatorname{Im} H_{g_2}$$

   and
$$\delta(g_1 + g_2) \leq \delta(g_1) + \delta(g_2).$$

2. *We have*
$$\delta(g_1 + g_2) = \delta(g_1) + \delta(g_2)$$

   if and only if q_1, q_2 are coprime.

Proof:

1. Obvious.

2. We have
$$g_1 + g_2 = \frac{q_1 p_2 + q_2 p_1}{q_1 q_2}.$$

Assume that q_1, q_2 are coprime. Then, by Proposition 5.3.2, we have $X^{q_1} + X^{q_2} = X^{q_1 q_2}$. On the other hand, the coprimeness of q_1, q_2 implies that of $q_1 q_2, q_1 p_2 + q_2 p_1$. Therefore,

$$\operatorname{Im} H_{g_1 + g_2} = X^{q_1 q_2} = X^{q_1} + X^{q_2} = \operatorname{Im} H_{g_1} + \operatorname{Im} H_{g_2}.$$

Conversely, assume that $\delta(g_1 + g_2) = \delta(g_1) + \delta(g_2)$. This implies that

$$\operatorname{Im} H_{g_1} \cap \operatorname{Im} H_{g_2} = X^{q_1} \cap X^{q_2} = \{0\}.$$

But, by invoking once again Proposition 5.3.2, this implies the coprimeness of q_1, q_2.   □

For the special case of the real field, the Hankel operator defines a quadratic form on $\mathbf{R}[z]$. Since $\mathbf{R}[z]^* = z^{-1}\mathbf{R}[[z^{-1}]]$, it follows from our definition of self-dual maps that H_g is self-dual. Thus, we also have an induced quadratic form on $\mathbf{R}[z]$ given by

$$[H_g f, f] = \sum_{i=0}^{\infty} \sum_{j=0}^{\infty} g_{i+j+1} f_i f_j. \tag{8.7}$$

Since f is a polynomial, the sum in Eq. (8.7) is well defined. Thus, with the Hankel map is associated the infinite Hankel matrix

$$H = \begin{pmatrix} g_1 & g_2 & g_3 & \cdot & \cdot \\ g_2 & g_3 & g_4 & \cdot & \cdot \\ g_3 & g_4 & g_5 & \cdot & \cdot \\ \cdot & \cdot & \cdot & \cdot & \cdot \\ \cdot & \cdot & \cdot & \cdot & \cdot \end{pmatrix}.$$

With g we also associate the finite Hankel forms H_k, $k = 1, 2, \ldots$ defined by the matrices

$$H_k = \begin{pmatrix} g_1 & \cdot & \cdot & \cdot & g_k \\ \cdot & \cdot & \cdot & \cdot & \cdot \\ \cdot & \cdot & \cdot & \cdot & \cdot \\ \cdot & \cdot & \cdot & \cdot & \cdot \\ g_k & \cdot & \cdot & \cdot & g_{2k-1} \end{pmatrix}.$$

In particular, assuming as we did that p and q are coprime, and that $\deg(q) = n$, then $\operatorname{rank}(H_k) = \deg(q) = n = \delta(g)$ for $k \geq n$.

Naturally, not only the rank information for H_n is derivable from g through its polynomial representation, but also the signature information. This is most conveniently done by the use of Bezoutians or the Cauchy index. We turn to these topics next.

8.4 Bezoutians

A quadratic form is associated with any symmetric matrix. We focus now on quadratic forms induced by polynomials and rational functions. In this section we present the basic theory of Bezout forms, or Bezoutians. Bezoutians are a special class of quadratic forms that were introduced to facilitate the study of root location of polynomials. Except for the study of signature information, we do not have to restrict the field. The study of Bezoutians is intimately connected to that of Hankel forms. The study of this connection will be undertaken in Section 8.7.

Let $p(z) = \sum_{i=0}^{n} p_i z^i$ and $q(z) = \sum_{i=0}^{n} q_i z^i$ be two polynomials, and let z and w be two, generally noncommuting variables. Then

$$
\begin{aligned}
p(z)q(w) - q(z)p(w) &= \sum_{i=0}^{n} \sum_{j=0}^{n} p_i q_j (z^i w^j - z^j w^i) \\
&= \sum_{0 \le i < j \le n} (p_i q_j - q_i p_j)(z^i w^j - z^j w^i).
\end{aligned}
$$

(8.8)

Observe now that

$$
z^i w^j - z^j w^i = \sum_{\nu=0}^{j-i-1} z^{i+\nu}(w - z)w^{j-\nu-1}.
$$

Thus, Eq. (8.8) can be rewritten as

$$
p(z)q(w) - q(z)p(w) = \sum_{0 \le i < j \le n} (p_i q_j - p_j q_i) \sum_{\nu=0}^{j-i-1} z^{i+\nu}(w - z)w^{j-\nu-1}
$$

or

$$
q(z)p(w) - p(z)q(w) = \sum_{i=1}^{n} \sum_{j=1}^{n} b_{ij} z^{i-1}(z - w)w^{j-1}.
$$

(8.9)

The last equality is obtained by changing the order of summation and properly defining the coefficients b_{ij}. Equation (8.9) also can be written as

$$
\frac{q(z)p(w) - p(z)q(w)}{z - w} = \sum_{i=1}^{n} \sum_{j=1}^{n} b_{ij} z^{i-1} w^{j-1}.
$$

Definition 8.4.1

1. The matrix (b_{ij}) defined by Eq. (8.9) is called the **Bezout form** associated with the polynomials q and p, or just the **Bezoutian of** q and p, and is denoted by $B(q, p)$. The function

$$
\frac{q(z)p(w) - p(z)q(w)}{z - w}
$$

is called the **generating function** of the Bezoutian form.

2. *If $g = p/q$ is the unique, irreducible representation of a rational function g, with q monic, then we write $B(g)$ for $B(q, p)$. $B(g)$ is called the Bezoutian of g.*

It should be noted that Eq. (8.9) holds even when z and w are a pair of noncommuting variables. This observation is extremely useful in the derivation of representation theorems for Bezoutians.

Note that $B(q, p)$ defines a bilinear form on F^n by

$$B(q, p)(\xi, \eta) = \sum_{i=1}^{n} \sum_{j=1}^{n} b_{ij} \xi_i \eta_j \qquad \text{for} \qquad \xi, \eta \in F^n.$$

No distinction will be made in the notation between the matrix and the bilinear form.

The following theorem summarizes the elementary properties of the Bezoutian:

Theorem 8.4.1 *Let $q, p \in F[z]$ with $\max(\deg q, \deg p) \leq n$. Then:*

1. *The Bezoutian matrix $B(q, p)$ is a symmetric matrix.*

2. *The Bezoutian $B(q, p)$ is linear in q and p.*

3. *$B(p, q) = -B(q, p)$.*

It is these basic properties of the Bezoutian, in particular the linearity in both variables, that turn the Bezoutian into such a powerful tool. In conjunction with the Euclidean algorithm, it is the starting point for many fast matrix inversion algorithms.

Lemma 8.4.1 *Let q and r be two coprime polynomials and p an arbitrary nonzero polynomial. Then*

$$\text{rank} \, (B(qp, rp)) = \text{rank} \, (B(q, r))$$

and

$$\sigma(B(qp, rp)) = \sigma(B(q, r)).$$

Proof: The Bezoutian $B(qp, rp)$ is determined by the polynomial expansion of

$$p(z) \frac{q(z)r(w) - r(z)q(w)}{z - w} p(w),$$

which implies that $B(qp, pr) = \tilde{X} B(q, r) X$ with

$$X = \begin{pmatrix} p_0 & \cdots & p_m & & \\ & \ddots & & \ddots & \\ & & \ddots & & \ddots \\ & & & p_0 & \cdots & p_m \end{pmatrix}.$$

The result follows now by an application of Theorem 8.2.3. $\square$

The following theorem is of central importance in the study of Bezout and Hankel forms. It gives a characterization of the Bezoutian as a matrix representation of an intertwining map. Thus, the study of Bezoutians is reduced to that of intertwining maps of the form $p(S_q)$ studied in Chapter 5. These are easier to handle and yield information on Bezoutians. This also leads to a conceptual theory of Bezoutians, in contrast to the purely computational approach that has been prevalent in their study for a long time.

Theorem 8.4.2 Let $p, q \in F[z]$, with $\deg p \leq \deg q$. Then the Bezoutian $B(q, p)$ of q and p satisfies

$$B(q, p) = [p(S_q)]_{co}^{st}. \tag{8.10}$$

Proof: Note that

$$\pi_+ w^{-k}(z - w)w^{j-1} = \begin{cases} 0 & \text{if } j < k \\ -1 & \text{if } j = k \\ (z - w)w^{j-k-1} & \text{if } j > k. \end{cases}$$

So from Eq. (8.9) it follows that

$$q(z)\pi_+ w^{-k}p(w) - p(z)\pi_+ w^{-k}q(w)|_{w=z} = \sum_{i=1}^{n} b_{ik}z^{i-1},$$

or, with $\{e_1, \ldots, e_n\}$ the control basis of X_q and p_k defined by

$$p_k(z) = \pi_+ z^{-k}p(z),$$

we have

$$q(z)p_k(z) - p(z)e_k(z) = \sum_{i=1}^{n} b_{ik}z^{i-1}.$$

Applying the projection π_q, we obtain

$$\pi_q p e_k = p(S_q)e_k = \sum_{i=1}^{n} b_{ik}z^{i-1}, \tag{8.11}$$

which, from the definition of a matrix representation, completes the proof.
    $\square$

Corollary 8.4.1 Let $p, q \in F[z]$, with $\deg p < \deg q$. Then the last row and column of the Bezoutian $B(q, p)$ consist of the coefficients of p.

Proof: By Eq. (8.11), and the fact that the last element of the control basis satisfies $e_n(z) = 1$, we have

$$\sum_{i=1}^{n} b_{in} z^{i-1} = (\pi_q p e_n)(z) = (\pi_q p)(z) = p(z) = \sum_{i=0}^{n-1} p_i z^i, \qquad (8.12)$$

that is,

$$b_{in} = p_{i-1} \qquad i = 1, \ldots, n. \qquad (8.13)$$

The statement for rows follows from the symmetry of the Bezoutian. □

In Section 3.3 we derived a determinantal test for the coprimeness of two polynomials. With the introduction of the Bezoutian, we have another such test. We can classify the resultant approach as geometric, for basically it gives a condition for a polynomial model being a direct sum of two shift-invariant subspaces. The Bezoutiant approach, on the other hand, is arithmetic inasmuch as it is based on the invertibility of intertwining maps. We shall see in Corollary 8.7.1 yet another such test, given in terms of Hankel matrices, and furthermore we shall clarify the connection between the various tests.

Theorem 8.4.3 *Given two polynomials $p, q \in F[z]$. Then*

1. $\dim(\operatorname{Ker} B(q, p))$ *is equal to the degree of the g.c.d. of q and p.*

2. $B(q, p)$ *is invertible if and only if q and p are coprime.*

3. *The polynomials p and q are coprime if and only if $\det B(p, q) \neq 0$.*

Proof: Part 1 follows from Proposition 5.1.9. Part 2 follows from Theorem 5.1.3 and Theorem 8.4.2. Part 3 follows from part 2. □

As an immediate consequence, we derive some well-known results concerning Bezoutians.

Corollary 8.4.2 *Let $p, q \in F[z]$, with $\deg p \leq \deg q$. Assume that p has a factorization $p = p_1 p_2$. Let $q(z) = \sum_{i=0}^{n} q_i z^i$ be monic, that is, $q_n = 1$. Let $C_q^{\sharp}$ and $C_q^{\flat}$ be the companion matrices of q, given by Eqs. (5.3) and (5.4), respectively, and let K be the Hankel matrix*

$$K = \begin{pmatrix} q_1 & \cdot & \cdot & \cdot & q_{n-1} & 1 \\ \cdot & & & \cdot & 1 & \\ \cdot & & \cdot & \cdot & & \\ \cdot & & \cdot & \cdot & & \\ q_{n-1} & 1 & & & & \\ 1 & & & & & \end{pmatrix}. \qquad (8.14)$$

Then

1. $B(q, p_1 p_2) = B(q, p_1) p_2(C_q^\flat) = p_1(C_q^\sharp) B(q, p_2)$ (8.15)

2. $B(q, p) = K p(C_q^\flat) = p(C_q^\sharp) K$ (8.16)

3. $B(q, p) C_q^\flat = C_q^\sharp B(p, q)$.

Note that the matrix K in Eq. (8.14) satisfies $K = B(q, 1)$ and is both a Bezoutian and a Hankel matrix.

Proof: Note that for the standard and control bases of X_q we have

$$C_q^\sharp = [S_q]_{st}^{st} \quad and \quad C_q^\flat = \tilde{C}_q^\sharp = [S_q]_{co}^{co}.$$

1. This follows from the equality

$$p(S_q) = p_1(S_q) p_2(S_q)$$

and the fact that

$$
\begin{aligned}
B(q, p_1 p_2) &= [(p_1 p_2)(S_q)]_{co}^{st} = [p_1(S_q) p_2(S_q)]_{co}^{st} \\
&= [p_1(S_q)]_{co}^{st} [p_2(S_q)]_{co}^{co} \\
&= [p_1(S_q)]_{st}^{st} [p_2(S_q)]_{co}^{st}.
\end{aligned}
$$

2. This follows from part 1 for the trivial factorization $p = p \cdot 1$.

3. From the commutativity of S_q and $p(S_q)$, it follows that

$$[p(S_q)]_{co}^{st} [S_q]_{co}^{co} = [S_q]_{st}^{st} [p(S_q)]_{co}^{st}.$$

We note that

$$[S_q]_{co}^{co} = (\widetilde{[S_q]_{st}^{st}}).$$ $\square$

Representation (2) for the Bezoutian sometimes is referred to as the **Barnett factorization**.

8.5   Representation of Bezoutians

Relation (8.9) easily leads to some interesting representation formulas for the Bezoutian. For a polynomial a of degree n we define the **reverse polynomial** $a^\sharp$ by

$$a^\sharp(z) = a(z^{-1}) z^n.$$ (8.17)

Theorem 8.5.1 Let $p(z) = \sum_{i=0}^n p_i z^i$ and $q(z) = \sum_{i=0}^n q_i z^i$ be polynomials of degree n. Then the Bezoutian has the following representations:

$$
\begin{aligned}
B(q,p) &= [p(\tilde{S})q^\sharp(S) - q(\tilde{S})p^\sharp(S)]J \\
&= J[p(S)q^\sharp(\tilde{S}) - q(S)p^\sharp(\tilde{S})] \\
&= -J[p^\sharp(\tilde{S})q(S) - q^\sharp(\tilde{S})p(S)] \\
&= -[p^\sharp(S)q(\tilde{S}) - q^\sharp(S)p(\tilde{S})]J.
\end{aligned}
\tag{8.18}
$$

Proof: We define the $n \times n$ shift matrix S by

$$
S = \begin{pmatrix}
0 & 1 & . & . & . & 0 \\
0 & . & 1 & . & . & 0 \\
0 & . & . & . & . & 0 \\
0 & . & . & . & 1 & 0 \\
0 & . & . & . & . & 1 \\
0 & . & . & . & . & 0
\end{pmatrix}
\tag{8.19}
$$

and the $n \times n$ transposition matrix J by

$$
J = \begin{pmatrix}
0 & . & . & . & 1 \\
0 & . & . & 1 & 0 \\
0 & . & . & . & 0 \\
0 & 1 & . & . & 0 \\
1 & . & . & . & 0
\end{pmatrix}.
\tag{8.20}
$$

Note also that, for an arbitrary polynomial a, we have $Ja(S)J = a(\tilde{S})$. From Eq. (8.9) we easily obtain

$$
q(z)p^\sharp(w) - p(z)q^\sharp(w) = [q(z)p(w^{-1}) - p(z)q(w^{-1})]w^n
$$

$$
= \sum_{i=1}^n \sum_{j=1}^n b_{ij} z^{i-1}(z - w^{-1})w^{-j+1}w^n
$$

$$
= \sum_{i=1}^n \sum_{j=1}^n b_{ij} z^{i-1}(zw - 1)w^{n-j}.
$$

In this identity we now substitute $z = \tilde{S}$ and $w = S$. Thus, we get for the central term

$$
(\tilde{S}S - I) = -\begin{pmatrix}
1 & 0 & . & . & . & 0 \\
0 & 0 & . & . & . & 0 \\
0 & 0 & . & . & . & 0 \\
0 & 0 & . & . & . & 0 \\
0 & 0 & . & . & . & 0
\end{pmatrix},
$$

and $\tilde{S}^{i-1}(\tilde{S}S - I)S^{n-j}$ is the matrix whose only nonzero term is -1 in the $i, n-j$ position. This implies the identity $B(q,p) = \{p(\tilde{S})q^\sharp(S) - q(\tilde{S})p^\sharp(S)\}J$. The other identities are derived similarly. □

From the representations (8.18) of the Bezoutian we obtain, by expanding the matrices, the **Gohberg–Semencul** formulas, of which we present one only:

$$B(q,p) = (p(\tilde{S})q^{\sharp}(S) - q(\tilde{S})p^{\sharp}(S))J$$

$$
= \left[\begin{pmatrix} p_0 & & & & \\ \cdot & p_0 & & & \\ \cdot & \cdot & p_0 & & \\ \cdot & \cdot & \cdot & p_0 & \\ p_{n-1} & \cdot & \cdot & \cdot & p_0 \end{pmatrix} \begin{pmatrix} q_n & q_{n-1} & \cdot & \cdot & q_1 \\ & \cdot & & \cdot & \cdot \\ & & & \cdot & \cdot \\ & & & & \cdot \\ & & & & q_n \end{pmatrix} \right.
$$

$$
\left. - \begin{pmatrix} q_0 & & & \\ \cdot & & & \\ \cdot & \cdot & & \\ \cdot & \cdot & \cdot & \\ q_{n-1} & \cdot & \cdot & \cdot & q_0 \end{pmatrix} \begin{pmatrix} p_n & p_{n-1} & \cdot & \cdot & p_1 \\ & \cdot & & \cdot & \cdot \\ & & & \cdot & \cdot \\ & & & & \cdot \\ & & & & p_n \end{pmatrix} \right] J
$$

$$
= \left[\begin{pmatrix} p_0 & & & & \\ \cdot & p_0 & & & \\ \cdot & \cdot & p_0 & & \\ \cdot & \cdot & \cdot & p_0 & \\ p_{n-1} & \cdot & \cdot & \cdot & p_0 \end{pmatrix} \begin{pmatrix} q_1 & \cdot & \cdot & q_{n-1} & q_n \\ \cdot & & \cdot & q_n & \\ \cdot & & \cdot & & \\ q_{n-1} & q_n & & & \\ q_n & & & & \end{pmatrix} \right.
$$

$$
\left. - \begin{pmatrix} q_0 & & & \\ \cdot & & & \\ \cdot & \cdot & & \\ \cdot & \cdot & \cdot & \\ q_{n-1} & \cdot & \cdot & \cdot & q_0 \end{pmatrix} \begin{pmatrix} p_1 & \cdot & \cdot & p_{n-1} & p_n \\ \cdot & & \cdot & p_n & \\ \cdot & & \cdot & & \\ p_{n-1} & p_n & & & \\ p_n & & & & \end{pmatrix} \right].
$$

$$\tag{8.21}$$

Given two polynomials p and q of degree n, we let their **Sylvester resultant**, $\operatorname{Res}(p,q)$, be defined by

$$
\operatorname{Res}(p,q) = \begin{pmatrix}
p_0 & \cdot & \cdot & \cdot & p_{n-1} & p_n & & & \\
 & \cdot & \cdot & \cdot & & \cdot & \cdot & & \\
 & & \cdot & \cdot & & & \cdot & \cdot & \\
 & & & \cdot & p_0 & p_1 & \cdot & \cdot & \cdot & p_n \\
 & & q_0 & \cdot & \cdot & \cdot & q_{n-1} & q_n & & \\
 & & & \cdot & \cdot & \cdot & & \cdot & \cdot & \\
 & & & & \cdot & \cdot & & & \cdot & \cdot \\
 & & & & & q_0 & q_1 & \cdot & \cdot & \cdot & q_n
\end{pmatrix}.
$$

$$\tag{8.22}$$

It has been proved in Theorem 3.3.1 that the resultant $\operatorname{Res}(p,q)$ is nonsingular if and only if p and q are coprime. Equation (8.22) can be rewritten as the 2×2 block matrix

$$\text{Res}\,(p, q) = \begin{pmatrix} p(S) & p^\sharp(\tilde{S}) \\ q(S) & q^\sharp(\tilde{S}) \end{pmatrix},$$

where the reverse polynomials $p^\sharp$ and $q^\sharp$ are defined in Eq. (8.17).

Based on the preceeding, we can state the following theorem:

Theorem 8.5.2 *Let* $p(z) = \sum_{i=0}^{n} p_i z^i$ *and* $q(z) = \sum_{i=0}^{n} q_i z^i$ *be polynomials of degree* n. *Then*

$$\widetilde{\text{Res}\,(p, q)} \begin{pmatrix} 0 & J \\ -J & 0 \end{pmatrix} \text{Res}\,(p, q) = \begin{pmatrix} 0 & B(p, q) \\ -B(p, q) & 0 \end{pmatrix}. \quad (8.23)$$

Proof: By expanding the left side of Eq. (8.23), we have

$$\begin{pmatrix} p(\tilde{S}) & q(\tilde{S}) \\ p^\sharp(S) & q^\sharp(S) \end{pmatrix} \begin{pmatrix} 0 & J \\ -J & 0 \end{pmatrix} \begin{pmatrix} p(S) & p^\sharp(\tilde{S}) \\ q(S) & q^\sharp(\tilde{S}) \end{pmatrix} =$$

$$\begin{pmatrix} (p(\tilde{S})Jq(S) - q(\tilde{S})Jp(S)) & (p(\tilde{S})Jq^\sharp(\tilde{S}) - q(\tilde{S})Jp^\sharp(\tilde{S})) \\ (p^\sharp(S)Jq(S) - q^\sharp(S)Jp(S)) & (p^\sharp(S)Jq^\sharp(\tilde{S}) - q^\sharp(S)Jp^\sharp(\tilde{S})) \end{pmatrix}.$$

Now,

$$p(\tilde{S})Jq(S) - q(\tilde{S})Jp(S) = J[p(S)q(S) - q(S)p(S)] = 0.$$

Similarly,

$$p^\sharp(S)Jq^\sharp(\tilde{S}) - q^\sharp(S)Jp^\sharp(\tilde{S}) = [p^\sharp(S)q^\sharp(S) - q^\sharp(S)p^\sharp(S)]J = 0.$$

In the same way,

$$p(\tilde{S})Jq^\sharp(\tilde{S}) - q(\tilde{S})Jp^\sharp(\tilde{S}) = J[p(S)q^\sharp(\tilde{S}) - q(S)p^\sharp(\tilde{S})] = B(q, p).$$

Finally,

$$p^\sharp(S)Jq(S) - q^\sharp(S)Jp(S) = J[p(\tilde{S})q^\sharp(S) - q^\sharp(\tilde{S})p(S)] = -B(q, p).$$

This proves the theorem. □

The nonsingularities of both the Bezoutiant and resultant matrices are equivalent to the coprimeness of the two underlying polynomials. As an immediate corollary to the previous theorem, we obtain the equality, up to a sign, of their determinants. Yet another determinantal test for coprimeness, given in terms of Hankel matrices, will be given in Corollary 8.7.1.

Corollary 8.5.1 *Let* p *and* q *be polynomials of degree* n. *Then*

$$|\det \text{Res}\,(p, q)| = |\det B(p, q)|. \quad (8.24)$$

In particular, the nonsingularity of either matrix implies that of the other. As we already know, these conditions are equivalent to the coprimeness of the polynomials p and q.

Actually, we can be more precise about the relation between the determinants of the resultant and Bezoutian matrices.

Corollary 8.5.2 *Let p and q be polynomials of degree n. Then:*

1.
$$\det B(q,p) = (-1)^{[n(n+1)]/2} \det \operatorname{Res}(q,p). \tag{8.25}$$

2.
$$\det p(S_q) = \det p(C_q^\sharp) = \det \operatorname{Res}(p,q). \tag{8.26}$$

Proof:

1. From the representation $B(q,p) = \{p(\tilde{S})q^\sharp(S) - q(\tilde{S})p^\sharp(S)\}J$ of the Bezoutian we get, by applying Lemma 3.2.1, $\det B(q,p) = \det \operatorname{Res}(q,p)\det J$. Since $B(q,p) = -B(p,q)$, it follows that $\det B(q,p) = (-1)^n \det B(p,q)$. On the other hand, for J, the $n \times n$ matrix defined by Eq. (8.20), we have $\det J = (-1)^{[n(n-1)]/2}$. Therefore, Eq. (1) follows.

2. We can compute the determinant of a linear transformation in any basis. Thus,
$$\det p(S_q) = \det p(C_q^\sharp) = \det p(C_q^\flat).$$

Also, by Theorem 8.4.2, $B(q,p) = [p(S_q)]_{co}^{st} = [I]_{co}^{st}[p(S_q)]_{co}^{co}$, and since $\det[I]_{co}^{st} = (-1)^{[n(n-1)]/2}$, we get

$$
\begin{aligned}
\det p(S_q) &= (-1)^{[n(n-1)]/2} \det B(q,p) \\
&= (-1)^{[n(n-1)]/2}(-1)^n \det B(p,q) \\
&= (-1)^{[n(n-1)]/2}(-1)^n(-1)^{[n(n+1)]/2} \det \operatorname{Res}(p,q) \\
&= \det \operatorname{Res}(p,q). \qquad \qquad \square
\end{aligned}
$$

8.6   Diagonalization of Bezoutians

Since the Bezoutian of a pair of polynomials is a symmetric matrix, it can be diagonalized by a congruence transformation. The diagonalization procedure described here is based on polynomial factorizations and related direct sum decompositions. We begin by analyzing a special case where the congruence transformation is given in terms of a Vandermonde matrix. This is related to Proposition 5.1.7.

Proposition 8.6.1 *Let $q(z)$ be a monic polynomial of degree n having distinct zeros $\lambda_1, \ldots, \lambda_n$, and let p be a polynomial of degree $\leq n$. Let $d_i(z) = \prod_{j \neq i}(z - \lambda_j)$ and V the Vandermonde matrix*

$$\tilde{V} = [I]_{st}^{in} = \begin{pmatrix} 1 & \lambda_1 & . & . & \lambda_1^{n-1} \\ . & . & . & . & . \\ . & . & . & . & . \\ . & . & . & . & . \\ . & . & . & . & . \\ 1 & \lambda_n & . & . & \lambda_n^{n-1} \end{pmatrix}.$$

Then the Bezoutian $B(q, p)$ satisfies the following identity:

$$\tilde{V} B(q, p) V = R,$$

where R is the diagonal matrix $\mathrm{diag}\,(r_1, \ldots, r_n)$ and

$$r_i = p(\lambda_i) d_i(\lambda_i) = p(\lambda_i) q'(\lambda_i).$$

Proof: The trivial operator identity $Ip(S_q)I = p(S_q)$ implies the matrix equality $[I]_{st}^{in}[p(S_q)]_{co}^{st}[I]_{sp}^{co} = [p(S_q)]_{sp}^{in}$. As $S_q q_i = \lambda_i q_i$, it follows that $p(S_q)p_i = p(\lambda_i)p_i = p(\lambda_i)d_i(\lambda_i)\pi_i$. Now, $d_i(\lambda_i) = \prod_{j \neq i}(\lambda_i - \lambda_j)$, but $q'(z) = \sum_{i=1}^n d_i(z)$ and hence $q'(\lambda_i) = \prod_{j \neq i}(\lambda_i - \lambda_j) = d_i(\lambda_i)$. Thus, $R = [p(S_q)]_{sp}^{in}$, and the result follows. $\qquad\square$

The previous proposition uses a factorization of $q(z)$ into distinct linear factors. Such a factorization does not always exist. If it does exist, then the polynomials $d_i(z) = \prod_{j \neq i}(z - \lambda_j)$, which, up to a constant factor, are equal to the Lagrange interpolation polynomials, form a basis for X_q. Moreover, we have $X_q = d_1 X_{z-\lambda_1} \oplus \cdots \oplus d_n X_{z-\lambda_n}$. The fact that the sum is direct is implied by the assumption that the zeros λ_i are distinct, or that the polynomials $z - \lambda_i$ are pairwise coprime. In this case, the internal direct sum decomposition is isomorphic to the external direct sum $X_{z-\lambda_1} \oplus \cdots \oplus X_{z-\lambda_n}$. This can be siginificantly extended and generalized.

To this end, let $q_1, \ldots, q_s$ be monic polynomials in $F[z]$. Let $X_{q_1} \oplus \cdots \oplus X_{q_s}$ be the external direct sum of the corresponding polynomial models. We would like to compare this space with $X_{q_1 \cdots q_s}$. We define the polynomials by $d_i(z) = \prod_{j \neq i} q_j(z)$ and the map $Z : X_{q_1} \oplus \cdots \oplus X_{q_s} \longrightarrow X_{q_1 \cdots q_s}$ by

$$Z \begin{pmatrix} f_1 \\ . \\ . \\ . \\ f_s \end{pmatrix} = \sum_{i=1}^s d_i f_i. \tag{8.27}$$

Then we have the following:

Proposition 8.6.2 *Let $Z : X_{q_1} \oplus \cdots \oplus X_{q_s} \longrightarrow X_{q_1 \cdots q_s}$ be defined by Eq. (8.27). Let $\mathcal{B}_{ST}$ be the standard basis in $X_{q_1 \cdots q_s}$, and let $\mathcal{B}_{st}$ be the basis of $X_{q_1} \oplus \cdots \oplus X_{q_s}$ constructed from the union of the standard bases in the X_{q_i}. Then the adjoint map to Z, that is, $Z^* : X_{q_1 \cdots q_s} \longrightarrow X_{q_1} \oplus \cdots \oplus X_{q_s}$, is given by:*

1.

$$Z^* f = \begin{pmatrix} \pi_{q_1} f \\ \cdot \\ \cdot \\ \cdot \\ \pi_{q_s} f \end{pmatrix}. \tag{8.28}$$

2. *The map Z defined by Eq. (8.27) is invertible if and only if the q_i are pairwise coprime.*

3. *For the case $s = 2$, we have*

$$[Z]_{st}^{ST} = \text{Res}\,(q_2, q_1), \tag{8.29}$$

where $\text{Res}\,(q_2, q_1)$ is the resultant matrix defined in Eq. (3.7).

4. *Assume that $q_i(z) = z - \lambda_i$. Then*

$$[Z^*]_{ST}^{st} = \begin{pmatrix} 1 & \lambda_1 & \cdot & \cdot & \lambda_1^{s-1} \\ \cdot & \cdot & \cdot & \cdot & \cdot \\ \cdot & \cdot & \cdot & \cdot & \cdot \\ \cdot & \cdot & \cdot & \cdot & \cdot \\ 1 & \lambda_n & \cdot & \cdot & \lambda_n^{s-1} \end{pmatrix}. \tag{8.30}$$

5. *We have*

$$[Z^*]_{ST}^{st} = [Z^*]_{in}^{st}[I]_{ST}^{in}$$

and $[Z^*]_{in}^{st} = I$.

Proof:

1. We compute

$$
\begin{aligned}
< Zf, g > \ &= \ < \sum_{i=1}^{s} d_i f_i, g >= \left[q^{-1} \sum_{i=1}^{s} d_i f_i, g \right] \\
&= \ \sum_{i=1}^{s} [q^{-1} d_i f_i, g] \\
&= \ \sum_{i=1}^{s} [q_i^{-1} f_i, g] = \sum_{i=1}^{s} < f_i, \pi_{q_i} g > \\
&= \ < f, Z^* g > .
\end{aligned}
$$

So Eq. (8.28) follows.

2. Clearly, $\dim(X_{q_1} \oplus \cdots \oplus X_{q_s}) = \dim X_{q_1 \cdots q_s} = \sum_{i=1}^{s} \deg q_i$. Therefore, to prove invertibility, it suffices to show that Z is surjective. From the definition of Z and Proposition 5.1.3, it follows that $\text{Im}\, Z = \sum_{i=1}^{s} d_i X_{q_i} = d_0 X_{q_0}$, where d_0 is the g.c.d. of the d_i and q_0 the l.c.m. of the q_i. Thus, $q_0 = q$ if and only if the q_i are mutually coprime.

3. In this case, $d_1 = q_2$ and $d_2 = q_1$.

4. This follows from the fact that, for an arbitrary polynomial p, we have $\pi_{z-\lambda_i} p = p(\lambda_i)$.

5. That $[I]_{ST}^{in}$ is the Vandermonde matrix has been proved in Section 2.10.1. That $[Z^*]_{in}^{st} = I$ follows from the fact that, for the Lagrange interpolation polynomials, we have $\pi_i(\lambda_j) = \delta_{ij}$.   □

Next we proceed to study a factorization of the map Z defined in Eq. (3.8). This leads to and generalizes, from a completely different perspective, the classical formula, given by Eq. (3.8), for the computation of the determinant of the Vandermonde matrix.

Proposition 8.6.3 *Let $q_1, \ldots, q_s$ be polynomials in $F[z]$. Then:*

1. *The map $Z : X_{q_1} \oplus \cdots \oplus X_{q_s} \longrightarrow X_{q_1 \cdots q_s}$, defined in Eq. (8.27), has the factorization*

$$Z = Z_{s-1} \cdots Z_1, \qquad (8.31)$$

where $Z_i : X_{q_1 \cdots q_i} \oplus X_{q_{i+1}} \oplus \cdots \oplus X_{q_s} \longrightarrow X_{q_1 \cdots q_{i+1}} \oplus X_{q_{i+2}} \oplus \cdots \oplus X_{q_s}$ is defined by

$$Z \begin{pmatrix} g \\ f_{i+1} \\ \cdot \\ \cdot \\ \cdot \\ f_s \end{pmatrix} = \begin{pmatrix} (q_1 \cdots q_i) f_{i+1} + q_{i+1} g \\ f_{i+2} \\ \cdot \\ \cdot \\ \cdot \\ f_s \end{pmatrix}.$$

2. *Choosing the standard basis in all spaces, we get*

$$[Z_i]_{st}^{st} = \begin{pmatrix} \mathrm{Res}\,(q_{i+1}, q_i \cdots q_1) & 0 \\ 0 & I \end{pmatrix}.$$

3. *For the determinant of $[Z_i]_{st}^{st}$, we have*

$$\det[Z_i]_{st}^{st} = \prod_{j=1}^{i} \mathrm{Res}\,(q_{i+1}, q_j).$$

4. *For the determinant of $[Z]_{st}^{st}$, we have*

$$\det[Z]_{st}^{st} = \prod_{1 \le i < j \le s} \mathrm{Res}\,(q_j, q_i). \qquad (8.32)$$

5. *For the special case of $q_i(z) = z - \lambda_i$, we have*

$$\det V(\lambda_1, \ldots, \lambda_s) = \prod_{1 \le i < j \le s} (\lambda_j - \lambda_i). \qquad (8.33)$$

Proof:

1. We prove this by induction. For $s = 2$, this is trivial. Assume that we proved it for all integers $\leq s$. Then we define the map $Z' : X_{q_1} \oplus \cdots \oplus X_{q_{s+1}} \longrightarrow X_{q_1 \cdots q_s} \oplus X_{q_{s+1}}$ by

$$
Z' \begin{pmatrix} f_1 \\ \cdot \\ \cdot \\ \cdot \\ f_s \\ f_{s+1} \end{pmatrix} = \begin{pmatrix} \sum_{j=1}^{s} d'_j f_j \\ f_{s+1} \end{pmatrix}.
$$

Here, $d'_i = \Pi_{j \neq i, s+1} q_j$.
Now $d_{s+1} = \Pi_{j \neq s+1} q_j$ and $q_{s+1} d'_i = q_{s+1} \Pi_{j \neq i, s+1} q_j = \Pi_{j \neq i} q_j = d_i$.
So

$$
Z_s Z' \begin{pmatrix} f_1 \\ \cdot \\ \cdot \\ \cdot \\ f_s \\ f_{s+1} \end{pmatrix} = (q_1 \cdots q_s) f_{s+1} + q_{s+1} \sum_{j=1}^{s} d'_j f_j = \sum_{j=1}^{s+1} d_j f_j
$$

$$
= Z \begin{pmatrix} f_1 \\ \cdot \\ \cdot \\ \cdot \\ f_s \\ f_{s+1} \end{pmatrix}.
$$

2. We use Proposition 8.6.2.

3. From Eq. (8.29) it follows that

$$
\det[Z_i]_{st}^{st} = \det \operatorname{Res}(q_{i+1}, q_i \cdots q_1).
$$

Now we use Eq. (8.26), that is, the equality $\det \operatorname{Res}(q, p) = \det p(S_q)$, to get

$$
\begin{aligned}
\det \operatorname{Res}(q_{i+1}, q_i \cdots q_1) &= (-1)^{(\sum_{j=1}^{i} m_j) m_{i+1}} \det \operatorname{Res}(q_i \cdots q_1, q_{i+1}) \\
&= (-1)^{(\sum_{j=1}^{i} m_j) m_{i+1}} \det(q_i \cdots q_1)(S_{q_{i+1}}) \\
&= \prod_{j=1}^{i} (-1)^{m_j m_{i+1}} \det(q_j)(S_{q_{i+1}}) \\
&= \prod_{j=1}^{i} (-1)^{m_j m_{i+1}} \det \operatorname{Res}(q_j, q_{i+1}) \\
&= \prod_{j=1}^{i} \det \operatorname{Res}(q_{i+1}, q_j).
\end{aligned}
$$

4. Clearly, by the factorization (8.31) and the product rule of determinants, we have

$$\det[Z]_{st}^{st} = \prod_{j=1}^{s-1} \det[Z_i]_{st}^{st} = \prod_{j=1}^{s-1} \prod_{j=1}^{i} \det \operatorname{Res}(q_{i+1}, q_j)$$

$$= \prod_{1 \le i < j \le s} \operatorname{Res}(q_j, q_i).$$

5. By Proposition 8.6.2, we have in this case $V(\lambda_1, \ldots, \lambda_s) = [Z^*]_{ST}^{st}$. Now $\det Z^* = \det Z$ implies that

$$\det V(\lambda_1, \ldots, \lambda_s) = \det[Z^*]_{ST}^{st} = \det[Z^*]_{co}^{CO} = \det \widetilde{[Z]_{st}^{ST}} = \det[Z]_{st}^{ST}.$$

Also, in this case,

$$\det \operatorname{Res}(z - \lambda_i, z - \lambda_j) = \begin{vmatrix} -\lambda_i & -\lambda_j \\ 1 & 1 \end{vmatrix}.$$

Thus, Eq. (8.33) follows from Eq. (8.30).

This agrees with the computation of the Vandermonde determinant given previously in Chapter 3. □

We now proceed with the block diagonalization of the Bezoutian. Here we assume an arbitrary factorization of q.

Proposition 8.6.4 *Let $q = q_1 \cdots q_s$, and let $Z : X_{q_1} \oplus \cdots \oplus X_{q_s} \longrightarrow X_{q_1 \cdots q_s}$ be the map defined by Eq. (8.27). Then:*

1. The following is a commutative diagram:

$$
\begin{array}{ccc}
X_{q_1} \oplus \cdots \oplus X_{q_s} & \xrightarrow{\quad Z \quad} & X_{q_1 \cdots q_s} \\
\Big\downarrow {\scriptstyle (pd_1)(S_{q_1}) \oplus \cdots \oplus (pd_s)(S_{q_k})} & & \Big\downarrow {\scriptstyle p(S_q)} \\
X_{q_1} \oplus \cdots \oplus X_{q_s} & \xleftarrow{\quad Z^* \quad} & X_{q_1 \cdots q_s}
\end{array}
$$

that is,

$$Z^* p(S_q) Z = (pd_1)(S_{q_1}) \oplus \cdots \oplus (pd_s)(S_{q_k}). \tag{8.34}$$

2. Let $\mathcal{B}_{CO}$ be the control basis in $X_{q_1 \cdots q_s}$, and let $\mathcal{B}_{co}$ be the basis of $X_{q_1} \oplus \cdots \oplus X_{q_s}$ constructed from the union of the control bases in the X_{q_i}. We have

$$\tilde{V}B(q,p)V = \mathrm{diag}\,(B(q_1,pd_1),\ldots,B(q_s,pd_s)),$$

where $d_i = \prod_{j \neq i} q_j$ and $V = [Z]_{co}^{CO}$.

Proof:

1. We compute

$$Z^*p(S_q)Z\begin{pmatrix} f_1 \\ \cdot \\ \cdot \\ \cdot \\ f_s \end{pmatrix} = Z^*p(S_q)\sum_{j=1}^{s}d_jf_j = Z^*\pi_q p\sum_{j=1}^{s}d_jf_j$$

$$= \begin{pmatrix} \pi_{q_1}\sum_{j=1}^{s}\pi_q pd_jf_j \\ \cdot \\ \cdot \\ \cdot \\ \pi_{q_s}\sum_{j=1}^{s}\pi_q pd_jf_j \end{pmatrix} = \begin{pmatrix} \pi_{q_1}\pi_q pd_1 f_1 \\ \cdot \\ \cdot \\ \cdot \\ \pi_{q_s}\pi_q pd_s f_s \end{pmatrix}$$

$$= \begin{pmatrix} (pd_1)(S_{q_1})f_1 \\ \cdot \\ \cdot \\ \cdot \\ (pd_s)(S_{q_s})f_s \end{pmatrix}$$

$$= ((pd_1)(S_{q_1}) \oplus \cdots \oplus (pd_s)(S_{q_k}))\begin{pmatrix} f_1 \\ \cdot \\ \cdot \\ \cdot \\ f_s \end{pmatrix}.$$

2. We start from Eq. (8.34) and take the following matrix representations:

$$[Z^*]_{ST}^{st}[p(S_q)]_{CO}^{ST}[Z]_{co}^{CO} = [(pd_1)(S_{q_1}) \oplus \cdots \oplus (pd_s)(S_{q_k})]_{co}^{st}.$$

The control basis in $X_{q_1} \oplus \cdots \oplus X_{q_s}$ refers to the ordered union of the control bases in the X_{q_i}. Of course we use Theorem 4.4.3 to infer the equality $[Z^*]_{ST}^{st} = \widetilde{[Z]_{co}^{CO}}$. $\square$

8.7   Bezout and Hankel Matrices

Given a strictly proper rational function g, we have considered two different representations of it. The first is the coprime factorization $g = p/q$, where we assume $\deg q = n$; whereas the second is the expansion at infinity, that is, the representation $g(z) = \sum_{i=1}^{\infty}(g_i/z^i)$. We saw that these two representations give rise to two different matrices, namely, the Bezoutian matrix and the Hankel matrix. Our next goal is to study the relation between these two objects. We already saw that the analysis of Bezoutians is facilitated by considering them as matrix representations of intertwining maps; see Theorem 5.1.3. The next theorem interpretes the finite Hankel matrix in the same spirit.

Theorem 8.7.1 *Let p, q be polynomials, with q monic and $\deg p < \deg q = n$. Let*

$$g(z) = \frac{p}{q} = \sum_{i=1}^{\infty} \frac{g_i}{z^i},$$

and let $\overline{H} : X_q \longrightarrow X^q$ be defined by

$$\overline{H}f = H_g f \qquad for \ f \in X_q.$$

*Let $\rho_q : X^q \longrightarrow X_q$ be the map defined by Eq. (5.21), that is, $\rho_q h = qh$. Let $\mathcal{B}_{st}, \mathcal{B}_{co}$ be the standard and control bases of X_q, and let $\mathcal{B}_{rc} = \{e_1/q, \ldots, e_n/q\}$ be the **rational control basis** of X^q defined as the image of the control basis in X_q under the map ρ_q^{-1}. Then:*

1.
$$H_n = [\overline{H}]_{st}^{rc} = \begin{pmatrix} g_1 & \cdot & \cdot & \cdot & g_n \\ \cdot & & \cdot & \cdot & \cdot \\ \cdot & & \cdot & \cdot & \cdot \\ \cdot & & \cdot & \cdot & \cdot \\ g_n & \cdot & \cdot & \cdot & g_{2n-1} \end{pmatrix}. \tag{8.35}$$

2. *H_n is invertible if and only if p and q are coprime.*

3. *If p and q are coprime, then $H_n^{-1} = B(q, a)$, where a arises out of any solution to the Bezout equation $ap + bq = 1$.*

Proof:

1. We compute, using the representation of π_q given in Eq. (5.19),

$$\rho_q^{-1}p(S_q)f = q^{-1}\pi_q(pf) = q^{-1}q\pi_-q^{-1}pf = \pi_-gf = H_g f = \overline{H}f.$$

This implies that the following diagram is commutative:

$$\begin{array}{ccc} X_q & \overset{p(S_q)}{\longrightarrow} & X_q \\ \overline{H} & \searrow & \downarrow \; \rho_q^{-1} \\ & & X^q \end{array} \qquad (8.36)$$

To compute the matrix representation of $\overline{H}$, let h_{ij} be the ij element of $[\overline{H}]_{st}^{rc}$. Then we have, by the definition of a matrix representation of a linear transformation, that

$$\overline{H}z^{j-1} = \pi_- pq^{-1}z^{j-1} = \sum_{i=1}^{n} h_{ij} e_i/q.$$

So

$$\sum_{i=1}^{n} h_{ij} e_i = q\pi_- q^{-1} pz^{j-1} = \pi_q pz^{j-1}.$$

Using the fact that B_{co} is the dual basis of B_{st} under the pairing (5.29), we have

$$\begin{aligned} h_{kj} &= \sum_{i=1}^{n} h_{ij} < e_i, z^{k-1} >=< \pi_q pz^{j-1}, z^{k-1} > \\ &= [q^{-1}q\pi_- q^{-1} pz^{j-1}, z^{k-1}] = [\pi_- q^{-1} pz^{j-1}, z^{k-1}] \\ &= [gz^{j-1}, z^{k-1}] = [g, z^{j+k-2}] = g_{j+k-1}. \end{aligned}$$

2. The invertibility of $\overline{H}$ and Eq. (1) imply the invertibility of H_n.

3. From the equality $\overline{H} = \rho_q^{-1} p(S_q)$, it is clear that $\overline{H}$ is invertible if and only if $p(S_q)$ is, and this, by Theorem 5.1.3, is the case if and only if p and q are coprime. $\qquad \square$

The Bezout and Hankel matrices corresponding to a rational function g are both symmetric matrices arising from the same data. It is therefore natural to conjecture that not only their ranks are equal, but also the signatures of the corresponding quadratic forms. The next theorem shows that indeed they are congruent; moreover, it provides the relevant congruence transformations.

Theorem 8.7.2 *Let $g = p/q$, with p and q, coprime polynomials satisfying $\deg p < \deg q = n$. Assume that $g(z) = \sum_{i=1}^{\infty} g_i z^{-i}$ and $q(z) = z^n + q_{n-1}z^{n-1} + \cdots + q_0$. Then the Hankel matrix H_n and the Bezoutian $B(q, p) = (B_{ij})$ are congruent. Specifically, we have:*

1.

$$
\begin{pmatrix}
g_1 & \cdot & \cdot & \cdot & g_n \\
\cdot & \cdot & \cdot & \cdot & \cdot \\
\cdot & \cdot & \cdot & \cdot & \cdot \\
\cdot & \cdot & \cdot & \cdot & \cdot \\
g_n & \cdot & \cdot & \cdot & g_{2n-1}
\end{pmatrix}
=
$$

$$
\begin{pmatrix}
 & & & & 1 \\
 & & & \cdot & \psi_1 \\
 & & \cdot & \cdot & \cdot \\
 & \cdot & \cdot & \cdot & \cdot \\
1 & \psi_1 & \cdot & \cdot & \psi_{n-1}
\end{pmatrix}
\begin{pmatrix}
b_{11} & \cdot & \cdot & \cdot & b_{1n} \\
\cdot & \cdot & \cdot & \cdot & \cdot \\
\cdot & \cdot & \cdot & \cdot & \cdot \\
\cdot & \cdot & \cdot & \cdot & \cdot \\
b_{n1} & \cdot & \cdot & \cdot & b_{nn}
\end{pmatrix}
\tag{8.37}
$$

$$
\times
\begin{pmatrix}
 & & & & 1 \\
 & & & \cdot & \psi_1 \\
 & & \cdot & \cdot & \cdot \\
 & \cdot & \cdot & \cdot & \cdot \\
1 & \psi_1 & \cdot & \cdot & \psi_{n-1}
\end{pmatrix}.
$$

Here, the ψ_i are given by the polynomial characterization of Proposition 5.6.4.

2. In the same way we have,

$$
\begin{pmatrix}
b_{11} & \cdot & \cdot & \cdot & b_{1n} \\
\cdot & \cdot & \cdot & \cdot & \cdot \\
\cdot & \cdot & \cdot & \cdot & \cdot \\
\cdot & \cdot & \cdot & \cdot & \cdot \\
b_{n1} & \cdot & \cdot & \cdot & b_{nn}
\end{pmatrix}
=
$$

$$
\begin{pmatrix}
q_1 & & \cdot & \cdot & q_{n-1} & 1 \\
\cdot & & \cdot & \cdot & 1 & \\
\cdot & & \cdot & \cdot & & \\
q_{n-1} & 1 & & & & \\
1 & & & & &
\end{pmatrix}
\begin{pmatrix}
g_1 & \cdot & \cdot & \cdot & g_n \\
\cdot & \cdot & \cdot & \cdot & \cdot \\
\cdot & \cdot & \cdot & \cdot & \cdot \\
\cdot & \cdot & \cdot & \cdot & \cdot \\
g_n & \cdot & \cdot & \cdot & g_{2n-1}
\end{pmatrix}
\tag{8.38}
$$

$$
\times
\begin{pmatrix}
q_1 & & \cdot & \cdot & q_{n-1} & 1 \\
\cdot & & \cdot & \cdot & 1 & \\
\cdot & & \cdot & \cdot & & \\
q_{n-1} & 1 & & & & \\
1 & & & & &
\end{pmatrix}.
$$

Note that Eq. (2) can be written more compactly as

$$
B(q,p) = B(q,1)H_{p/q}B(q,1). \tag{8.39}
$$

3. *We have*
$$\text{rank}\,(H_n) = \text{rank}\,(B(q,p))$$
and
$$\sigma(H_n) = \sigma(B(q,p)).$$

Proof:

1. To this end recall diagram (8.36) and also the definition of the bases B_{st}, B_{co}, and B_{rc}. Clearly, the operator equality
$$\overline{H} = \rho_q^{-1} p(S_q)$$
implies a variety of matrix representations, depending on the choice of bases. In particular, it implies
$$[\overline{H}]_{st}^{rc} = [\rho_q^{-1}]_{st}^{rc}[p(S_q)]_{co}^{st}[I]_{st}^{co}. = [\rho_q^{-1}]_{co}^{rc}[I]_{st}^{co}[p(S_q)]_{co}^{st}[I]_{st}^{co}.$$

Now it was proved in Theorem 8.7.1 that
$$[\overline{H}]_{st}^{rc} = H_n = \begin{pmatrix} g_1 & \cdot & \cdot & \cdot & g_n \\ \cdot & \cdot & \cdot & \cdot & \cdot \\ & \cdot & \cdot & \cdot & \\ \cdot & \cdot & \cdot & \cdot & \cdot \\ g_n & \cdot & \cdot & \cdot & g_{2n-1} \end{pmatrix}.$$

Obviously, $[\rho_q^{-1}]_{co}^{rc} = I$ and, using matrix representations for dual maps, $[I]_{co}^{st} = \widetilde{[I^*]_{co}^{st}} = \widetilde{[I]_{co}^{st}}$. Here we used the fact that B_{st} and B_{co} are dual bases. So we see that $[I]_{co}^{st}$ is symmetric. In fact, it is a Hankel matrix, easily computed to be
$$[I]_{co}^{st} = \begin{pmatrix} q_1 & \cdot & \cdot & \cdot & q_{n-1} & 1 \\ \cdot & & \cdot & \cdot & \cdot & \\ \cdot & & & \cdot & \cdot & \\ \cdot & & \cdot & \cdot & & \\ q_{n-1} & \cdot & & & & \\ 1 & & & & & \end{pmatrix}.$$

Finally, from the identification of the Bezoutian as the matrix representation of an intertwining map for S_q, that is, from $B(q,p) = [p(S_q)]_{co}^{st}$, and defining $R = [I]_{st}^{co} = \widetilde{[I]_{st}^{co}}$, we obtain the equality
$$H_n = \tilde{R}B(q,p)R. \tag{8.40}$$

2. Follows from part 1 by inverting the congruence matrix; see Proposition 5.6.4.

3. Follows from the congruence relations.   □

The following corollary gives another determinantal test for the coprimeness of two polynomials, this time in terms of a Hankel matrix:

Corollary 8.7.1 *Let $g = p/q$, with p and q polynomials satisfying $\deg p < \deg q = n$. If $g(z) = \sum_{i=1}^{\infty} g_i z^{-i}$ and*

$$
H_n = \begin{pmatrix} g_1 & \cdot & \cdot & \cdot & g_n \\ \cdot & \cdot & \cdot & \cdot & \cdot \\ \cdot & \cdot & \cdot & \cdot & \cdot \\ \cdot & \cdot & \cdot & \cdot & \cdot \\ g_n & \cdot & \cdot & \cdot & g_{2n-1} \end{pmatrix},
$$

then

$$
\det H_n = \det B(q, p). \tag{8.41}
$$

As a consequence, p and q are coprime if and only if $\det H_n \neq 0$.

Proof: Clearly, $\det R = (-1)^{n(n-1)/2}$ and so $(\det R)^2 = (-1)^{n(n-1)} = 1$, and hence Eq. (8.41) follows from Eq. (8.40). □

The matrix identities appearing in Theorem 8.7.2 are also interesting for other reasons. They include in them some other identities for the principal minors of H_n. We state it in the following way.

Corollary 8.7.2 *Under the previous assumption we have, for $k \leq n$,*

$$
\begin{pmatrix} g_1 & \cdot & \cdot & \cdot & g_k \\ \cdot & \cdot & \cdot & \cdot & \cdot \\ \cdot & \cdot & \cdot & \cdot & \cdot \\ \cdot & \cdot & \cdot & \cdot & \cdot \\ g_k & \cdot & \cdot & \cdot & g_{2k-1} \end{pmatrix} =
$$

$$
\begin{pmatrix} & & & 1 \\ & & \cdot & \psi_1 \\ & \cdot & \cdot & \cdot \\ 1 & \psi_1 & \cdot & \psi_{k-1} \end{pmatrix} \begin{pmatrix} b_{(n-k+1)(n-k+1)} & \cdot & \cdot & \cdot & b_{(n-k+1)n} \\ \cdot & & & & \cdot \\ \cdot & & & & \cdot \\ \cdot & & & & \cdot \\ b_{n(n-k+1)} & \cdot & \cdot & \cdot & b_{nn} \end{pmatrix}
$$

$$
\times \begin{pmatrix} & & & 1 \\ & & \cdot & \psi_1 \\ & \cdot & \cdot & \cdot \\ 1 & \psi_1 & \cdot & \psi_{k-1} \end{pmatrix} \tag{8.42}
$$

□

In particular, we obtain the following.

Corollary 8.7.3 *The rank of the Bezoutian of the polynomials q and p is equal to the rank of the Hankel matrix $H_{p/q}$ and hence to the order of the* *largest nonsingular principal minor, when starting from the lower right-hand corner.*

Proof: Equation (8.7.2) implies the following equality:

$$\det(g_{i+j-1})_{i,j=1}^{k} = \det(b_{ij})_{i,j=n-k+1}^{n}. \tag{8.43}$$

$\square$

At this point we recall the connection, given in Eq. (8.41), between the determinant of the Bezoutian and that of the resultant of the polynomials q and p. In fact, this cannot be made more precise, but we can find relations between the minors of the resultant and those of the corresponding Hankel and Bezout matrices. We state this as follows:

Theorem 8.7.3 *Let q be of degree n and p of degree $\leq n$. Then the lower $k \times k$ right-hand corner of the Bezoutian $B(q,p)$ is given by*

$$
\begin{pmatrix}
b_{(n-k+1)(n-k+1)} & \cdots & b_{(n-k+1)n} \\
& & \\
& & \\
b_{n(n-k+1)} & \cdots & b_{nn}
\end{pmatrix}
$$

$$
= \begin{pmatrix}
p_{n-k} & \cdots & p_{n-2k+1} \\
& & \\
& & \\
p_{n-1} & \cdots & p_{n-k}
\end{pmatrix}
\begin{pmatrix}
q_{n-k+1} & \cdots & q_{n-1} & q_n \\
& & \\
q_{n-1} & q_n & \\
q_n & &
\end{pmatrix}
$$

$$
- \begin{pmatrix}
q_{n-k} & \cdots & q_{n-2k+1} \\
& & \\
& & \\
q_{n-1} & \cdots & q_{n-k}
\end{pmatrix}
\begin{pmatrix}
p_{n-k+1} & \cdots & p_{n-1} & p_n \\
& & \\
p_{n-1} & p_n & \\
p_n & &
\end{pmatrix}
$$

$$
= \left[\begin{pmatrix}
p_{n-k} & \cdots & p_{n-2k+1} \\
& & \\
& & \\
p_{n-1} & \cdots & p_{n-k}
\end{pmatrix}
\begin{pmatrix}
q_n & q_{n-1} & \cdots & q_{n-k+1} \\
& & \\
& & \\
q_n & &
\end{pmatrix} \right.
$$

$$
\left. - \begin{pmatrix}
q_{n-k} & \cdots & q_{n-2k+1} \\
& & \\
& & \\
q_{n-1} & \cdots & q_{n-k}
\end{pmatrix}
\begin{pmatrix}
p_n & p_{n-1} & \cdots & p_{n-k+1} \\
& & \\
& & \\
p_n & &
\end{pmatrix} \right] J_k,
$$

$$\tag{8.44}$$

where we take $p_j = q_j = 0$ for $j < 0$. Moreover, we have

$$\det(g_{i+j})_{i,j=1}^{k} = \det(b_{ij})_{i,j=n-k+1}^{n} =$$

$$\begin{vmatrix} p_{n-k} & q_{n-k} & p_{n-k-1} & q_{n-k-1} & & & q_{n-2k-1} \\ \cdot & \cdot & p_{n-k} & q_{n-k} & & & \\ \cdot & \cdot & \cdot & \cdot & & & \\ \cdot & \cdot & \cdot & \cdot & \cdot & \cdot & p_{n-k} & q_{n-k} \\ p_{n-1} & q_{n-1} & \cdot & \cdot & \cdot & \cdot & \cdot & \cdot \\ p_n & q_n & p_{n-1} & \cdot & \cdot & \cdot & \cdot & \cdot \\ & & p_n & q_n & \cdot & \cdot & \cdot & \cdot \\ & & & & \cdot & \cdot & \cdot & \cdot \\ & & & & & p_{n-1} & q_{n-1} \\ & & & & & p_n & q_n \end{vmatrix}. \tag{8.45}$$

Proof: We use Eq. (8.21) to infer Eq. (8.44). Using Lemma 3.2.1, we have

$$\det(b_{ij})_{i,j=n-k+1}^{n} = \det J_k$$

$$= \begin{vmatrix} p_{n-k} & \cdot & \cdot & \cdot & p_{n-2k+1} & q_{n-k} & \cdot & \cdot & \cdot & q_{n-2k+1} \\ \cdot & \cdot & \cdot & \cdot & \cdot & \cdot & \cdot & \cdot & \cdot & \cdot \\ \cdot & \cdot & \cdot & \cdot & \cdot & \cdot & \cdot & \cdot & \cdot & \cdot \\ \cdot & \cdot & \cdot & \cdot & \cdot & \cdot & \cdot & \cdot & \cdot & \cdot \\ p_{n-1} & \cdot & \cdot & \cdot & p_{n-k} & q_{n-1} & \cdot & \cdot & \cdot & q_{n-k} \\ p_n & p_{n-1} & \cdot & \cdot & p_{n-k+1} & q_n & & \cdot & \cdot & q_{n-k+1} \\ & & \cdot & \cdot & \cdot & & & \cdot & \cdot & \cdot \\ & & & \cdot & \cdot & & & & \cdot & \cdot \\ & & & \cdot & p_{n-1} & & & & \cdot & q_{n-1} \\ & & & & p_n & & & & & q_n \end{vmatrix}.$$

Since $\det J_k = (-1)^{[k(k-1)]/2}$, we can use this factor to rearrange the columns of the determinant on the right to get Eq. (8.45).

An application of Eq. (8.7) yields equality (8.45). □

Equation (8.45) will be useful in the derivation of the Hurwitz stability test.

8.8 Inversion of Hankel Matrices

In this section we study the problem of inverting a finite, nonsingular Hankel matrix

$$H_n = \begin{pmatrix} g_1 & \cdot & \cdot & \cdot & g_n \\ \cdot & \cdot & \cdot & \cdot & \cdot \\ \cdot & \cdot & \cdot & \cdot & \cdot \\ \cdot & \cdot & \cdot & \cdot & \cdot \\ g_n & \cdot & \cdot & \cdot & g_{2n-1} \end{pmatrix}. \tag{8.46}$$

We already saw, in Theorem 8.7.1, that if the g_i arise out of a rational function $g = p/q$, with p, q coprime, via the expansion $g(z) = \sum_{i=1}^{\infty} g_i z^{-i}$, then the corresponding Hankel matrix is invertible. Our aim here is to utilize the special structure of Hankel matrices in order to facilitate the computation of the inverse. In addition, we would like to provide a recursive algorithm, one that uses the information in the nth step for the computation in the next one. This is a natural entry point to the general area of fast algorithms for computing with specially structured matrices, of which the Hankel matrices are just an example.

The method we shall employ is to use the coprime factorization of g as the data of the problem. This leads quickly to concrete representations of the inverse. Going back to the data given in the form of the coefficients $g_1, \ldots, g_{2n-1}$, we construct all possible rational functions with rank equal to n and whose first $2n - 1$ coefficients are prescribed by the g_i. Such a rational function is called a **minimal rational extension** of the sequence $g_1, \ldots, g_{2n-1}$. We use these rational functions to obtain inversion formulas. It turns out, as is to be expected, that the nature of a specific minimal extension is irrelevant as far as the inversion formulas are concerned. Finally, we will study the recursive structure of the solution, that is, how to use the solution to the inversion problem for a given n when the amount of data is increased.

Suppose now that, rather than having a rational function g as our data, we are given only numbers $g_1, \ldots, g_{2n-1}$ for which the matrix H_n is invertible. We consider this to correspond to the rational function $\sum_{i=1}^{2n-1} g_i/z^i$. We now look for all other possible minimal rational extensions. It turns out that, preserving the rank condition, there is relatively little freedom in choosing the coefficients g_i, $i > 2n - 1$. In fact, we have the following.

Proposition 8.8.1 *Given $g_1, \ldots, g_{2n-1}$ for which the Hankel matrix H_n is invertible, then any minimal rational extension is completely determined by the choice of $\xi = g_{2n}$.*

Proof: Since H_n is invertible, there exists a unique solution to the equation

$$
\begin{pmatrix} g_1 & \cdot & \cdot & \cdot & g_n \\ \cdot & \cdot & \cdot & \cdot & \cdot \\ \cdot & \cdot & \cdot & \cdot & \cdot \\ \cdot & \cdot & \cdot & \cdot & \cdot \\ g_n & \cdot & \cdot & \cdot & g_{2n-1} \end{pmatrix} \begin{pmatrix} x_0 \\ \cdot \\ \cdot \\ \cdot \\ x_{n-1} \end{pmatrix} = \begin{pmatrix} g_{n+1} \\ \cdot \\ \cdot \\ \cdot \\ g_{2n} \end{pmatrix}.
$$

Since the rank of the infinite Hankel matrix does not increase, we have

$$
\begin{pmatrix}
g_1 & \cdot & \cdot & \cdot & g_n \\
\cdot & & \cdot & \cdot & \cdot \\
\cdot & & \cdot & \cdot & \cdot \\
\cdot & & \cdot & \cdot & \cdot \\
g_n & \cdot & \cdot & \cdot & g_{2n-1} \\
g_{n+1} & \cdot & \cdot & \cdot & g_{2n} \\
\cdot & & \cdot & \cdot & \cdot \\
\cdot & & \cdot & \cdot & \cdot \\
\cdot & & \cdot & \cdot & \cdot
\end{pmatrix}
\begin{pmatrix}
x_0 \\
\cdot \\
\cdot \\
\cdot \\
x_{n-1}
\end{pmatrix}
=
\begin{pmatrix}
g_{n+1} \\
\cdot \\
\cdot \\
\cdot \\
g_{2n} \\
g_{2n+1} \\
\cdot
\end{pmatrix},
$$

which shows that the g_i with $i > 2n$ are completely determined by the choice of g_{2n} via the recursion relation $g_k = \sum_{i=0}^{n-1} x_i g_{k-n+i}$. $\square$

Incidentally, it is quite easy to obtain the monic denominator polynomial of a minimal rational extension.

Theorem 8.8.1 *Given $g_1, \ldots, g_{2n-1}$ for which the Hankel matrix H_n is invertible, let $g_\xi = p_\xi / q_\xi$ be any minimal rational extension of the sequence $g_1, \ldots, g_{2n-1}$ determined by $\xi = g_{2n}$, p_ξ and q_ξ coprime, and q_ξ monic. Let $g = p/q$ be the one corresponding to $\xi = 0$. Also, let a and b be the unique polynomials of degree $< n$ and $< \deg p$, respectively, that solve the Bezout equation*

$$a(z)p(z) + b(z)q(z) = 1. \tag{8.47}$$

Then:

1. *The polynomial a is independent of ξ, and its coefficients are given as a solution of the system of linear equations*

$$
\begin{pmatrix}
g_1 & \cdot & \cdot & \cdot & g_n \\
\cdot & \cdot & \cdot & \cdot & \cdot \\
\cdot & \cdot & \cdot & \cdot & \cdot \\
\cdot & \cdot & \cdot & \cdot & \cdot \\
g_n & \cdot & \cdot & \cdot & g_{2n-1}
\end{pmatrix}
\begin{pmatrix}
a_0 \\
\cdot \\
\cdot \\
\cdot \\
a_{n-1}
\end{pmatrix}
= -
\begin{pmatrix}
0 \\
\cdot \\
\cdot \\
0 \\
1
\end{pmatrix}.
\tag{8.48}
$$

2. *The polynomial b is independent of ξ and has the representation*

$$b(z) = -\sum_{i=1}^{n} g_i a_i(z),$$

where $a_i(z)$, $i = 1, \ldots, n-2$ are the polynomials defined by

$$a_i(z) = \pi_+ z^{-i} a.$$

3. *The coefficients of the polynomial* $q_\xi(z) = z^n + q_{n-1}(\xi)z^{n-1} + \cdots + q_0(\xi)$ *are solutions of the system of linear equations*

$$
\begin{pmatrix} g_1 & \cdot & \cdot & \cdot & g_n \\ \cdot & \cdot & \cdot & \cdot & \cdot \\ \cdot & \cdot & \cdot & \cdot & \cdot \\ \cdot & \cdot & \cdot & \cdot & \cdot \\ g_n & \cdot & \cdot & \cdot & g_{2n-1} \end{pmatrix} \begin{pmatrix} q_0(\xi) \\ \cdot \\ \cdot \\ \cdot \\ q_{n-1}(\xi) \end{pmatrix} = - \begin{pmatrix} g_{n+1} \\ \cdot \\ \cdot \\ g_{2n-1} \\ \xi \end{pmatrix}, \qquad (8.49)
$$

The polynomial q_ξ *also can be given in determinantal form as*

$$
q_\xi(z) = (\det H_n)^{-1} \begin{vmatrix} g_1 & \cdot & \cdot & g_n & 1 \\ \cdot & \cdot & \cdot & \cdot & z \\ \cdot & \cdot & \cdot & \cdot & \cdot \\ \cdot & \cdot & \cdot & \cdot & \cdot \\ g_n & \cdot & \cdot & g_{2n-1} & z^{n-1} \\ g_{n+1} & \cdot & \cdot & \xi & z^n \end{vmatrix}. \qquad (8.50)
$$

4. *We have* $q_\xi(z) = q(z) - \xi a(z)$.

5. *We have* $p_\xi(z) = p(z) + \xi b(z)$.

6. *With respect to the control basis polynomials corresponding to* q_ξ, *that is,*

$$
e_i(\xi, z) = \pi_+ z^{-i} q_\xi = q_i(\xi) + q_{i+1}(\xi)z + \cdots + z^{n-i}
$$

$i = 1, \ldots, n$, *the polynomial* p_ξ *has the representation*

$$
p_\xi(z) = \sum_{i=1}^{n} g_i e_i(\xi, z). \qquad (8.51)
$$

Proof:

1. Let p_ξ, q_ξ be the polynomials in the coprime representation $g_\xi = p_\xi / q_\xi$ corresponding to the choice $g_{2n} = \xi$. Let a and b be the unique polynomials, assuming $\deg a < \deg q_\xi$, that solve the Bezout equation $ap_\xi + bq_\xi = 1$. We rewrite this equation as

$$
a \frac{p_\xi}{q_\xi} + b = \frac{1}{q_\xi}, \qquad (8.52)
$$

and note that the right-hand side has an expansion of the form

$$
\frac{1}{q_\xi(z)} = \frac{1}{z^n} + \frac{\sigma_{n+1}}{z^{n+1}} + \cdots.
$$

Equating coefficients of $z^{-i}, i = 1, \ldots, n$ in Eq. (8.52), we have

$$
\begin{cases}
a_0 g_1 + \cdots + a_{n-1} g_n & = & 0 \\
a_0 g_2 + \cdots + a_{n-1} g_{n+1} & = & 0 \\
\quad \cdot \\
\quad \cdot \\
\quad \cdot \\
a_0 g_{n-1} + \cdots + a_{n-1} g_{2n-2} & = & 0 \\
a_0 g_n + \cdots + a_{n-1} g_{2n-1} & = & -1,
\end{cases}
$$

which is equivalent to Eq. (8.48). Since $g_1, \ldots, g_{2n-1}$ do not depend on ξ, this shows that the polynomial $a(z)$ is also independent of ξ.

2. Equating nonnegative indexed coefficients in Eq. (8.52), we obtain

$$
\begin{cases}
a_{n-1} g_1 & = & -b_{n-2} \\
a_{n-1} g_2 + a_{n-2} g_1 & = & -b_{n-3} \\
\quad \cdot \\
\quad \cdot \\
\quad \cdot \\
a_{n-1} g_{n-1} + \cdots + a_1 g_1 & = & -b_0
\end{cases}
$$

or, in matrix form,

$$
\begin{pmatrix}
a_{n-1} & & \\
\cdot & \cdot & \\
\cdot & & \cdot \\
\cdot & & & \cdot \\
a_1 & \cdots & a_{n-1}
\end{pmatrix}
\begin{pmatrix}
g_0 \\
\cdot \\
\cdot \\
\cdot \\
g_{n-1}
\end{pmatrix}
= -
\begin{pmatrix}
b_{n-2} \\
\cdot \\
\cdot \\
\cdot \\
b_0
\end{pmatrix}
\qquad (8.53)
$$

This means that the polynomial $b(z)$ is independent of ξ and has the representation $b(z) = -\sum_{i=1}^{n} g_i a_i(z)$.

3. From the representation $g_\xi = p_\xi / q_\xi$ we get $p_\xi = q_\xi g_\xi$. In the last equality we equate coefficients of $z^{-1}, \ldots, z^{-n}$ to get the system of equations (8.49). Since H_n is assumed to be invertible, this system has a unique solution, and the solution obviously depends linearly on the parameter ξ. This system can be solved by inverting H_n to get

$$
\begin{pmatrix}
q_0(\xi) \\
\cdot \\
\cdot \\
\cdot \\
q_{n-1}(\xi)
\end{pmatrix}
= -H_n^{-1}
\begin{pmatrix}
g_{n+1} \\
\cdot \\
\cdot \\
g_{2n-1} \\
\xi
\end{pmatrix}
= -\frac{\mathrm{adj}\, H_n}{\det H_n}
\begin{pmatrix}
g_{n+1} \\
\cdot \\
\cdot \\
g_{2n-1} \\
\xi
\end{pmatrix}.
$$

Multiplying on the left by the polynomial row vector $(1 \ z \ \cdot \ \cdot \ z^{n-1})$, and using Eq. (3.6), we get

$$q_\xi(z) = z^n + q_{n-1}(\xi)z^{n-1} + \cdots + q_0(\xi)$$

$$= z^n - \begin{pmatrix} 1 & z & \cdot & \cdot & \cdot z^{n-1} \end{pmatrix} \frac{\mathrm{adj}\, H_n}{\det H_n} \begin{pmatrix} g_{n+1} \\ \cdot \\ \cdot \\ g_{2n-1} \\ \xi \end{pmatrix}$$

$$= (\det H_n)^{-1} \left[\det H_n z^n - \begin{pmatrix} 1 & z & \cdot & \cdot & \cdot & z^{n-1} \end{pmatrix} \right.$$

$$\left. \times \mathrm{adj}\, H_n \begin{pmatrix} g_{n+1} \\ \cdot \\ \cdot \\ g_{2n-1} \\ \xi \end{pmatrix} \right]$$

$$= (\det H_n)^{-1} \begin{vmatrix} g_1 & \cdot & \cdot & \cdot & g_n & g_{n+1} \\ \cdot & \cdot & \cdot & \cdot & \cdot & \cdot \\ \cdot & \cdot & \cdot & \cdot & \cdot & \cdot \\ \cdot & \cdot & \cdot & \cdot & \cdot & \cdot \\ g_n & \cdot & \cdot & \cdot & g_{2n-1} & \xi \\ 1 & z & \cdot & \cdot & z^{n-1} & z^n \end{vmatrix}.$$

This is equivalent to Eq. (8.50).

4. By writing the last column of the determinant in Eq. (8.50) as

$$(g_{n+1} \ldots \xi\ z^n) = (g_{n+1} \ldots 0\ z^n) + (0 \ldots 0\ \xi\ 0),$$

and using the linearity of the determinant as a function of each of its columns, we get

$$q_\xi(z) = (det H_n)^{-1} \begin{vmatrix} g_1 & \cdot & \cdot & \cdot & g_n & 1 \\ \cdot & \cdot & \cdot & \cdot & \cdot & z \\ \cdot & \cdot & \cdot & \cdot & \cdot & \cdot \\ \cdot & \cdot & \cdot & \cdot & \cdot & \cdot \\ g_n & \cdot & \cdot & \cdot & g_{2n-1} & z^{n-1} \\ g_{n+1} & \cdot & \cdot & \cdot & \xi & z^n \end{vmatrix}$$

$$= (det H_n)^{-1} \begin{vmatrix} g_1 & \cdot & \cdot & \cdot & g_n & 1 \\ \cdot & \cdot & \cdot & \cdot & \cdot & z \\ \cdot & \cdot & \cdot & \cdot & \cdot & \cdot \\ \cdot & \cdot & \cdot & \cdot & \cdot & \cdot \\ g_n & \cdot & \cdot & \cdot & g_{2n-1} & z^{n-1} \\ g_{n+1} & \cdot & \cdot & \cdot & 0 & z^n \end{vmatrix}$$

$$+ (det H_n)^{-1} \begin{vmatrix} g_1 & \cdot & \cdot & \cdot & g_n & 1 \\ \cdot & \cdot & \cdot & \cdot & \cdot & z \\ \cdot & \cdot & \cdot & \cdot & \cdot & \cdot \\ \cdot & \cdot & \cdot & \cdot & \cdot & \cdot \\ g_n & \cdot & \cdot & \cdot & g_{2n-1} & z^{n-1} \\ 0 & \cdot & \cdot & 0 & \xi & 0 \end{vmatrix}$$

$$= q(z) - \xi r(z)$$

$$r(z) = \begin{vmatrix} g_1 & \cdot & \cdot & \cdot & g_n & 1 \\ \cdot & \cdot & \cdot & \cdot & \cdot & z \\ \cdot & \cdot & \cdot & \cdot & \cdot & \cdot \\ \cdot & \cdot & \cdot & \cdot & \cdot & \cdot \\ g_n & \cdot & \cdot & \cdot & g_{2n-2} & z^{n-1} \end{vmatrix}.$$

We will show that $r = a$. From the Bezout equations $ap + bq = 1$ and $ap_\xi + bq_\xi = 1$, it follows by subtraction that $a(p_\xi - p) + b(q_\xi - q) = 0$. Since $q_\xi = q - \xi r$, we must have $a(p_\xi - p) = \xi br$. The coprimeness of a and b implies that a is a divisor of r. Setting $r = ad$, it follows that $p_\xi = p + \xi bd$ and $q_\xi = q - \xi ad$. Now,

$$g_\xi - g = \frac{p + \xi bd}{q - \xi ad} - \frac{p}{q} = \frac{\xi(ap + bq)d}{q(q - \xi ad)} = \frac{\xi d}{q(q - \xi ad)}.$$

Since g_ξ and g have expansions in powers of z^{-1} agreeing for the first $2n - 1$ terms, the next term has to be ξz^{-2n}. This implies that $d = 1$.

5. Follows from $p_\xi = p + \xi bd$, by substituting $d = 1$.

6. Equating positively indexed coefficients in the equality $p_\xi(z) = q_\xi(z) \times g_\xi(z)$, we have

$$\begin{cases} p_{n-1}(\xi) &= g_1 \\ p_{n-2}(\xi) &= g_2 + q_{n-1}(\xi)g_1 \\ & \quad \cdot \\ & \quad \cdot \\ & \quad \cdot \\ p_0(\xi) &= g_n + q_{n-1}(\xi)g_{n-1} + \cdots + q_1(\xi)g_1 \end{cases}$$

or

$$
\begin{pmatrix} p_0(\xi) \\ \cdot \\ \cdot \\ \cdot \\ p_{n-1}(\xi) \end{pmatrix} = \begin{pmatrix} q_1(\xi) & & \cdot & \cdot & \cdot & q_{n-1}(\xi) & 1 \\ & & \cdot & \cdot & \cdot & \cdot & \\ & & \cdot & \cdot & \cdot & & \\ & & \cdot & \cdot & \cdot & & \\ q_{n-1}(\xi) & 1 & & & & & \\ 1 & & & & & & \end{pmatrix} \begin{pmatrix} g_1 \\ \cdot \\ \cdot \\ \cdot \\ g_n \end{pmatrix}.
$$

This is equivalent to

$$
p_\xi(z) = g_1 e_1(\xi, z) + \cdots + g_n e_n(\xi, z),
$$

which proves Eq. (8.51).   □

Incidentally, parts 4 and 5 of the theorem provide a nice parametrization of all solutions to the minimal partial realization problem arising out of a nonsingular Hankel matrix.

Corollary 8.8.1 *Given the nonsingular Hankel matrix H_n. If $g = p/q$ is the minimal rational extension corresponding to the choice $g_{2n} = 0$, then the minimal rational extension corresponding to $g_{2n} = \xi$ is given by*

$$
g_\xi = \frac{p + \xi b}{q - \xi a},
$$

where a, b are the polynomials arising from the solution of the Bezout equation $ap + bq = 1$.

This theorem explains the mechanism of how the particular minimal rational extension chosen does not influence the computation of the inverse, as of course it should not. Indeed, we see that

$$
B(q_\xi, a) = B(q - \xi a, a) = B(q, a) - \xi B(a, a) = B(q, a),
$$

and so $H_n^{-1} = B(q_\xi, a) = B(q, a)$.

8.9   Continued Fractions and Orthogonal Polynomials

The Euclidean algorithm provides a tool that is not only for the computation of a g.c.d. of a pair of polynomials, but also, if coprimeness is assumed, for the solution of the corresponding Bezout equation. The application of the Euclidean algorithm yields a representation of a rational function in the form of a continued fraction. It should be noted that a slight modification of the process makes the method applicable to an arbitrary (formal) power series. This is the theme of the present section.

Given an irreducible representation p/q of a strictly proper rational function g, we define, using the division rule for polynomials, a sequence of polynomials q_i, nonzero constants β_i, and monic polynomials $a_i(z)$ by

$$\begin{cases} q_{-1} = q, \quad q_0 = p \\ q_{i+1}(z) = a_{i+1}(z)q_i(z) - \beta_i q_{i-1}(z) \end{cases} \tag{8.54}$$

with $\deg q_{i+1} < \deg q_i$. The procedure ends when $q_n = 0$, and then q_{n-1} is the g.c.d. of p and q. Since p and q are assumed coprime, q_{n-1} is a nonzero constant. The pairs $\{\beta_{i-1}, a_i(z)\}$ are called the **atoms** of g. We define $\alpha_i = \deg a_i$.

For $i = 0$, we rewrite Eq. (8.54) as

$$\frac{p}{q} = \frac{q_0}{q_{-1}} = \frac{\beta_0}{a_1 - \dfrac{q_1}{q_0}}.$$

Thus, $g_1 = q_1/q_0$ is also a strictly proper rational function and, furthermore, this is an irreducible representation. By an inductive argument, this leads to the **continued fraction representation** of g in terms of the atoms β_{i-1} and $a_i(z)$:

$$g(z) = \cfrac{\beta_0}{a_1(z) - \cfrac{\beta_1}{a_2(z) - \cfrac{\beta_2}{a_3(z) - \cfrac{\ddots}{\quad a_{n-1}(z) - \cfrac{\beta_{n-1}}{a_n(z)}}^{\beta_{n-2}}}}}. \tag{8.55}$$

There are two ways to truncate the continued fraction. The first one is to consider the top part, that is,

$$g_i(z) = \cfrac{\beta_0}{a_1(z) - \cfrac{\beta_1}{a_2(z) - \cfrac{\beta_2}{a_3(z) - \cfrac{\ddots}{\quad a_{i-1}(z) - \cfrac{\beta_{i-1}}{a_i(z)}}^{\beta_{i-2}}}}}. \tag{8.56}$$

The other option is to consider the rational functions γ_i defined by the bottom part, namely,

$$\gamma_i(z) = \cfrac{\beta_i}{a_{i+1}(z) - \cfrac{\beta_{i+1}}{a_{i+2}(z) - \cfrac{}{\ddots \cfrac{\beta_{n-2}}{a_{n-1}(z) - \cfrac{\beta_{n-1}}{a_n(z)}}}}}. \tag{8.57}$$

Clearly, we have $g_n = \gamma_0 = g = p/q$. We also set $\gamma_n = 0$.

Just as a rational function determines a unique continued fraction, it is also completely determined by it.

Proposition 8.9.1 *Let g be a rational function having the continued fraction representation (8.55). Then $\{\beta_i, a_{i+1}\}_{i=0}^{n-1}$ are the atoms of g.*

Proof: Assume that $g = p/q$. Then

$$g = \frac{p}{q} = \frac{\beta_0}{a_1 - \gamma_1},$$

with γ_i strictly proper. This implies $p\gamma_1 = a_1 p - \beta_0 q$. This shows that $r = p\gamma_1$ is a polynomial and, since γ_1 is strictly proper, we have $\deg r < \deg p$. It follows that $\{\beta_0, a_1\}$ is the first atom of g. The rest follows by induction.
□

Fractional linear transformations are crucial to the analysis of continued fractions and play a central role in a wide variety of parametrization problems.

Definition 8.9.1 *A **fractional linear transformation**, or a **Moebius transformation**, is a function of the form*

$$f(w) = \frac{aw + b}{cw + d},$$

under the extra condition $ad - bc \neq 0$. With this fractional linear transformation we associate the matrix $\begin{pmatrix} a & b \\ c & d \end{pmatrix}$.

The following lemma relates the composition of fractional linear transformations to matrix multiplication:

Lemma 8.9.1 *Let $f(w) = (aw+b)/(cw+d)$, $i = 1, 2$ be two fractional linear transformations. Then their composition $(f_1 \circ f_2)(w) = (aw+b)/(cw+d)$, where*

$$\begin{pmatrix} a & b \\ c & d \end{pmatrix} = \begin{pmatrix} a_1 & b_1 \\ c_1 & d_1 \end{pmatrix} \begin{pmatrix} a_2 & b_2 \\ c_2 & d_2 \end{pmatrix}.$$

Proof: By computation.
□

We now define two sequences of polynomials, $\{Q_k\}$ and $\{P_k\}$, by the three-term recursion formulas

$$\begin{cases} Q_{-1} = 0, \quad Q_0 = 1 \\ Q_{k+1}(z) = a_{k+1}(z)Q_k(z) - \beta_k Q_{k-1}(z) \end{cases} \tag{8.58}$$

and

$$\begin{cases} P_{-1} = -1, \quad P_0 = 0 \\ P_{k+1}(z) = a_{k+1}(z)P_k(z) - \beta_k P_{k-1}(z). \end{cases} \tag{8.59}$$

The polynomials Q_i and P_i so defined are called the **Lanczos polynomials** of the first and second kind, respectively.

Theorem 8.9.1 *Let g be a strictly proper rational function having the continued fraction representation (8.55). Let the Lanczos polynomials be defined by Eqs. (8.58) and (8.59), and define the polynomials A_k, B_k by*

$$A_k(z) = \frac{Q_{k-1}(z)}{\beta_0 \cdots \beta_{k-1}}$$

$$\tag{8.60}$$

$$B_k(z) = \frac{-P_{k-1}(z)}{\beta_0 \cdots \beta_{k-1}}.$$

Then:

1. *For the rational functions γ_i, $i = 1, \ldots, n$, we have*

$$a_{i+1}(z)\gamma_i(z) - \beta_i = \gamma_i(z)\gamma_{i+1}(z). \tag{8.61}$$

2. *Defining the rational functions E_i by*

$$E_i = Q_i P_n / Q_n - P_i = (Q_i P_n - P_i Q_n)/Q_n, \tag{8.62}$$

    they satisfy the following recursion:

$$\begin{cases} E_{-1} = 1, \quad E_0 = g_0 = g \\ E_{k+1}(z) = a_{k+1}(z)E_k(z) - \beta_k E_{k-1}(z). \end{cases} \tag{8.63}$$

3. *We have*

$$E_i(z) = \gamma_0(z) \cdots \gamma_i(z). \tag{8.64}$$

4. *In the expansion of E_i in powers of z^{-1}, the leading terms is $\beta_0 \cdots \beta_i / z^{\alpha_1 + \cdots + \alpha_{i+1}}$.*

5. *With $\alpha_i = \deg a_i$, we have*

$$\deg Q_k = \sum_{i=1}^{k} \alpha_i$$

    and

$$\deg P_k = \sum_{i=2}^{k} \alpha_i.$$

6. *The polynomials A_k, B_k are coprime and solve the Bezout equation*

$$P_k(z)A_k(z) + Q_k(z)B_k(z) = 1. \tag{8.65}$$

7. *The rational function P_k/Q_k is strictly proper, and the expansions of p/q and P_k/Q_k in powers of z^{-1} agree up to order $2\sum_{i=1}^{k} \deg a_i + \deg a_{k+1} = 2\sum_{i=1}^{k} \alpha_i + \alpha_{i+1}$.*

8. *For the unimodular matrices $\begin{pmatrix} 0 & \beta_i \\ -1 & a_{i+1} \end{pmatrix}$, $i = 0, \ldots, n-1$, we have*

$$\begin{pmatrix} 0 & \beta_0 \\ -1 & a_1 \end{pmatrix} \cdots \begin{pmatrix} 0 & \beta_{i-1} \\ -1 & a_i \end{pmatrix} = \begin{pmatrix} -P_{i-1} & P_i \\ -Q_{i-1} & Q_i \end{pmatrix}.$$

9. *Let T_i be the fractional linear transformation corresponding to the matrix $\begin{pmatrix} 0 & \beta_i \\ -1 & a_{i+1} \end{pmatrix}$. Then*

$$g = (T_0 \circ \cdots \circ T_{i-1})(\gamma_i) = \frac{P_i - P_{i-1}\gamma_i}{Q_i - Q_{i-1}\gamma_i}. \tag{8.66}$$

In particular,

$$g = (T_0 \circ \cdots \circ T_{n-1})(\gamma_n) = P_n/Q_n,$$

that is, we have
$$P_n = p, \qquad Q_n = q. \tag{8.67}$$

10. *The atoms of P_k/Q_k are $\{\beta_{i-1}, a_i(z)\}_{i=1}^{k}$.*

11. *The generating function of the Bezoutian of Q_k and A_k satisfies*

$$\frac{Q_k(z)A_k(w) - A_k(z)Q_k(w)}{z - w}$$
$$= \sum (\beta_0 \cdots \beta_{k-1})^{-1} Q_j(z) \frac{a_{j+1}(z) - a_{j+1}(w)}{z - w} Q_j(w). \tag{8.68}$$

The formula for the expansion of the generating function is called the **generalized Christoffel–Darboux formula**. *The regular Christoffel–Darboux formula is the generic case, namely, when all of the atoms a_j have degree 1.*

12. *The Bezoutian matrix $B(Q_k, A_k)$ can be written as*

$$B(Q_k, A_k) = \sum_{j=1}^{k-1} (\beta_0 \cdots \beta_{j-1})^{-1} \tilde{R}_j B(a_{j+1}, 1) R_j,$$

where R_j is the $n_j \times n$ Toeplitz matrix

$$R_j = \begin{pmatrix} \sigma_0^{(j)} & \sigma_1^{(j)} & \cdot & \cdot & \cdot & \sigma_{n_j}^{(j)} \\ & & & & & \\ & & & & & \\ & & & & & \\ & & & & & \\ & & \sigma_0^{(j)} & \sigma_1^{(j)} & \cdot & \cdot & \cdot & \sigma_{n_j}^{(j)} \end{pmatrix}$$ (8.69)

and $Q_j(z) = \sum \sigma_\nu^{(j)} z^\nu$.

13. *Assume that the Hankel matrix H_n in Eq. (8.46) to be nonsingular. Let $g = p/q$ be any minimal rational extension of the sequence $g_1, \ldots, g_{2n-1}$. Then*

$$H_n^{-1} = \sum_{j=1}^{n} \frac{1}{\beta_0 \cdots \beta_{j-1}} \tilde{R}_j B(1, a_{j+1}) R_j.$$

Proof:

1. Clearly, by the definition of the γ_i, we have, for $i = 1, \ldots, n-1$,

$$\gamma_i = \frac{\beta_i}{a_{i+1} - \gamma_{i+1}},$$ (8.70)

which can be rewritten as Eq. (8.61). Since $\gamma_{n-1} = \beta_{n-1}/a_n$ and $\gamma_n = 0$, Eq. (8.61) also holds for $i = n$.

2. We compute first the initial conditions, using the initial conditions of the Lanczos polynomials. Thus,

$$\begin{aligned} E_{-1} &= Q_{-1}P_n/Q_n - P_{-1} = 1 \\ E_0 &= Q_0 P_n/Q_n - P_0 = P_n/Q_n = \gamma_0 = g. \end{aligned}$$

Next,

$$\begin{aligned} E_{i+1} &= Q_{i+1}P_n/Q_n - P_{i+1} = (Q_{i+1}P_n - P_{i+1}Q_n)/Q_n \\ &= [(a_{i+1}Q_i - \beta_i Q_{i-1}(z))P_n - (a_{i+1}P_i - \beta_i P_{i-1}(z))]Q_n/Q_n \\ &= a_{i+1}(Q_i P_n - P_i Q_n)/Q_n - \beta_i(Q_{i-1}P_n - P_{i-1}Q_n)/Q_n \\ &= a_{i+1}E_i - \beta_i E_{i-1}. \end{aligned}$$

3. We use the recursion formula (8.63) and Eq. (8.61). The proof goes by induction. For $i = 1$, we have

$$E_1 = a_1 E_0 - \beta_0 = a_1\gamma_0 - \beta_0 = \gamma_0\gamma_1.$$

Assuming that $E_i = \gamma_0 \cdots \gamma_i$, we compute

$$\begin{aligned} E_{i+1} &= a_{i+1}E_i - \beta_i E_{i-1} \\ &= a_{i+1}\gamma_0 \cdots \gamma_i - \beta_i\gamma_0 \cdots \gamma_{i-1} \\ &= \gamma_0 \cdots \gamma_{i-1}(a_{i+1}\gamma_i - \beta_i) \\ &= \gamma_0 \cdots \gamma_{i-1}(\gamma_i\gamma_{i+1}). \end{aligned}$$

4. From Eq. (8.70) we have $\beta_i/z^{\alpha_{i+1}}$ for the leading term of the expansion of γ_i. Since $E_i = \gamma_0 \cdots \gamma_i$, we have $\beta_0 \cdots \beta_i/z^{\alpha_1+\cdots+\alpha_{i+1}} = \beta_0 \cdots \beta_i/z^{\nu_{i+1}}$ for the leading term of the expansion of E_i.

5. By induction, using the recursion formulas as well as the initial conditions.

6. We compute the expression $Q_{k+1}P_k - Q_kP_{k+1}$ using the defining recursion formulas. So

$$
\begin{aligned}
Q_{k+1}P_k - Q_kP_{k+1} &= (a_{k+1}Q_k - \beta_kQ_{k-1})P_k \\
&\quad - (a_{k+1}P_k - \beta_kP_{k-1})Q_k \\
&= \beta_k(Q_kP_{k-1} - Q_{k-1}P_k)
\end{aligned}
$$

and, by induction, this implies

$$
\begin{aligned}
Q_{k+1}P_k - Q_kP_{k+1} &= \beta_k \cdots \beta_0(Q_0P_{-1} - Q_{-1}P_0) \\
&= -\beta_k \cdots \beta_0.
\end{aligned}
$$

Therefore, for each k, Q_k and P_k are coprime. Moreover, dividing by $-\beta_k \cdots \beta_0$, we see that the Bezout identities

$$
P_k(z)A_k(z) + Q_k(z)B_k(z) = 1.
$$

7. With the E_i defined by Eq. (8.62), we compute

$$
P_n/Q_n - P_i/Q_i = (P_nQ_i - Q_nP_i)/Q_nQ_i = E_i/Q_i.
$$

Since $\deg Q_i = \sum_{j=1}^{i} \alpha_j$ and using part 4, it follows that the leading term of E_i/Q_i is $\beta_0 \cdots \beta_i/z^{2\sum_{j=1}^{i}\alpha_j+\alpha_{i+1}}$. This is in agreement with the statement of the theorem as $\deg a_i = \sum_{j=1}^{i} \alpha_j$.

8. We will show it by induction. Indeed, for $i = 0$, we have

$$
\begin{pmatrix} 0 & \beta_0 \\ -1 & a_1 \end{pmatrix} = \begin{pmatrix} -P_0 & P_1 \\ -Q_0 & Q_1 \end{pmatrix}.
$$

Next, using the induction hypothesis, we compute

$$
\begin{aligned}
\begin{pmatrix} -P_{i-1} & P_i \\ -Q_{i-1} & Q_i \end{pmatrix} \begin{pmatrix} 0 & \beta_i \\ -1 & a_{i+1} \end{pmatrix} &= \begin{pmatrix} -P_i & a_{i+1}P_i - \beta_iP_{i-1} \\ -Q_i & a_{i+1}Q_i - \beta_iQ_{i-1} \end{pmatrix} \\
&= \begin{pmatrix} -P_i & P_{i+1} \\ -Q_i & Q_{i+1} \end{pmatrix}.
\end{aligned}
$$

9. We clearly have

$$
\begin{aligned}
g = \gamma_0 = T_0(\gamma_1) &= T_0(T_1(\gamma_2)) = (T_0 \circ T_1)(\gamma_2) \\
&= \cdots = (T_0 \circ \cdots \circ T_{i-1})(\gamma_i).
\end{aligned}
$$

To prove (8.67) we use the fact that $\gamma_n = 0$.

10. From Eq. (8.56), and using Eq. (8.66), we get $g_k = (T_0 \circ \cdots \circ T_{k-1})(0) = P_k/Q_k$.

11. Obviously, the Bezout identity (8.65) also implies the coprimeness of the polynomials A_k and Q_k. Since the Bezoutiant is linear in each of its arguments, we can get a recursive formula for $B(A_k, Q_k)$:

$$
\begin{aligned}
B(Q_k, A_k) &= B(Q_k, (\beta_0 \cdots \beta_{k-1})^{-1} Q_{k-1}) \\
&= B(a_k Q_{k-1} - \beta_{k-1} Q_{k-2}, (\beta_0 \cdots \beta_{k-1})^{-1} Q_{k-1}) \\
&= (\beta_0 \cdots \beta_{k-1})^{-1} Q_{k-1}(z) B(a_k, 1) Q_{k-1}(w) \\
&\quad + B(Q_{k-1}, (\beta_0 \cdots \beta_{k-2})^{-1} Q_{k-2}) \\
&= \cdots = \sum (\beta_0 \cdots \beta_{k-1})^{-1} Q_j(z) B(a_{j+1}, 1) Q_j(w).
\end{aligned}
$$

12. This is just the matrix representation of Eq. (8.68).

13. We use Theorem 8.7.1, the equalities $Q_n = q$, $P_n = p$, and the Bezout equation (8.65) for $k = n$. □

A continued fraction representation is an extremely convenient tool for the computation of the signature of the Hankel or Bezout quadratic forms.

Theorem 8.9.2 (Frobenius) *Let g be rational, having the continued fraction representation (8.55). Then*

$$
\sigma(H_g) = \sum_{i=1}^{r} \left(\operatorname{sign} \prod_{j=0}^{i-1} \beta_j \right) \frac{1 + (-1)^{\deg a_i - 1}}{2}. \tag{8.71}
$$

Proof: It suffices to compute the signature of the Bezoutian $B(q, p)$. Using the equations of the Euclidean algorithm, we have

$$
\begin{aligned}
B(q, p) &= B(q_{-1}, q_0) = B\left(\frac{a_1 q_0 - q_1}{\beta_0}, q_0 \right) \\
&= \frac{1}{\beta_0} q_0(z) B(a_1, 1) q_0(w) + \frac{1}{\beta_0} B(q_0, q_1).
\end{aligned}
$$

Here we used the fact that the Bezoutian is alternating. Since p and q are coprime, we have

$$
\begin{aligned}
\operatorname{rank} B(a_1, 1) + \frac{1}{\beta_0} \operatorname{rank} B(q_0, q_1) &= \deg a_1 + \deg q_0 = \deg q \\
&= \operatorname{rank} B(q, p).
\end{aligned}
$$

The additivity of the ranks implies, by Theorem 8.2.4, the additivity of the signatures, and therefore

$$
\sigma(B(q, p)) = \sigma\left(\frac{1}{\beta_0} B(a_1, 1) \right) + \sigma\left(\frac{1}{\beta_0} B(q_0, q_1) \right)
$$

We proceed by induction and readily derive

$$\sigma(B(p,q)) = \sum_{i=1}^{r} \text{sign}\left(\frac{1}{\beta_0 \cdots \beta_i}\right) \sigma(B(a_{i+1}, 1)). \tag{8.72}$$

Now, given a polynomial $c(z) = z^k + c_{k-1}z^{k-1} + \cdots + c_0$, we have

$$B(c,1) = \frac{c(z) - c(w)}{z - w} = \sum_{i=1}^{k} c_i \frac{z^i - w^i}{z - w}$$

with $c_k = 1$. Using the equality

$$\frac{z^i - w^i}{z - w} = \sum_{j=1}^{i} z^j w^{i-j-1},$$

we have

$$B(c,1) = \begin{pmatrix} c_1 & c_2 & . & . & c_{i-1} & 1 \\ c_2 & . & . & . & . & \\ . & . & . & . & & \\ . & . & . & . & & \\ c_{i-1} & 1 & & & & \\ 1 & & & & & \end{pmatrix}.$$

Hence,

$$\sigma(B(c,1)) = \begin{cases} 1 & \text{if } \deg c \text{ is odd} \\ 0 & \text{if } \deg c \text{ is even.} \end{cases}$$

Equality (8.72) now can be rewritten as Eq. (8.71).   □

The Lanczos polynomials Q_i generated by the recursion relation (8.58) satisfy certain orthogonality conditions. These orthogonality conditions depend of course on the rational function g from which the polynomials were derived. These orthogonality relations are related to an inner product induced by the Hankel operator H_g. We now explain this relation. Starting from the strictly proper rational function g, we have the bilinear form defined on the space of polynomials by

$$\{x, y\} = [H_g x, y] = \sum_i \sum_j g_{i+j} \xi_i \eta_j, \tag{8.73}$$

where $x(z) = \sum_i \xi_i z^i$ and $y(z) = \sum_j \eta_j z^j$. If g has the irreducible representation $g = p/q$, then $\text{Ker } H_g = qF[z]$, and hence, we may as well restrict ourselves to the study of H_g as a map from X_q to $X^q = X_q^*$. However, we can view all of the action as taking place in X_q, using the pairing $< \cdot, \cdot >$ defined in Eq. (5.29). Thus, for $a, b \in X_q$, we have

$$\begin{aligned} [H_g x, y] &= [\pi_- q^{-1} p x, y] \\ &= [q^{-1} q \pi_- q^{-1} p x, y] = < \pi_q p x, y > \\ &= < p(S_q) x, y >. \end{aligned}$$

In fact, $p(S_q)$ is, as a result of Theorem 5.6.5, a self-adjoint operator in X_q. The inner product

$$\{x, y\} = < p(S_q)x, y >$$

of course can be evaluated by choosing an appropriate basis. In fact, we have $< p(S_q)x, y > = ([p(S_q)]_{st}^{co}[x]^{st}, [y]^{st})$. As $[p(S_q)]_{st}^{co} = H_n$, we recover Eq. (8.73).

Proposition 8.9.2 *Let P_i, Q_i be the Lanczos polynomials associated with the rational function g. Relative to the inner product*

$$\{x, y\} = < p(S_q)x, y > = [H_g x, y] = \sum_i \sum_j g_{i+j} \xi_i \eta_j, \qquad (8.74)$$

the polynomial Q_i is orthogonal to all polynomials of lower degree. Specifically,

$$\{Q_i, z^j\} = [H_g Q_i, z^j] = \left\{ \begin{array}{ll} 0 & j < \alpha_1 + \cdots + \alpha_i \\ \beta_0 \cdots \beta_i & j = \alpha_1 + \cdots + \alpha_i. \end{array} \right. \qquad (8.75)$$

Proof: Recalling that the rational functions E_i are defined in Eq. (8.62) by $E_i = gQ_i - P_i$, this means that

$$\pi_- E_i = \pi_- gQ_i = H_g Q_i.$$

Since the leading term of E_i is $\beta_0 \cdots \beta_i / z^{\alpha_1 + \cdots + \alpha_{i+1}}$, this implies Eq. (8.75).
$\qquad \square$

The set

$$B_{or} = \{1, z, \ldots, z^{n_1 - 1}, Q_1, zQ_1, \ldots, z^{n_2 - 1}Q_1, \ldots, Q_{r-1}, \ldots, z^{n_{r-1} - 1}Q_{r-1}\}$$

is clearly a basis for X_q as it contains one polynomial for each degree between 0 and $n - 1$. We call this basis the **orthogonal basis** because of its relation to orthogonal polynomials. The properties of the Lanczos polynomials can be used to compute a matrix representation of the shift S_q relative to this basis.

Theorem 8.9.3 *Let the strictly proper transfer function $g = p/q$ have the sequence of atoms $\{a_i(z), \beta_{i-1}\}_{i=1}^r$, and assume that $\alpha_1, \ldots, \alpha_r$ are the degrees of the atoms and*

$$a_k(z) = z^{\alpha_k} + \sum_{i=0}^{\alpha_k - 1} a_i^{(k)} z^i.$$

Then the shift operator S_q has the following block tridiagonal matrix representation with respect to the orthogonal basis:

$$[S_q]_{or}^{or} = \left(\begin{array}{cccccc} A_{11} & A_{12} & & & \\ A_{21} & A_{22} & . & & \\ & . & . & . & \\ & & . & . & A_{r-1\,r} \\ & & & A_{r\,r-1} & A_{rr} \end{array} \right),$$

where

$$A_{ii} = \begin{pmatrix} 0 & & \cdots & & -a_0^{(i)} \\ 1 & & & & \cdot \\ & \cdot & & & \cdot \\ & & \cdot & & \cdot \\ & & & 1 & -a_{\alpha_i-1}^{(i)} \end{pmatrix}, \; i = 1, \ldots r,$$

$$A_{i+1\,i} = \begin{pmatrix} 0 & \cdot & \cdot & 0 & 1 \\ & & & & 0 \\ \cdot & & & & \cdot \\ \cdot & & & & \cdot \\ 0 & \cdot & \cdot & \cdot & 0 \end{pmatrix},$$

and $A_{i\,i+1} = \beta_{i-1}A_{i+1\,i}$.

Proof: On the basis elements $z^i Q_{j-1}$, $i = 0, \ldots, n_j - 2$, $j = 1, \ldots, r-1$, the map S_q indeed acts as the shift to the next basis element. For $i = n_j - 1$, we compute

$$\begin{aligned}
S_q z^{n_j-1} Q_{j-1} &= \pi_q z^{n_j} Q_{j-1} \\
&= \pi_q \left(z^{n_j} + \sum_{k=0}^{n_j-1} a_k^{(j)} z^k - \sum_{k=0}^{n_j-1} a_k^{(j)} z^k \right) Q_{j-1} \\
&= a_j Q_{j-1} - \sum_{k=0}^{n_j-1} a_k^{(j)} z^k \cdot Q_{j-1} \\
&= Q_j + \beta_{j-1} Q_{j-2} - \sum_{k=0}^{n_j-1} a_k^{(j)} z^k \cdot Q_{j-1}. \qquad \Box
\end{aligned}$$

In the generic case, when all of the Hankel matrices H_i, $i = 1, \ldots, n$, are nonsingular, all polynomials a_i are linear and we write $a_i(z) = z - \theta_i$. The recursion equation for the orthogonal polynomials become $Q_{i+1}(z) = (z - \theta_{i+1})Q_i(z) - \beta_i Q_{i-1}(z)$ and we obtain the following:

Corollary 8.9.1 *Assuming that all Hankel matrices H_i, $i = 1, \ldots, n$, are nonsingular, then with respect to the orthogonal basis $\{Q_0, \ldots, Q_{n-1}\}$ of $X_q = X_{Q_n}$, the shift operator S_q has the tridiagonal matrix representation*

$$[S_q]_{or}^{or} = \begin{pmatrix} \theta_1 & \beta_1 & & & \\ 1 & \theta_2 & \cdot & & \\ & \cdot & \cdot & \cdot & \\ & & \cdot & \cdot & \cdot \\ & & & \cdot & \beta_{n-1} \\ & & & 1 & \theta_n \end{pmatrix}.$$

The Hankel matrix H_n is symmetric and hence can be diagonalized by a congruence transformation. In fact, we can exhibit explicitly a diagonalizing congruence. Using the same method, we can clarify the connection between $C_q^\sharp$, the companion matrix of q, and the tridiagonal representation.

Proposition 8.9.3 *Let $g = p/q$ be a rational of McMillan degree $n = \deg q$. Assume that all of the Hankel matrices H_i, $i = 1, \ldots, n$, are nonsingular, and let $\{Q_0, \ldots, Q_{n-1}\}$ be the orthogonal basis of X_q. Then:*

1.

$$
\begin{pmatrix}
q_{0,0} & & & \\
q_{1,0} & q_{1,1} & & \\
\cdot & & \cdot & \\
\cdot & & & \\
q_{n-1,0} & \cdot & \cdot & \cdot q_{n-1,n-1}
\end{pmatrix}
\begin{pmatrix}
g_1 & \cdot & \cdot & \cdot & g_n \\
\cdot & & \cdot & \cdot & \cdot \\
\cdot & & \cdot & \cdot & \cdot \\
\cdot & & \cdot & \cdot & \cdot \\
g_n & \cdot & \cdot & \cdot & g_{2n-1}
\end{pmatrix}
$$

$$
\times
\begin{pmatrix}
q_{0,0} & q_{1,0} & \cdot & \cdot & q_{n-1,0} \\
 & q_{1,1} & & & \cdot \\
 & & \cdot & & \cdot \\
 & & & \cdot & \cdot \\
 & & & & q_{n-1,n-1}
\end{pmatrix}
$$

$$
=
\begin{pmatrix}
\beta_0 & & & \\
 & \beta_0\beta_1 & & \\
 & & \cdot & \\
 & & & \beta_0 \cdots \beta_{n-1}
\end{pmatrix}.
$$

$$(8.76)$$

2.

$$
\begin{pmatrix}
0 & & & -q_0 \\
1 & & & \cdot \\
\cdot & & \cdot & \cdot \\
 & \cdot & \cdot & \cdot \\
 & & 1 & -q_{n-1}
\end{pmatrix}
\begin{pmatrix}
q_{0,0} & q_{1,0} & \cdot & \cdot & q_{n-1,0} \\
 & q_{1,1} & & & \cdot \\
 & & \cdot & & \cdot \\
 & & & \cdot & \cdot \\
 & & & & q_{n-1,n-1}
\end{pmatrix}
$$

$$
=
\begin{pmatrix}
q_{0,0} & q_{1,0} & \cdot & \cdot & q_{n-1,0} \\
 & q_{1,1} & & & \cdot \\
 & & \cdot & & \cdot \\
 & & & \cdot & \cdot \\
 & & & & q_{n-1,n-1}
\end{pmatrix}
\begin{pmatrix}
\theta_1 & \beta_1 & & & \\
1 & \theta_2 & \cdot & & \\
 & \cdot & \cdot & \cdot & \\
 & & \cdot & \cdot & \beta_{n-1} \\
 & & & 1 & \theta_n
\end{pmatrix}.
$$

$$(8.77)$$

Proof:

1. This is the matrix representation of the orthogonality relations given in Eq. (8.75).

2. We use the trivial operator identity $S_q I = I S_q$ and take the matrix representation $[S_q]^{st}_{st}[I]^{st}_{or} = [I]^{st}_{or}[S_q]^{or}_{or}$. This is equivalent to Eq. (8.77). $\qquad\square$

It is easy to read off from the diagonal representation (8.76) of the Hankel matrix the relevant signature information. A special case is the following:

Corollary 8.9.2 *Let g be a rational function with the continued fraction representation (8.55). Then H_n is positive definite if and only if all a_i are linear and all β_i are positive.*

8.10   The Cauchy Index

Let g be a rational transfer function with real coefficients having the coprime representation $g = p/q$. The **Cauchy index** of g, denoted by I_g, is defined as the number of jumps of g from $-\infty$ to $+\infty$ minus the number of jumps from $+\infty$ to $-\infty$. Thus, the Cauchy index is related to discontinuities of g on the real axis. These discontinuities arise from the real zeros of q. However, the contributions to the Cauchy index come only from zeros of q of odd order.

In this section we establish some connections between the Cauchy index, the signature of the Hankel map induced by g, and the existence of signature-symmetric realizations of g. The central result is the classical result of Hermite and Hurwitz. However, before proving it we state and prove an auxiliary result that is of independent interest.

Proposition 8.10.1 *Let g be a real rational function. Then the following scaling operations:*

$$
\begin{array}{llll}
(1) & g(z) & \longrightarrow & mg(z) & m > 0 \\
(2) & g(z) & \longrightarrow & g(z - a) & a \in \mathbf{R} \\
(3) & g(z) & \longrightarrow & g(rz) & r > 0
\end{array}
$$

leave the rank and signature of the Hankel map as well as the Cauchy index invariant.

Proof: (1) is obvious. To prove the rank invariance, let $g = p/q$ with p and q coprime. By the Euclidean algorithm there exist polynomials a and b such that $ap + bq = 1$. This implies that

$$a(z - a)p(z - a) + b(z - a)q(z - a) = 1$$

as well as

$$a(rz)p(rz) + b(rz)q(rz) = 1,$$

that is, $p(z-a), q(z-a)$ are coprime and so are $p(rz), q(rz)$. Now, $g(z-a) = p(z - a)/q(z - a)$ and $g(rz) = p(rz)/q(rz)$, which proves the invariance of the McMillan degree, which is the same as the rank of the Hankel map. Now it is easy to check that, given any polynomial u, we have

$$[H_g u, u] = [H_{g_a} u_a, u_a],$$

where $g_a(z) = g(z - a)$. If we define a map $R_a : \mathbf{R}[z] \longrightarrow \mathbf{R}[z]$ by

$$(R_a u)(z) = u(z - a) = u_a(z),$$

then R_a is invertible, $R_a^{-1} = R_{-a}$ and

$$[H_g u, u] = [H_{g_a} u_a, u_a] = [H_{g_a} R_a u, R_a u] = [R_a * H_{g_a} R_a u, u],$$

which shows that

$$H_g = R_a^* H_{g_a} R_a$$

and hence that

$$\sigma(H_g) = \sigma(H_{g_a}),$$

which proves (2). To prove (3) define, for $r > 0$, a map $P_r : \mathbf{R}[z] \longrightarrow \mathbf{R}[z]$ by

$$(P_r u)(z) = u(rz).$$

Clearly, P_r is invertible and $P_r^{-1} = P_{1/r}$. Letting $u_r = P_r u$, we have

$$
\begin{aligned}
[H_{g_r} u, u] &= [\pi_- g(rz)u, u] = \left[\pi_- \sum (g_k / r^{k+1} z^{k+1}) u, u\right] \\
&= \sum \sum (g_{i+j} r^{-i-j} u_i) u_j = \sum \sum g_{i+j} (u_i r^{-i})(u_j r^{-j}) \\
&= [H_g P_r u, P_r u] = [P_r^* H_g P_r u, u]
\end{aligned}
$$

and hence $H_g = P_r^* H_g P_r$, which implies that $\sigma(H_{g_r}) = \sigma(H_g)$. The invariance of the Cauchy index under these scaling operations is obvious. $\qquad\square$

Theorem 8.10.1 (Hermite–Hurwitz) *Let $g = p/q$ be a strictly proper, real rational function with p and q coprime. Then*

$$I_g = \sigma(H_g) = \sigma(B(q,p)). \qquad (8.78)$$

Proof: That $I_g = \sigma(H_g) = \sigma(B(q,p))$ has been proved in Section 8.7.2. So it suffices to prove the equality $I_g = \sigma(H_g)$.

Let us analyze first the case where q is a polynomial with simple real zeros, that is, $q(z) = \prod_{j=1}^n (z - a_j)$ and $a_i \neq a_j$ for $i \neq j$. Let $d_i(z) = q(z)/(z - a_i)$. Given any polynomial $u \in X_q$, it has a unique expansion $u = \sum_{i=1}^n u_i d_i$. Then

$$
\begin{aligned}
[H_g u, u] &= [\pi_- gu, u] = [\pi_- q^{-1} pu, u] = [q^{-1} q \pi_- q^{-1} pu, u] \\
&= <\pi_q pu, u> = <p(S_q)u, u> = \sum_{i=1}^n \sum_{j=1}^n <p(S_q) d_i, d_j> u_i u_j \\
&= \sum_{i=1}^n \sum_{j=1}^n <p(a_i) d_i, d_j> u_i u_j = \sum_{i=1}^n p(a_i) d_i(a_i) u_i^2,
\end{aligned}
$$

as d_i are eigenfunctions of S_q corresponding to the eigenvalues a_i and as, by Proposition (5.6.3), $< d_i, d_j >= d_i(a_i)\delta_{ij}$. From this computation it follows, since $[p(S_q)]_{co}^{st} = B(q, p)$, that

$$\sigma(H_g) = \sigma(B(q, p)) = \sum_{i=1}^{n} \text{sign} \, [p(a_i)d_i(a_i)].$$

On the other hand, we have the partial fraction decomposition

$$g(z) = \frac{p(z)}{q(z)} = \sum_{i=1}^{n} \frac{c_i}{z - a_i}$$

or

$$p(z) = \sum_{i=1}^{n} c_i \frac{q(z)}{z - a_i} = \sum_{i=1}^{n} c_i d_i(z),$$

which implies $p(a_i) = c_i d_i(a_i)$ or, equivalently, that $c_i = p(a_i)/d_i(a_i)$. Now obviously

$$I_g = \sum_{i=1}^{n} \text{sign} \, (c_i) = \sum_{i=1}^{n} \text{sign} \, (\frac{p(a_i)}{d_i(a_i)}),$$

and as $\text{sign} \, (p(a_i)d_i(a_i)) = \text{sign} \, (p(a_i)/d_i(a_i))$, the equality (8.78) is proved in this case.

We pass now to the general case. Let $q = q_1 \cdots q_s$ be the unique factorization of q into powers of relatively prime irreducible monic polynomials. As before, we define polynomials d_i by

$$d_i(z) = \frac{q(z)}{q_i(z)}.$$

Since we have the direct sum decomposition $X_q = d_1 X_{q_1} \oplus \cdots \oplus d_s X_{q_s}$, it follows that each $f \in X_q$ has a unique representation of the form $f = \sum_{i=1}^{s} d_i u_i$ with $u_i \in X_{q_i}$. Relative to the indefinite metric of X_q, introduced in Eq. (5.29), this is an orthogonal direct sum decomposition, that is,

$$< d_i X_{q_i}, d_j X_{q_j} >= 0 \qquad \text{for} \qquad i \neq j.$$

Indeed, if $f_i \in X_{q_i}$ and $g_j \in X_{q_j}$, then

$$< d_i f_i, d_j f_j >= [q^{-1} d_i f_i, d_j f_j] = [d_j q^{-1} d_i f_i, f_j] = 0$$

as $d_i d_j$ is divisible by q and $F[z]^{\perp} = F[z]$.

Let $g = \sum_{i=1}^{s} p_i/q_i$ be the partial fraction decomposition of g. Since the zeros of the q_i are distinct, it is clear that

$$I_g = I_{\frac{p}{q}} = \sum_{i=1}^{s} I_{p_i/q_i}.$$

Also, as a consequence of Proposition 8.3.3, it is clear that for the McMillan degree $\delta(g)$ of g we have $\delta(g) = \sum_{i=1}^{s} \delta(p_i/q_i)$; and hence, by Theorem 8.2.4, the signatures of the Hankel forms are additive, namely, $\sigma(H_g) = \sum_{i=1}^{s} \sigma(H_{p_i/q_i})$. Therefore, to prove the Hermite–Hurwitz theorem, it suffices to prove it in the case of q being the power of a monic prime. Since we discuss the real case, the primes are of the form $z - a$ or $((z-a)^2 + b^2)$, with $a, b \in \mathbf{R}$. By applying the previous scaling result, the proof of the Hermite–Hurwitz theorem reduces to the two cases $q(z) = z^m$ and $q(z) = (z^2 + 1)^m$.

Case 1: $q(z) = z^m$.

Assume that $p(z) = p_0 + p_1 z + \cdots + p_{m-1} z^{m-1}$. Then the coprimeness of p and q is equivalent to $p_0 \neq 0$. Therefore, we have

$$g(z) = p_{m-1} z^{-1} + \cdots + p_0 z^{-m},$$

which shows that

$$I_g = I_{(p/z^m)} = \begin{cases} 0 & \text{if} \quad m \text{ is even} \\ \text{sign}(p_0) & \text{if} \quad m \text{ is odd.} \end{cases}$$

On the other hand, $\operatorname{Ker} H_g = z^{m+1} \mathbf{R}[z]$, and so

$$\sigma(H_g) = \sigma(H_g | X_{z^m}).$$

Relative to the standard basis, the truncated Hankel map has the matrix representation

$$\begin{pmatrix} p_{m-1} & \cdot & \cdot & \cdot & p_1 & p_0 \\ \cdot & & \cdot & \cdot & \cdot & \\ \cdot & & & \cdot & \cdot & \\ \cdot & & & & \cdot & \\ p_1 & & \cdot & & & \\ p_0 & & & & & \end{pmatrix}.$$

Now, clearly, the previous matrix has the same signature as

$$\begin{pmatrix} 0 & \cdot & \cdot & \cdot & 0 & p_0 \\ \cdot & & \cdot & \cdot & \cdot & \\ \cdot & & & \cdot & \cdot & \\ \cdot & & & & \cdot & \\ 0 & & \cdot & & & \\ p_0 & & & & & \end{pmatrix},$$

and hence

$$\sigma(H_g) = \begin{cases} \text{sign}(p_0) & \text{if} \quad m \text{ is odd} \\ 0 & \text{if} \quad m \text{ is even.} \end{cases}$$

Case 2: $q(z) = (z^2 + 1)^m$.

Since q has no real zeros, it follows that in this case $I_g = I_{p/q} = 0$. So it suffices to prove that $\sigma(H_g) = 0$ also. Let $g(z) = p(z)/(z^2 + 1)^m$ with $\deg p < 2m$. Let us expand p in the form

$$p(z) = \sum_{k=0}^{m-1} (p_k + q_k z)(z^2 + 1)^k,$$

with the p_k and q_k uniquely determined. The coprimeness condition is equivalent to p_0 and q_0 not being zero together. The transfer function g therefore has the representation

$$g(z) = \sum_{k=0}^{m-1} \frac{p_k + q_k z}{(z^2 + 1)^{m-k}}.$$

In much the same way, every polynomial u can be written in a unique way as

$$u(z) = \sum_{i=0}^{m-1} (u_i + v_i z)(z^2 + 1)^i.$$

We now compute the matrix representation of the Hankel form with respect to the basis

$$\mathcal{B} = \{1, z, (z^2 + 1), z(z^2 + 1), \ldots, (z^2 + 1)^{m-1}, z(z^2 + 1)^{m-1}\}$$

of $X_{(z^2+1)^m}$. Thus, we need to compute

$$[H_g z^\alpha (z^2 + 1)^\lambda, z^\beta (z^2 + 1)^\mu][H_g z^{\alpha+\beta}(z^2 + 1)^{\lambda+\mu}, 1].$$

Now, with $0 \leq \gamma \leq 2$ and $0 \leq \nu \leq 2m - 2$, we compute

$$[H_g z^\gamma (z^2 + 1)^\nu, 1] = \left[\frac{\displaystyle\sum_{i=0}^{m-1}(p_i + q_i z)(z^2 + 1)^i}{(z^2 + 1)^m} z^\gamma (z^2 + 1)^\nu, 1 \right]$$

$$= \left[\sum_{i=0}^{m-1} \frac{(p_i + q_i z) z^\gamma (z^2 + 1)^{\nu+i}}{(z^2 + 1)^m}, 1 \right].$$

The only nonzero contributions come from the terms where $\nu + i = m - 2, m - 1$, or, equivalently, when $i = m - \lambda - \mu - 2, m - \lambda - \mu - 1$. Now

$$\left[\frac{(p_i + q_i z) z^\gamma (z^2 + 1)^{\nu+i}}{(z^2 + 1)^m}, 1 \right] = \begin{cases} \begin{cases} q_i & \gamma = 0 \\ p_i & \gamma = 1 \quad i = m - \nu - 1 \\ -q_i & \gamma = 2 \end{cases} \\[2em] \begin{cases} 0 & \gamma = 0 \\ 0 & \gamma = 1 \quad i = m - \nu - 2 \\ q_i & \gamma = 2. \end{cases} \end{cases}$$

Thus, the matrix of the Hankel form in this basis has the following block triangular form:

$$
M = \begin{pmatrix}
q_{m-1} & p_{m-1} & & & q_1 & p_1 & & q_0 & p_0 \\
p_{m-1} & q_{m-2}-q_{m-1} & & & p_1 & q_0-q_1 & & p_0 & -q_0 \\
& & & & q_0 & p_0 & & & \\
& & & & p_0 & -q_0 & & & \\
& & & & & & & & \\
& & & & & & & & \\
q_0 & p_0 & & & & & & & \\
p_0 & -q_0 & & & & & & &
\end{pmatrix}.
$$

Now, by our assumption on coprimeness, the matrix $\begin{pmatrix} q_0 & p_0 \\ p_0 & -q_0 \end{pmatrix}$ is nonsingular and its determinant is negative. Hence, it has signature equal to zero. This implies that the signature of M also is zero. With this the proof of the theorem is complete. $\qquad\square$

8.11   Exercises

1. **Jacobi's signature rule.** Let $A(x,x)$ be a Hermitian form, and let Δ_i be the determinants of the principal minors. If the rank of A is r and $\Delta_1, \ldots, \Delta_r$ are nonzero, then

$$
\pi = P(1, \Delta_1, \ldots, \Delta_r), \qquad \nu = V(1, \Delta_1, \ldots, \Delta_r),
$$

where P and V denote, respectively, the number of sign permanences and the number of sign changes in the sequence $1, \Delta_1, \ldots, \Delta_r$.

2. Let $p(z) = \sum_{i=0}^{m} p_i z^i = p_m \prod_{i=1}^{m}(z - \alpha_i)$ and $q(z) = \sum_{i=0}^{n} q_i z^i = q_n \prod_{i=1}^{m}(z - \beta_i)$. Show that $\det Res(p,q) = p_m^n q_n^m \prod_{i,j}(\beta_i - \alpha_j)$.

3. Show that a real Hankel form

$$
H_n(x,x) = \sum_{i=1}^{n} \sum_{j=1}^{n} s_{i+j-1} \xi_i \xi_j
$$

is positive if and only if the parameters s_k allow a representation of the form

$$
s_k = \sum_{j=1}^{n} \rho_j \theta_j^k,
$$

with $\rho_j > 0$ and θ_j distinct real numbers.

4. Given $g_1, \ldots, g_{2n-1}$, consider the Hankel matrix H_n in Eq. (8.46). Let $a(z) = \sum_{i=0}^{n-1} a_i z^i$ and $x(z) = \sum_{i=0}^{n-1} x_i z^i$ be the polynomials arising out of the solutions of the system of linear equations

$$
\begin{pmatrix}
g_1 & \cdot & \cdot & \cdot & g_n \\
\cdot & & & & \cdot \\
\cdot & & & & \cdot \\
\cdot & & & & \cdot \\
g_n & \cdot & \cdot & \cdot & g_{2n-1}
\end{pmatrix}
\begin{pmatrix}
x_0 \\
\cdot \\
\cdot \\
\cdot \\
x_{n-1}
\end{pmatrix}
=
\begin{pmatrix}
r_1 \\
\cdot \\
\cdot \\
\cdot \\
r_n
\end{pmatrix}
$$

with the right-hand side being given, respectively, by

$$
r_i = \begin{cases} 0 & i = 1, \ldots, n-1 \\ 1 & i = n \end{cases}
$$

and

$$
r_i = \begin{cases} 1 & i = 1 \\ 0 & i = 2, \ldots, n. \end{cases}
$$

Show that if $a_0 \neq 0$, we have $x_{n-1} = a_0$, the Hankel matrix H_n is invertible, and its inverse is given by $H_n^{-1} = B(y, a)$, where the polynomial y is defined by $y(z) = a(0)^{-1} z x(z) = q(z) - a(0)^{-1} q(0) a(z)$.

5. Show that a Hermitian **Toeplitz form**

$$
T_n(x, x) = \sum_{i=1}^{n} \sum_{j=1}^{n} c_{i-j} \xi_i \bar{\xi}_j
$$

with the matrix

$$
T_n = \begin{pmatrix}
c_0 & \cdot & \cdot & c_{-n+1} & c_{-n} \\
\cdot & \cdot & \cdot & \cdot & \cdot \\
\cdot & \cdot & \cdot & \cdot & \cdot \\
\cdot & \cdot & \cdot & \cdot & \cdot \\
c_n & \cdot & \cdot & & c_0
\end{pmatrix}
$$

is positive definite if and only if the parameters $c_k = \bar{c}_{-k}$ allow a representation of the form

$$
c_k = \sum_{j=1}^{n} \rho_j \theta_j^k,
$$

with $\rho_j > 0$, $|\theta_j| = 1$ and the θ_j distinct.

6. We say that a sequence $c_0, c_1, \ldots$ of complex numbers is a **positive sequence** if, for all $n \geq 0$, the corresponding **Toeplitz form** is positive definite. An inner product on $\mathbf{C}[z]$ is defined by

$$< a, b >_C \;\; = (\, a_0 \;\; \cdots \;\; a_n \,) \begin{pmatrix} c_0 & \cdots & & c_n \\ & & & & \\ & & & & \\ & & & & \\ c_n & \cdots & & c_0 \end{pmatrix} \begin{pmatrix} \bar{b}_0 \\ \cdot \\ \cdot \\ \bar{b}_n \end{pmatrix}$$

$$= \sum_{i=0}^{n} \sum_{j=0}^{n} c_{(i-j)} a_i \bar{b}_j .$$

Under our assumption of positivity, this is a definite inner product on $\mathbf{C}[z]$.

(a) Show that the inner product defined above has the following properties:

$$< z^i, z^j >_C = c_{(i-j)}$$

and

$$< za, zb >_C = < a, b >_C .$$

(b) Let ϕ_n be the sequence of polynomials obtained from the polynomials $1, z, z^2, \ldots$ by an application of the Gram–Schmidt orthogonalization procedure. Assume that all of the ϕ_n to be monic. Show that

$$\begin{cases} \phi_0(z) = 1 \\ \\ \phi_n(z) = \dfrac{1}{\det T_{n-1}} \begin{vmatrix} c_0 & \cdots & & c_{-n} \\ & & & \cdot \\ & & & \cdot \\ c_{n-1} & \cdots & c_0 & c_{-1} \\ 1 & \cdots & z^{n-1} & z^n \end{vmatrix} . \end{cases}$$

(c) Show that

$$< \phi_n^\sharp, z^{n-i} >_C = 0, \qquad i = 0, \ldots, n-1.$$

(d) Show that

$$\phi_n^\sharp(z) = 1 - z \sum_{i=0}^{n-1} \gamma_{i+1} \phi_i(z),$$

where the γ_i are defined by

$$\gamma_{i+1} = \frac{< 1, z\phi_i >}{\|\phi_i\|^2}.$$

The γ_i are called the **Schur parameters** of the sequence $c_0, c_1, \ldots$.

(e) Show that the orthogonal polynomials satisfy the following recursions:

$$\begin{cases} \phi_n &= z\phi_{n-1} - \gamma_n \phi_{n-1}^\sharp \\[2mm] \phi_n^\sharp &= \phi_{n-1}^\sharp - z\gamma_n \phi_{n-1} \end{cases}$$

with the initial conditions for the recursions $\phi_0(z) = \phi_0^\sharp(z) = 1$. Equivalently,

$$\begin{pmatrix} \phi_n \\ \phi_n^\sharp \end{pmatrix} = \begin{pmatrix} z & -\gamma_n \\ -z\gamma_n & 1 \end{pmatrix} \begin{pmatrix} \phi_{n-1} \\ \phi_{n-1}^\sharp \end{pmatrix}, \begin{pmatrix} \phi_0(0) \\ \phi_0^\sharp(0) \end{pmatrix} = \begin{pmatrix} 1 \\ 1 \end{pmatrix}.$$

(f) Show that

$$\gamma_n = -\phi_n(0).$$

(g) Show that

$$\|\phi_n\|^2 = (1 - \gamma_n^2)\|\phi_{n-1}\|^2.$$

(h) Show that the Schur parameters γ_n satisfy

$$|\gamma_n| < 1.$$

(i) The **Levinson algorithm.** Let $\{\phi_n(z)\}$ be the monic orthogonal polynomials associated with the positive sequence $\{c_n\}$. Assume that $\phi_n(z) = z^n + \sum_{i=0}^{n-1} \phi_{n,i} z^i$. Show that they can be computed recursively by

$$\begin{cases} r_{n+1} &= (1 - \gamma_n^2)r_n, & r_0 = c_0 \\[2mm] \gamma_{n+1} &= \dfrac{1}{r_n}\sum_{i=0}^{n} c_{i+1}\phi_{n,i} \\[2mm] \phi_{n+1}(z) &= z\phi_n(z) - \gamma_{n+1}\phi_n^\sharp(z), & \phi_0(z) = 1. \end{cases}$$

(j) Show that the shift operator S_{ϕ_n} acting in X_{ϕ_n} satisfies

$$S_{\phi_n}\phi_i = \phi_{i+1} - \gamma_{i+1}\sum_{j=1}^{i}\gamma_j \prod_{\nu=j+1}^{i}(1 - \gamma_\nu^2)\phi_j + \gamma_{i+1}\prod_{\nu=1}^{i}(1 - \gamma_\nu^2)\phi_0.$$

(k) Let $c_0, c_1, \ldots$ be a positive sequence. Define a new sequence $\hat{c}_0, \hat{c}_1, \ldots$ through the infinite system of equations given by

$$\begin{pmatrix} c_0/2 & 0 & \cdot & \cdot & \cdot \\ c_1 & c_0/2 & 0 & \cdot & \cdot \\ c_2 & c_1 & \cdot & \cdot & \cdot \\ \cdot & \cdot & \cdot & \cdot & \cdot \end{pmatrix} \begin{pmatrix} \hat{c}_0/2 \\ \hat{c}_1 \\ \cdot \\ \cdot \end{pmatrix} = \begin{pmatrix} 1 \\ 0 \\ \cdot \\ \cdot \end{pmatrix}.$$

Show that $\hat{c}_0, \hat{c}_1, \ldots$ is also a positive sequence.

(l) Show that the Schur parameters for the orthogonal polynomials ψ_n, corresponding to the new sequence, are given by $-\gamma_n$.

(m) Show that ψ_n satisfy the recurrence relation

$$\begin{cases} \psi_n & = & z\psi_{n-1} + \gamma_n \psi_{n-1}^\sharp \\ \\ \psi_n^\sharp & = & \psi_{n-1}^\sharp + z\gamma_n \psi_{n-1}. \end{cases}$$

The initial conditions for the recursions are $\psi_0(z) = \psi_0^\sharp(z) = 1$.

8.12 Notes and Remarks

The study of quadratic forms owes much to the work of Sylvester and Cayley. Hermite studied both quadratic and Hermitian forms, arriving independently, and more or less simultaneously, at Sylvester's law of inertia. Bezout forms and matrices were studied already by Sylvester and Cayley. The connection of Hankel and Bezout forms was known to Jacobi. For some of the history, we refer to Krein and Naimark [1936] as well as to Gantmacher [1959]. Both sources contain very extensive bibliographies.

Continued fraction representations need not be restricted to rational functions. In fact, most applications of continued fractions are in the area of rational approximations of functions, or of numbers. Here is the Magnus' **generalized Euclidean algorithm**. Given $g \in F((z^{-1}))$, we set $g = a_0 + g_0$ with $a_0 = \pi_+ g$. Define recursively

$$\frac{\beta_{n-1}}{g_{n-1}(z)} = a_n(z) - g_n(z),$$

where β_{n-1} is a normalizing constant, $a_n(z)$ is a monic polynomial, and $g_n \in z^{-1} F[[z^{-1}]]$. This leads, in the nonrational case, to an infinite continued fraction.

9
Stability

9.1 Root Location Using Quadratic Forms

The problem of root location of algebraic equations has served to motivate the introduction and study of quadratic and Hermitian forms.

The first result in this direction seems to be that of Borchardt. This in turn motivated Jacobi as well as Hermite [1856], who no doubt made the greatest contribution to this subject. We follow, as much as possible, the masterful exposition of Krein and Naimark [1936], which every reader is advised to consult. It contains also a carefully prepared list of references to the historically important literature on the subject.

The following theorem summarizes the first application of quadratic forms to the problem of zero location:

Theorem 9.1.1 *Let $q(z) = \sum_{i=0}^{n} q_i z^i$ be a real polynomial and $g = q'/q$. Then:*

1. *The number of distinct roots of q is equal to* rank (H_g).

2. *The number of distinct real roots of q is equal to $\sigma(H_g)$ or, equivalently, to I_g, the Cauchy index of g.*

Proof:

1. Let $\alpha_1, \ldots, \alpha_m$ be the distinct roots of $q(z) = \sum_{i=0}^{m} q_i z^i$, whether real or complex. So

$$q(z) = q_m \prod_{i=1}^{m} (z - \alpha_i)^{\nu_i},$$

with $\nu_1 + \cdots + \nu_m = n$. Hence the rational function g defined below satisfies

$$g(z) = \frac{q'(z)}{q(z)} = \sum_{i=1}^{m} \frac{\nu_i}{(z - \alpha_i)}.$$

From this it is clear that m is equal to $\delta(g)$, the McMillan degree of g, and so in turn to rank (H_g).

2. On the other hand, the Cauchy index of g obviously is equal to the number of real zeroes, as all residues ν_i are positive. But, by the Hermite–Hurwitz theorem, we have $I_g = \sigma(H_g)$.   □

The same problem can be approached using Bezoutians.

Theorem 9.1.2 *Let q be as in the previous theorem. Then:*

1. *The number of distinct roots of q is equal to* codim Ker $B(q, q')$.

2. *The number of distinct real roots of q is equal to* $\sigma(B(q, q'))$.

3. *All roots of q are real if and only if $B(q, q') \geq 0$.*

Proof:

1. From our study of Bezoutians and intertwining maps, we know that $\dim \mathrm{Ker}\, B(q, q') = \deg(r)$, where r is the g.c.d. of q and q'. It is easy to check that the g.c.d. of q and q' is equal to $r(z) = q_n \prod_{i=1}^{m} (z - \alpha_i)^{\nu_i - 1}$; hence its degree is $n - m$.

2. Follows as before.

3. Note that if In $(B(q, q')) = (\pi, \nu, \delta)$, then $\nu = (\mathrm{rank}\,(B(q, q')) - \sigma(B(q, q')))/2$. So $\nu = 0$ if and only if rank $(B(q, q')) = \sigma(B(q, q'))$, that is, if and only if all roots of q are real.   □

Naturally, in Theorem 9.1.1, the infinite quadratic form H_g can be replaced by H_n, the $n \times n$ truncated form. In fact, it is this form, in various guises, that appears in the early studies.

We digress a little on this point. If we expand $1/(z - \alpha_i)$ in powers of z^{-1}, we obtain

$$g(z) = \frac{q'(z)}{q(z)} = \sum_{i=1}^{m} \frac{\nu_i}{z - \alpha_i} = \sum_{i=1}^{m} \nu_i \sum_{k=1}^{\infty} \frac{\alpha_i^k}{z^{k+1}} = \sum_{k=1}^{\infty} \frac{s_k}{z^{k+1}},$$

with $s_k = \sum_{i=1}^{m} \nu_i \alpha_i^k$. In case all the zeroes are simple, the numbers s_k are called the **Newton sums** of q. The finite Hankel quadratic form $\sum_{i=0}^{n-1} s_{i+j} \xi_i \xi_j$ can easily be seen to be a different representation of $\sum_{i=1}^{n} \times (\sum_{j=0}^{n-1} \xi_j \alpha_i^j)^2$.

The result stated in Theorem 9.1.1 is far from exhausting the power of the method of quadratic forms. We address ourselves now to the problem of determining, for an arbitrary complex polynomial q, the number of its zeroes in an open half-plane. We begin with the problem of determining the number of zeroes of $q(z) = \sum_{k=0}^{n} q_k z^k$ in the open upper half-plane. Once this is accomplished, we can apply the result to other half-planes, most importantly to the left half-plane. To solve this problem, Hermite introduced the notion of Hermitian forms, which had far-reaching consequences in mathematics, Hilbert space theory being one offshoot.

Theorem 9.1.3 *Given a complex polynomial q of degree n, we define*

$$\tilde{q}(z) = \overline{q(\bar{z})}$$

and a polynomial $Q(z, w)$ in two variables by

$$Q(z, w) = -i\frac{q(z)\tilde{q}(w) - \tilde{q}(z)q(w)}{z - w} = \sum_{j=1}^{n}\sum_{k=1}^{n} Q_{jk} z^{j-1} w^{k-1}. \qquad (9.1)$$

The polynomial $Q(z, w)$ is called the **generating function** *of the form*

$$Q = \sum_{j=1}^{n}\sum_{k=1}^{n} Q_{jk}\xi_j\overline{\xi_k}.$$

Then:

1. *The form Q defined above is Hermitian, that is, we have $Q_{jk} = \overline{Q}_{kj}$.*

2. *Let (π, ν, δ) be the inertia of the form Q. Then the number of real zeroes of q together with the number of zeroes of q arising from complex conjugate pairs is equal to δ. There are π more zeroes of q in the upper half-plane and ν more in the lower half-plane. In particular, all zeroes of q are in the upper half-plane if and only if Q is positive definite.*

Proof:

1. We compute

$$
\begin{aligned}
\overline{Q(\bar{w}, \bar{z})} &= \overline{-i\Big[\frac{q(\bar{w})\tilde{q}(\bar{z}) - \tilde{q}(\bar{w})q(\bar{z})}{\bar{w} - \bar{z}}\Big]} \\
&= i\left[\frac{\tilde{q}(w)q(z) - q(w)\tilde{q}(z)}{w - z}\right] = -i\left[\frac{\tilde{q}(z)q(w) - q(z)\tilde{q}(w)}{z - w}\right] \\
&= Q(z, w).
\end{aligned}
$$

This in turn implies

$$\sum_{j=1}^{n}\sum_{k=1}^{n}Q_{jk}z^{j-1}w^{k-1} = \overline{\sum_{j=1}^{n}\sum_{k=1}^{n}\overline{Q}_{jk}\overline{w}^{j-1}\overline{z}^{k-1}}$$

$$= \sum_{j=1}^{n}\sum_{k=1}^{n}\overline{Q}_{jk}z^{k-1}w^{j-1}$$

$$= \sum_{j=1}^{n}\sum_{k=1}^{n}\overline{Q}_{kj}z^{j-1}w^{k-1}.$$

So, comparing coefficients, the equality $Q_{jk} = \overline{Q}_{kj}$ is obtained.

2. Let d be the g.c.d. of q and $\tilde{q}$. Clearly, the zeroes of d are the real zeroes of q and all pairs of complex conjugate zeroes of q. Without loss of generality, we may assume that d is a real polynomial. Write $q = dq'$ and $\tilde{q} = d\tilde{q}'$. Using Lemma 8.4.1, we have $B(q, \tilde{q}) = B(dq', d\tilde{q}')$, and hence rank $B(q, \tilde{q}) = \operatorname{rank} B(q', \tilde{q}')$ and $\sigma(B(q, \tilde{q})) = \sigma(B(q', \tilde{q}'))$. This shows that δ is the number of real zeroes added to the number of zeroes arising from complex conjugate zeroes of q. Therefore, we may as well assume that q and $\tilde{q}$ are coprime. Let $q = q_1 q_2$ be any factorization of q into real or complex factors. We do not assume that q_1 and q_2 are coprime. We compute

$$-i\frac{q(z)\tilde{q}(w) - \tilde{q}(z)q(w)}{z - w}$$

$$= -iq_2(z)\frac{q_1(z)\tilde{q}_1(w) - \tilde{q}_1(z)q_1(w)}{z - w}\tilde{q}_2(w)$$

$$= -i\tilde{q}_1(z)\frac{q_2(z)\tilde{q}_2(w) - \tilde{q}_2(z)q_2(w)}{z - w}q_1(w).$$

The ranks of the two Hermitian forms on the right are $\deg(q_1)$ and $\deg(q_2)$, respectively. By Theorem 8.2.4, the signature of Q is equal to the sum of the signatures of the forms

$$Q_1 = -i\frac{q_1(z)\tilde{q}_1(w) - \tilde{q}_1(z)q_1(w)}{z - w}$$

and

$$Q_2 = -i\frac{q_2(z)\tilde{q}_2(w) - \tilde{q}_2(z)q_2(w)}{z - w}.$$

Let $\alpha_1, \ldots, \alpha_n$ be the zeroes of q. Then, by an induction argument, it follows that

$$\sigma(-iB(q, \tilde{q})) = \sum_{k=1}^{n}\sigma(-iB(z - \alpha_k, z - \overline{\alpha}_k)) = \sum_{k=1}^{n}\operatorname{sign}(2\operatorname{Im}(\alpha_k)).$$

So obviously π is the number of zeroes of q in the upper half-plane and ν the number in the lower half-plane.   □

We can restate the previous theorem in terms of either the Cauchy index or the Hankel form.

Theorem 9.1.4 *Let $q(z)$ be a complex polynomial of degree n. Define its real and imaginary parts by*

$$\begin{cases} q_r(z) & = & \dfrac{q(z) + \tilde{q}(z)}{2} \\[2mm] q_i(z) & = & \dfrac{q(z) - \tilde{q}(z)}{2i}. \end{cases}$$

We assume that $\deg(q_r) \geq \deg(q_i)$. Define the proper rational function $g = q_i/q_r$. Then:

1. *The Cauchy index of g, $I_g = \sigma(H_g)$, is the difference between the number of zeroes of q in the lower half-plane and the upper half-plane.*

2. *We have $I_g = \sigma(H_g) = -n$, that is, H_g or, equivalently, $B(q_r, q_i)$ are negative definite, if and only if all the zeroes of q lie in the open upper half-plane.*

Proof:

1. Note that the assumption that $\deg(q_r) \geq \deg(q_i)$ does not limit the generality. For, if this condition is not satisfied, we consider instead the polynomial $iq(z)$. The Hermitian form $-iB(q, \tilde{q})$ also can be represented as the Bezoutian of two real polynomials. Indeed, if $q(z) = q_r(z) + iq_i(z)$ with q_r and q_i real polynomials, then $\tilde{q}(z) = q_r(z) - iq_i(z)$ and

$$\begin{aligned} -iB(q, \tilde{q}) & = & -iB(q_r + iq_i, q_r - iq_i) \\ & = & -i[B(q_r, q_r) + iB(q_i, q_r) - iB(q_r, q_i) + B(q_i, q_i)] \\ & = & 2B(q_i, q_r) \end{aligned}$$

or

$$-iB(q, \tilde{q}) = 2B(q_i, q_r). \tag{9.2}$$

From equality (9.2) it follows that $\sigma(-iB(q, \tilde{q})) = \sigma(B(q_i, q_r))$ and, by the Hermite–Hurwitz theorem, that

$$\sigma(B(q_i, q_r)) = -\sigma(B(q_r, q_i)) = -\sigma(H_g) = -I_g.$$

2. Follows from part 1.   □

Definition 9.1.1 *A polynomial $q(z)$ will be called **stable**, or a **Hurwitz** polynomial, if all of its zeroes lie in the open left half-plane Π_-.*

To obtain stability characterizations we need to replace the upper half-plane by the left half-plane, and this can be done easily by a change of variable.

Theorem 9.1.5 *Let $q(z)$ be a complex polynomial. Then a necessary and sufficient condition for q to be a Hurwitz polynomial is that the* **Hermite–Fujiwara quadratic form** *with generating function*

$$H = \frac{q(z)\tilde{q}(w) - \tilde{q}(-z)q(-w)}{z + w} = \sum_{j=1}^{n} \sum_{k=1}^{n} H_{jk} z^{j-1} w^{k-1} \qquad (9.3)$$

be positive definite.

Proof: Obviously, a zero α of q is in the open left half-plane if and only if $-i\alpha$, which is in the upper half-plane, is a zero of f. Thus, by Theorem 9.1.4, all of the zeroes of q are in the left half-plane if and only if the Hermitian form $-iB(f, \tilde{f})$ is positive definite.

Now, as $\tilde{f}(z) = \overline{f(\bar{z})} = \overline{q(i\bar{z})} = \overline{q(\overline{(-iz)})} = \tilde{q}(-iz)$, it follows that

$$-iB(f, \tilde{f}) = -i\frac{f(z)\tilde{f}(w) - \tilde{f}(z)f(w)}{z - w} = \frac{q(iz)\tilde{q}(-iw) - \tilde{q}(-iz)q(iw)}{i(z - w)}.$$

If we substitute $\zeta = iz$ and $\omega = -iw$, then we get the Hermite–Fujiwara generating function (9.3), and this form has to be positive for all of the zeroes of q to be in the upper half-plane.   □

Definition 9.1.2 *Let $q(z)$ be a real monic polynomial of degree m with real simple zeroes*

$$\alpha_1 < \alpha_2 < \cdots < \alpha_m,$$

and let $p(z)$ be a real polynomial with positive leading coefficients and zeroes

$$\beta_1 < \beta_2 < \cdots < \beta_{m-1} \qquad if \qquad \deg p = m - 1$$

and

$$\beta_1 < \beta_2 < \cdots < \beta_m \qquad if \qquad \deg p = m.$$

Then:

1. *We say that q and p are a* **real pair** *if the zeroes satisfy*

$$\alpha_1 < \beta_1 < \alpha_2 < \beta_2 < \cdots < \beta_{m-1} < \alpha_m \qquad if \qquad \deg p = m - 1$$

and

$$\beta_1 < \alpha_1 < \beta_2 < \cdots < \beta_m < \alpha_m \qquad if \qquad \deg p = m.$$

2. *We say that q and p form a* **positive pair** *if they form a real pair and $\alpha_m < 0$.*

Theorem 9.1.6 *Let q and p be a pair of real polynomials as above. Then:*

1. *q and p form a real pair if and only if $B(q,p) > 0$.*

2. *q and p form a positive pair if and only if $B(q,p) > 0$ and $B(zp,q) > 0$.*

Proof:

1. Assume that q and p form a real pair. By our assumption, all zeroes α_i of q are real and simple. Define a rational function by $g = p/q$. Clearly, $q(z) = \prod_{j=1}^{m}(z - \alpha_j)$. Hence, $g(z) = \sum_{j=1}^{m} c_j/(z - \alpha_j)$, and it is easily checked that $c_j = p(\alpha_i)/q'(\alpha_i)$. Using the Hermite–Hurwitz theorem, to show that $B(q,p) > 0$ it suffices to show that $\sigma(B(q,p)) = I_g = m$ or, equivalently, that $p(\alpha_i)/q'(\alpha_i) > 0$, for all i. Since both p and q have positive leading coefficients, we have, except for the trivial case, where $m = 1$ and $\deg p = 0$, $\lim_{x \to \infty} p(x) = \lim_{x \to \infty} q(x) = +\infty$. This implies that $q'(\alpha_m) > 0$ and $p(\alpha_m) > 0$. As a result, we have $p(\alpha_m)/q'(\alpha_m) > 0$. The simplicity of the zeroes of q and p and the interlacing property force the signs of the $q'(\alpha_i)$ and $p(\alpha_i)$ to alternate, and this forces all other residues $p(\alpha_i)/q'(\alpha_i)$ to be positive.

   Conversely, assume that $B(q,p) > 0$. Then, by the Hermite–Hurwitz theorem $I_{p/q} = m$, which means that all of the zeroes of q have to be real and simple, and the residues have to satisfy $p(\alpha_i)/q'(\alpha_i) > 0$. As all of the zeroes of q are simple, necessarily the signs of the $q'(\alpha_i)$ alternate. Thus, p has values of different signs at neighboring zeroes of q, and hence, its zeroes are located between zeroes of q. If $\deg p = m - 1$, we are done. If $\deg p = m$, there is an extra zero of p. Since $q'(\alpha_m) > 0$, necessarily $p(\alpha_m) > 0$. Now $\beta_m > \alpha_m$ would imply that $\lim_{x \to \infty} p(x) = -\infty$, contrary to the assumption that p has a positive leading coefficient. So necessarily we have $\beta_1 < \alpha_1 < \beta_2 < \cdots < \beta_m < \alpha_m$, that is, q and p form a real pair.

2. Assume now that q and p are a positive pair. By our assumption, all α_i are negative. Moreover, as q and p in particular are a real pair, it follows by part 1 that $B(q,p) > 0$. This means that $p(\alpha_i)/q'(\alpha_i) > 0$ for all $i = 1, \ldots, m$, and hence that $\alpha_i p(\alpha_i)/q'(\alpha_i) < 0$. But these are the residues of $zp(z)/q(z)$; so $I_{zp/q} = \sigma(B(q,zp)) = -m$. So $B(q,zp)$ is negative definite and $B(zp,q)$ is positive definite.

   Conversely, assume that the two Bezoutians $B(q,p)$ and $B(zp,q)$ are positive definite. By part 1, the polynomials q and p form a real pair. Thus the residues satisfy $p(\alpha_i)/q'(\alpha_i) > 0$ and $\alpha_i p(\alpha_i)/q'(\alpha_i) < 0$. So, in particular, all zeroes of q are negative. In particular, $\alpha_m < 0$, and so q and p form a positive pair. $\qquad\square$

As a corollary we obtain the following characterization:

Theorem 9.1.7 *Let q be a complex polynomial, and let q_+, q_- be its real and imaginary parts. Then all of the zeroes of $q(z) = q_+(z) + iq_-(z)$ are in the open upper half-plane if and only if all of the zeroes of q_+ and q_- are simple, real, and separate each other.*

Proof: Follows from Theorems 9.1.4 and 9.1.6.   □

In most applications, stability criterias are needed for real polynomials. The assumption of realness of the polynomial q leads to further simplifications. These simplifications arise out of the decomposition of a real polynomial into its even and odd parts, defined, with $q(z) = \sum q_j z^j$, by

$$
\begin{cases}
q_+(z) &= \displaystyle\sum_{j \geq 0} q_{2j} z^j \\[2mm]
q_-(z) &= \displaystyle\sum_{j \geq 0} q_{2j+1} z^j.
\end{cases}
\tag{9.4}
$$

This is equivalent to writing

$$
q(z) = q_+(z^2) + zq_-(z^2).
\tag{9.5}
$$

The following theorem sums up the central results on the characterization of stability of real polynomials:

Theorem 9.1.8 *Let $q(z)$ be a real monic polynomial of degree n, and let q_+, q_- be its even and odd parts, respectively, as defined in Eq. (9.4). Then the following statements are equivalent:*

1. *$q(z)$ is a stable, or Hurwitz, polynomial.*

2. *The Hermite–Fujiwara form is positive definite.*

3. *The two Bezoutians $B(q_+, q_-)$ and $B(zq_-, q_+)$ are positive definite.*

4. *The polynomials q_+ and q_- form a positive pair.*

5. *The Bezoutian $B(q_+, q_-)$ is positive definite and all q_i are positive.*

Proof: (1) $\Leftrightarrow$ (2) By Theorem 9.1.5, the stability of q is equivalent to the positive definiteness of the Hermite–Fujiwara form.

(2) $\Leftrightarrow$ (3) Since the polynomial q is real, we have in this case $\tilde{q}(z) = q(z)$. From Eq. (9.5) it follows that $q(-z) = q_+(z^2) - zq_-(z^2)$. Therefore,

$$\frac{q(z)q(w)-q(-z)q(-w)}{(z+w)}$$

$$= \frac{(q_+(z^2)+zq_-(z^2))(q_+(w^2)+wq_-(w^2))-(q_+(z^2)-zq_-(z^2))(q_+(w^2)-wq_-(w^2))}{z+w}$$

$$= 2\frac{zq_-(z^2)q_+(w^2)+q_+(z^2)wq_-(w^2)}{z+w}$$

$$= 2\frac{z^2q_-(z^2)q_+(w^2)-q_+(z^2)w^2q_-(w^2)}{z^2-w^2}$$

$$+ 2zw\frac{q_+(z^2)q_-(w^2)-q_-(z^2)q_+(w^2)}{z^2-w^2}.$$

One form contains only even terms and the other only odd ones. So the positive definiteness of the form (9.3) is equivalent to the positive definiteness of the two Bezoutians $B(q_+, q_-)$ and $B(zq_-, q_+)$.

(3) $\Leftrightarrow$ (4) This is the content of Theorem 9.1.6.

(4) $\Leftrightarrow$ (5) Let C_{q_+}, C_{q_-} be the companion matrices associated with q_+ and q_-, respectively. Note that when $n = 2m+1$, that is, n is odd, we have $\deg q_+ = \deg q_- = m$, and when $n = 2m$, that is, n is even, then $\deg q_+ = m$ and $\deg q_- = m - 1$. We also observe that the following relation holds:

$$B(q_+, q_-)C_{q_+} = \tilde{C}_{q_+}B(q_+, q_-). \qquad (9.6)$$

This follows trivially from $p(S_{q_+})S_{q_+} = S_{q_+}p(S_{q_+})$ and taking matrix representations as follows:

$$[p(S_{q_+})]_{co}^{st}[S_{q_+}]_{co}^{co} = [S_{q_+}]_{st}^{st}[p(S_{q_+})]_{co}^{st}.$$

Let (ξ, η) be the standard inner product on $\mathbf{R}^m$, or $\mathbf{C}^m$. Since $B(q_+, q_-)$ is symmetric, we can introduce a new inner product in $\mathbf{R}^m$ by letting

$$[\xi, \eta]_B = (B(q_+, q_-)\xi, \eta).$$

This is in general an indefinite inner product and is definite if and only if the Bezoutian $B(q_+, q_-)$ is positive definite. Clearly, relation (9.6) means that

$$[C_{q_+}\xi, \eta]_B = [\xi, C_{q_+}\eta]_B,$$

that is, that C_{q_+} is a self-adjoint map in the $B(q_+, q_-)$ metric. In particular, if $B(q_+, q_-) > 0$, it follows that the spectrum of C_{q_+}, $\sigma(C_{q_+})$, is real. This means that all of the zeroes of $q_+(z)$ are real. If $n = 2m + 1$, the same holds true for $q_-(z)$. Moreover, the zeroes of C_{q_+} and C_{q_-} are simple as both C_{q_+} and C_{q_-} are cyclic matrices.

Assume now that part 3 holds, that is, $B(q_+, q_-)$ and $B(zq_-, q_+)$ are both positive definite. By our previous remarks, the zeroes of q_+ and q_- are all real and simple. Let the zeroes of q_+ be ordered as

$$\lambda_1 < \cdots < \lambda_m$$

and the zeroes of q_- as

$$\mu_1 < \mu_2 < \cdots < \mu_{m-1} < \mu_m),$$

the last inequality holding if $n = 2m + 1$.

Now $B(q_+, q_-)$ is, by Proposition 8.6.1, congruent to the diagonal matrix diag $(r_1, \ldots, r_m)$ with

$$r_i = q_-(\lambda_i) q_+'(\lambda_i). \tag{9.7}$$

In the same way, $B(zq_-, q_+) = -B(q_+, zq_-)$ is congruent to diag $(s_1, \ldots, s_m)$ with

$$s_i = -\lambda_i q_-(\lambda_i) q_+'(\lambda_i). \tag{9.8}$$

We now consider two cases.

Case I: $n = 2m + 1$.

In this case, q_+ and q_- are both of degree m. By our assumption, $B(q_+, q_-)$ and $B(zq_-, q_+)$ are both positive definite. This means that, for all $i = 1, \ldots, m$, we have $r_i > 0$ and $s_i > 0$. Comparing Eqs. (9.7) and (9.8), we get $\lambda_i < 0$. Moreover, from Eq. (9.7) it follows that $q_-(z)$ has different signs at neighboring zeroes λ_i of q_+. So the zeroes of q_+ and q_- interlace. Now $\mu_m > \lambda_m$ is impossible, as

$$r_m = (\lambda_m - \mu_1) \cdots (\lambda_m - \mu_m)(\lambda_m - \lambda_1) \cdots (\lambda_m - \lambda_{m-1})$$

and the condition $r_m > 0$ would be violated for all of the factors. But $(\lambda_m - \mu_m)$ are positive, so we get

$$\lambda_1 < \mu_1 < \lambda_2 < \mu_2 < \cdots < \mu_{m-1} < \lambda_m < 0.$$

Case II: $n = 2m$.

In this case, $\deg q_+ = m$ and $\deg q_- = m - 1$. By the same reasoning as before we get $\lambda_i < 0$, and the zeroes of q_- interlace those of q_+. Since q_- has only $m - 1$ zeroes, we must have

$$\lambda_1 < \mu_1 < \lambda_2 < \mu_2 < \cdots < \mu_{m-1} < \lambda_m < 0.$$

Thus, part 3 implies part 4.

Conversely, assume that q_+ and q_- form a positive pair. In particular, the leading coefficient of q_- is positive. Let

$$r_i = q_-(\lambda_i) q_+'(\lambda_i).$$

We distinguish between two cases.

Case I: $n = 2m + 1$.

In this case,

$$r_i = \prod_{j=1}^{i}(\lambda_i - \mu_j) \cdot \prod_{k=1}^{i-1}(\lambda_i - \lambda_k) \cdot \prod_{j=i+1}^{m}(\lambda_i - \mu_j) \cdot \prod_{k=i+1}^{m}(\lambda_i - \lambda_k).$$

The factors in the first bracket are all positive, and in the second they are all negative. However, there is an even number of negative factors, which implies that $r_i > 0$.

Case II: $n = 2m$.

In the same way,

$$r_i = \prod_{j=1}^{i-1}(\lambda_i - \mu_j) \cdot \prod_{k=1}^{i-1}(\lambda_i - \lambda_k) \cdots \prod_{j=i}^{i-1}(\lambda_i - \mu_j) \cdot \prod_{k=i+1}^{m}(\lambda_i - \lambda_k).$$

The number of negative factors is now $(m - 1 - i) + (m - (i + 1)) = 2(m - i - 1)$. Hence once again we have $r_i > 0$. By our congruence result we get $B(q_+, q_-) > 0$. In the same way, we get $s_i = -\lambda_i r_i > 0$, and so also $B(zq_-, q_+) > 0$, that is, part 4 implies part 3.

That part 3 implies part 5 is trivial, for we assume q to be monic, and, since all λ_i are negative, clearly all coefficients q_i of q are positive.

Assume now that $B(q_+, q_-) > 0$ and all q_i are positive. By our identification of the Bezoutian as a matrix representation we have

$$B(q_+, zq_-) = [S_{q_+}q_-(S_{q_+})]_{co}^{st} = [S_{q_+}]_{st}^{st}[q_-(S_{q_+})]_{co}^{st} = [q_-(S_{q_+})]_{co}^{st}[S_{q_+}]_{co}^{co}$$

or

$$B(q_+, zq_-) = B(q_+, q_-)C_{q_+} = \tilde{C}_{q_+}B(q_+, q_-).$$

This equality implies first that C_{q_+} is self-adjoint in the $B(q_+, q_-)$ metric. Hence, all zeroes of q_+ are real. But, by our assumption, all of the coefficients of q, and therefore also all of the coefficients of q_+, are positive. In particular, $q_+(z) > 0$ for $z \geq 0$. Thus all zeroes of q_+ are real and negative, and so C_{q_+} is a negative operator in the $B(q_+, q_-)$ metric.

Let $\xi \in \mathbf{R}^m$ be an arbitrary nonzero constant vector. Then

$$[B(zq_-, zq_+)\xi, \xi] = -(\tilde{C}_{q_+}B(q_+, q_-)\xi, \xi) = -[\xi, C_{q_+}\xi] = -[C_{q_+}\xi, \xi] > 0.$$

Thus, $B(zq_-, zq_+)$ is positive definite and we proved that part 5 implies part 3. $\square$

Corollary 9.1.1 *Let q be a real monic polynomial of degree n, having the expansion $q(z) = z^n + q_{n-1}z^{n-1} + \cdots + q_0$, and let*

$$q(z) = q_+(z^2) + zq_-(z^2)$$

be its decomposition into its even and odd parts. Then q is a Hurwitz polynomial if and only if the Cauchy index of the function g defined by

$$g(z) = \frac{q_{n-1}z^{n-1} - q_{n-3}z^{n-3} + \cdots}{z^n - q_{n-2}z^{n-2} + \cdots}$$

is equal to n.

Proof: We distinguish between two cases.

Case I: n is even.

Let $n = 2m$. Clearly, in this case,

$$\begin{cases} \begin{aligned} q_+(-z^2) &= (-1)^m(q_{2m}z^{2m} - q_{2m-2}z^{2m-2} + \cdots) \\ &= (-1)^m(z^n - q_{n-2}z^{n-2} + \cdots) \\[1em] -zq_-(-z^2) &= (-1)^m(q_{2m-1}z^{2m-1} - q_{2m-3}z^{2m-3} + \cdots) \\ &= (-1)^m(q_{n-1}z^{n-1} - q_{n-3}z^{n-3} + \cdots). \end{aligned} \end{cases}$$

So we have, in this case, $g(z) = [-zq_-(-z^2)]/[q_+(-z^2)]$. By the Hermite–Hurwitz theorem, the Cauchy index of g is equal to the signature of the Bezoutian of the polynomials $q_+(-z^2), -zq_-(-z^2)$, and, moreover, $I_g = n$ if and only if that Bezoutian is positive definite. We compute

$$\frac{-q_+(-z^2)wq_-(-w^2) + zq_-(-z^2)q_+(-w^2)}{z - w}$$

$$= \frac{[-q_+(-z^2)wq_-(-w^2) + zq_-(-z^2)q_+(-w^2)](z + w)}{z^2 - w^2}$$

$$= \frac{z^2 q_-(-z^2)q_+(-w^2) - q_+(-z^2)w^2 q_-(-w^2)}{z^2 - w^2}$$

$$+ zw\frac{q_-(-z^2)q_+(-w^2) - q_+(-z^2)q_-(-w^2)}{z^2 - w^2}.$$

So the Bezoutian of $-zq_-(-z^2)$ and $q_+(-z^2)$ is isomorphic to $B(zq_-, q_+) \oplus B(q_+, q_-)$. In particular, the Cauchy index of g is equal to n if and only both Bezoutians are positive definite. This, by Theorem 9.1.8, is equivalent to the stability of q.

Case II: $n = 2m + 1$.

In this case,

$$\begin{cases} \begin{aligned} q_+(-z^2) &= (-1)^m(q_{2m}z^{2m} - q_{2m-2}z^{2m-2} + \cdots) \\ &= (-1)^m(q_{n-1}z^{n-1} - q_{n-3}z^{n-3} + \cdots) \\[1em] -zq_-(-z^2) &= (-1)^{m+1}(q_{2m+1}z^{2m+1} - q_{2m-1}z^{2m-1} + \cdots) \\ &= (-1)^{m+1}(z^n - q_{n-2}z^{n-2} + \cdots). \end{aligned} \end{cases}$$

So we have, in this case, $g(z) = [q_+(-z^2)]/[zq_-(-z^2)]$. We conclude the proof as before. $\qquad\square$

Since Bezout and Hankel forms are related by congruence, we expect that stability criterias also can be given in terms of Hankel forms. We present next such a result.

As before, let $q(z) = q_+(z^2) + zq_-(z^2)$. Since for $n = 2m$ we have $\deg q_+ = m$ and $\deg q_- = m - 1$, whereas for $n = 2m + 1$ we have $\deg q_+ = \deg q_- = m$, the rational function g defined by

$$g(z) = \frac{q_-(z)}{q_+(z)}$$

is proper for odd n and strictly proper for even n. Thus, g can be expanded in a power series in z^{-1} to

$$g(z) = g_0 + g_1 z^{-1} + \cdots$$

with $g_0 = 0$ in case n is even.

Theorem 9.1.9 *Let q be a real monic polynomial of degree n, and let q_+, q_- be its even and odd parts, respectively. Let $m = [n/2]$. Define the rational function g by*

$$g(z) = \frac{-q_-(-z)}{q_+(-z)} = -g_0 + \sum_{i=1}^{\infty} \frac{g_i}{z^i}.$$

Then q is stable if and only if the two Hankel forms

$$H_m = \begin{pmatrix} g_1 & \cdot & \cdot & \cdot & g_m \\ \cdot & \cdot & \cdot & \cdot & \cdot \\ \cdot & \cdot & \cdot & \cdot & \cdot \\ \cdot & \cdot & \cdot & \cdot & \cdot \\ g_m & \cdot & \cdot & \cdot & g_{2m-1} \end{pmatrix}$$

and

$$(\sigma H)_m = \begin{pmatrix} g_2 & \cdot & \cdot & \cdot & g_{m+1} \\ \cdot & \cdot & \cdot & \cdot & \cdot \\ \cdot & \cdot & \cdot & \cdot & \cdot \\ \cdot & \cdot & \cdot & \cdot & \cdot \\ g_{m+1} & \cdot & \cdot & \cdot & g_{2m} \end{pmatrix}$$

are positive definite.

Proof: For the purpose of the proof we let $\hat{p}(z) = p(-z)$. We clearly have $\deg q_+ = m$, and $\deg q_-$ is equal to m when n is odd and to $m - 1$ when n is even. We utilize the polynomial model $X_{\hat{q}_+}$. We also use the identification of the Bezoutian and the Hankel matrix as matrix representations of intertwining maps with respect to the standard and control bases. Now we compute

$$\begin{aligned} (H_m[f]^{st}, [f]^{st}) &= [\pi_- g f, f] = [-\pi_- \hat{q}_+^{-1} \hat{q}_- f, f] \\ &= -[\hat{q}_+^{-1} \hat{q}_+ \pi_- \hat{q}_+^{-1} \hat{q}_- f, f] = - < \hat{q}_-(S_{\hat{q}_+}) f, f > \\ &= (-B(\hat{q}_+, \hat{q}_-)[f]^{co}, [f]^{co}). \end{aligned}$$

We proceed to compute the Bezoutian:

$$-\frac{\hat{q}_+(z)\hat{q}_-(w) - \hat{q}_-(z)\hat{q}_+(w)}{z - w} = \frac{q_+(-z)q_-(-w) - q_-(-z)q_+(-w)}{(-z) - (-w)}.$$

This, by a change of variable, implies that $-B(\hat{q}_+, \hat{q}_-) = B(q_+, q_-)$.
Similarly,

$$((\sigma H_m)[f]^{st}, [f]^{st}) = [\pi_-(-g)f, f] = [\pi_-\hat{q}_+^{-1}\widehat{zq_-}f, f]$$

$$= (B(\hat{q}_+, \widehat{zq_-})[f]^{co}, [f]^{co}).$$

Computing the appropriate Bezoutian,

$$\frac{\hat{q}_+(z)\widehat{zq_-}(w) - \widehat{zq_-}(z)\hat{q}_+(w)}{z - w} = \frac{-q_+(-z)wq_-(-w) + zq_-(z)q_+(-w)}{z - w}.$$

Again, by a change of variable, we conclude that $B(\hat{q}_+, z\hat{q}_-) = B(zq_-, q_+)$.
Now we know that the stability of q is equivalent to the positive definiteness of the Bezoutians $B(q_+, q_-)$ and $B(zq_-, q_+)$. By the computations performed above, this is equivalent to the positive definiteness of the two Hankel forms.   □

We wish to remark that, if we do not assume q to be monic, or to have its highest coefficient positive, we need to impose the extra condition that $g_0 > 0$ if n is odd.

We conclude our discussion with the Hurwitz determinantal stability criterion.

Theorem 9.1.10 *All of the zeroes of the real polynomial $q(z) = \sum_{i=0}^{n} q_i z^i$ lie in the open left half-plane if and only if, under the assumption $q_n > 0$, the n determinantal conditions,*

$$\mathbf{H}_1 = |q_{n-1}| > 0, \qquad \mathbf{H}_2 = \begin{vmatrix} q_{n-2} & q_{n-3} \\ q_n & q_{n-1} \end{vmatrix} > 0,$$

$$\mathbf{H}_3 = \begin{vmatrix} q_{n-3} & q_{n-4} & q_{n-5} \\ q_{n-1} & q_{n-2} & q_{n-3} \\ 0 & q_n & q_{n-1} \end{vmatrix} > 0, \ldots,$$

$$\mathbf{H}_n = \begin{vmatrix} q_0 & 0 & . & . & . & 0 \\ . & . & . & . & . & . \\ . & . & . & . & . & . \\ . & . & . & q_{n-3} & q_{n-4} & q_{n-5} \\ . & . & . & q_{n-1} & q_{n-2} & q_{n-3} \\ 0 & . & . & 0 & q_n & q_{n-1} \end{vmatrix} > 0,$$

*are satisfied. We interprete $q_{n-k} = 0$ for $k > n$. These determinants are called the **Hurwitz determinants**.*

Proof: Clearly, the assumption $q_n > 0$ involves no loss of generality; otherwise, we consider the polynomial $-q$. By Theorem 9.1.6, the polynomial q is stable if and only if the two Bezoutians $B(q_+, q_-)$ and $B(zq_-, q_+)$ are positive definite. This, by Theorem 8.2.5, is equivalent to the positive definiteness of all principal minors of both Bezoutians. We find it convenient to check for positivity all of the lower right-hand corner minors rather than all upper left-hand corner minors. In the proof we will use the Gohberg–Semencul formula (8.21) for the representation of the Bezoutian. The proof will be split into two parts, according to $n = \deg q$ being even or odd.

Case I: Assume that n is even and $n = 2m$. The even and odd parts of q are given by

$$\begin{cases} q_+(z) = q_0 + \cdots + q_{2m}z^m \\ q_-(z) = q_1 + \cdots + q_{2m-1}z^{m-1}. \end{cases}$$

The Bezoutian $B(q_+, q_-)$ therefore has the representation

$$\left[\begin{pmatrix} q_1 & & \\ & \ddots & \\ q_{2m-1} & \cdots & q_1 \end{pmatrix} \begin{pmatrix} q_{2m} & q_{2m-2} & \cdot & \cdot & q_2 \\ & & & & \\ & & & & q_{2m} \end{pmatrix} \right.$$

$$\left. - \begin{pmatrix} q_0 & & \\ & \ddots & \\ q_{2m-2} & \cdots & q_0 \end{pmatrix} \begin{pmatrix} 0 & q_{2m-1} & \cdot & \cdot & q_3 \\ & & & & \cdot \\ & & & q_{2m-1} & \\ & & & & 0 \end{pmatrix} \right] J_m.$$

We now consider the lower right-hand $k \times k$ submatrix. This is given by

$$\left[\begin{pmatrix} q_{2m-2k+1} & & \\ & \ddots & \\ q_{2m-1} & \cdots & q_{2m-2k+1} \end{pmatrix} \begin{pmatrix} q_{2m} & \cdot & \cdot & q_{2m-2k+2} \\ & & & \\ & & & q_{2m} \end{pmatrix} \right.$$

$$\left. - \begin{pmatrix} q_{2m-2k} & & \\ & \ddots & \\ q_{2m-2} & \cdots & q_{2m-2k} \end{pmatrix} \begin{pmatrix} 0 & q_{2m-1} & \cdot & \cdot & q_{2m-2k+3} \\ & & & & \cdot \\ & & & q_{2m-1} & \\ & & & & 0 \end{pmatrix} \right] J_k.$$

Applying Lemma 3.2.1 we have

$$\det(B(q_+, q_-))_{i,j=m-k+1}^m$$

$$= \begin{vmatrix} q_{2m-2k+1} & & & & & & q_{2m-2k} & & & \\ \cdot & & \cdot & & & & \cdot & & \cdot & \\ \cdot & & & \cdot & & & \cdot & & & \cdot \\ \cdot & & & & \cdot & & \cdot & & & \\ q_{2m-1} & \cdot & \cdot & \cdot & q_{2m-2k+1} & q_{2m-2} & \cdot & \cdot & \cdot & q_{2m-2k} \\ 0 & \cdot & \cdot & \cdot & q_{2m-2k+3} & q_{2m} & \cdot & \cdot & \cdot & q_{2m-2k+2} \\ & \cdot & \cdot & \cdot & \cdot & & \cdot & \cdot & \cdot & \cdot \\ & & \cdot & \cdot & \cdot & & & \cdot & \cdot & \cdot \\ & & & \cdot & q_{2m-1} & & & & \cdot & \cdot \\ & & & & 0 & & & & & q_{2m} \end{vmatrix} \cdot \det J_k,$$

which, using the factor $\det J_k = (-1)^{[k(k-1)]/2}$, can be rearranged in the form

$$\begin{vmatrix} q_{2m-2k+1} & q_{2m-2k} & 0 & & & & & & & \cdot \\ \cdot & q_{2m-2k+2} & q_{2m-2k+1} & & & & & & & \cdot \\ \cdot & & \cdot & & & & & & & \cdot \\ \cdot & & \cdot & \cdot & & & & & & 0 \\ q_{2m-1} & q_{2m-2} & \cdot & & \cdot & & \cdot & q_{2m-2k+1} & q_{2m-2k} & \\ 0 & q_{2m} & \cdot & & \cdot & & \cdot & q_{2m-2k+3} & q_{2m-2k+2} & \\ \cdot & & 0 & & \cdot & \cdot & & & \cdot & \cdot \\ \cdot & & & \cdot & \cdot & \cdot & & \cdot & & \cdot \\ \cdot & & & & & 0 & q_{2m} & q_{2m-1} & q_{2m-2} & \\ \cdot & & & & & & \cdot & 0 & q_{2m} & \end{vmatrix} \cdot$$

Since $q_{2m} = q_n > 0$, it follows that the following determinantal conditions hold:

$$\begin{vmatrix} q_{2m-2k+1} & q_{2m-2k} & 0 & & & & & \\ \cdot & q_{2m-2k+2} & q_{2m-2k+1} & & & & & \\ \cdot & & \cdot & & & & & \\ \cdot & & \cdot & \cdot & & & & \\ q_{2m-1} & q_{2m-2} & \cdot & & \cdot & & \cdot & q_{2m-2k+1} \\ 0 & q_{2m} & \cdot & & \cdot & & \cdot & q_{2m-2k+3} \\ \cdot & & 0 & & \cdot & \cdot & & \cdot \\ \cdot & & & & 0 & q_{2m} & q_{2m-1} & \end{vmatrix} \cdot$$

Using the fact that $n = 2m$, we can write the last determinantal condition as

$$
\begin{vmatrix}
q_{n-2k+1} & q_{n-2k} & 0 & & & & & \\
\cdot & q_{n-2k+2} & q_{n-2k+1} & & & & & \\
\cdot & & & \cdot & & & & \\
\cdot & & & & \cdot & \cdot & & \\
q_{n-1} & q_{n-2} & \cdot & & \cdot & \cdot & \cdot & q_{n-2k+1} \\
0 & q_n & \cdot & \cdot & & \cdot & \cdot & q_{n-2k+3} \\
\cdot & & 0 & \cdot & \cdot & & & \\
\cdot & & & \cdot & \cdot & \cdot & \cdot & \\
\cdot & & & & 0 & q_n & q_{n-1}
\end{vmatrix} > 0.
$$

Note that the order of these determinants is $1, \ldots, 2m - 1$.

Next we consider the Bezoutian $B(zq_-, q_+)$, which has the representation

$$
\left[\begin{pmatrix} q_0 & & \\ \cdot & \cdot & \\ \cdot & \cdot & \cdot \\ \cdot & \cdot & \cdot \\ q_{2m-2} & \cdots & q_0 \end{pmatrix} \begin{pmatrix} q_{2m-1} & q_{2m-3} & \cdot & \cdot & q_1 \\ & \cdot & \cdot & \cdot & \\ & & \cdot & \cdot & \\ & & & \cdot & \\ & & & & q_{2m-1} \end{pmatrix} \right.
$$

$$
\left. - \begin{pmatrix} 0 & & & \\ q_1 & \cdot & & \\ \cdot & \cdot & \cdot & \\ \cdot & \cdot & \cdot & \\ q_{2m-3} & \cdot & \cdot & q_1 & 0 \end{pmatrix} \begin{pmatrix} q_{2m} & & \cdot & \cdot & q_2 \\ & \cdot & \cdot & \cdot & \cdot \\ & & \cdot & \cdot & \cdot \\ & & & & \cdot \\ & & & & q_{2m} \end{pmatrix} \right] J_m.
$$

The lower right-hand $k \times k$ submatrix is given by

$$
\left[\begin{pmatrix} q_{2m-2k} & & \\ \cdot & \cdot & \\ \cdot & \cdot & \cdot \\ \cdot & \cdot & \cdot \\ q_{2m-2} & \cdots & q_{2m-2k} \end{pmatrix} \begin{pmatrix} q_{2m-1} & q_{2m-3} & \cdot & \cdot & q_{2m-2k+1} \\ & \cdot & \cdot & \cdot & \\ & & \cdot & \cdot & \\ & & & \cdot & \\ & & & & q_{2m-1} \end{pmatrix} \right.
$$

$$
\left. - \begin{pmatrix} q_{2m-2k-1} & & \\ \cdot & \cdot & \\ \cdot & \cdot & \cdot \\ \cdot & \cdot & \cdot \\ q_{2m-3} & \cdots & q_{2m-2k-1} \end{pmatrix} \begin{pmatrix} q_{2m} & \cdot & \cdot & q_{2m-2k+2} \\ & \cdot & \cdot & \cdot \\ & & \cdot & \cdot \\ & & & \cdot \\ & & & q_{2m} \end{pmatrix} \right] J_k.
$$

Again, by the use of Lemma 3.2.1, we have

$$\det(B(zq_-, q_+))_{i,j=m-k+1}^m$$

$$
= \begin{vmatrix}
q_{2m-2k} & & & & q_{2m-2k-1} & & \\
\cdot & \cdot & & & & \cdot & \cdot \\
\cdot & & \cdot & & & \cdot & & \cdot \\
\cdot & & & \cdot & & \cdot & & & \cdot \\
q_{2m-2} & \cdots & q_{2m-2k} & q_{2m-3} & \cdots & q_{2m-2k-1} & \\
q_{2m} & \cdots & q_{2m-2k+2} & q_{2m-1} & \cdots & q_{2m-2k+1} & \\
& \cdot & \cdot & \cdot & & \cdot & \\
& & \cdot & \cdot & & & \cdot & \cdot & \\
& & \cdot & q_{2m-1} & & & & \cdot & & \cdot \\
& & & q_{2m} & & & q_{2m-1} &
\end{vmatrix} \cdot \det J_k
$$

$$
= \begin{vmatrix}
q_{2m-2k} & q_{2m-2k-1} & 0 & & & & & & \cdot \\
\cdot & q_{2m-2k+1} & q_{2m-2k} & & & & & & \cdot \\
\cdot & & \cdot & & & & & & \cdot \\
\cdot & & & \cdot & \cdot & & & & 0 \\
q_{2m-1} & q_{2m-2} & \cdot & \cdot & \cdot & \cdot & q_{2m-2k+1} & q_{2m-2k} & \\
q_{2m} & q_{2m-1} & \cdot & \cdot & \cdot & \cdot & q_{2m-2k+3} & q_{2m-2k+2} & \\
& & \cdot & \cdot & \cdot & \cdot & & & \\
& & & \cdot & \cdot & \cdot & \cdot & & \cdot & \\
\cdot & & & & q_{2m} & q_{2m-1} & q_{2m-2} & q_{2m-3} \\
\cdot & & & & & \cdot & q_{2m} & q_{2m-1}
\end{vmatrix}
$$

Again, using $n = 2m$, this determinant can be rewritten as

$$
\begin{vmatrix}
q_{n-2k} & q_{n-2k-1} & 0 & & & & & & \cdot \\
\cdot & q_{n-2k+1} & q_{n-2k} & & & & & & \cdot \\
\cdot & & \cdot & & & & & & \cdot \\
\cdot & & & \cdot & \cdot & & & & 0 \\
q_{n-1} & q_{n-2} & \cdot & \cdot & \cdot & \cdot & q_{n-2k+1} & q_{n-2k} & \\
q_n & q_{n-1} & \cdot & \cdot & \cdot & \cdot & q_{n-2k+3} & q_{n-2k+2} & \\
& & \cdot & \cdot & \cdot & \cdot & & & \\
& & & \cdot & \cdot & \cdot & \cdot & & \cdot & \\
\cdot & & & & q_n & q_{n-1} & q_{n-2} & q_{n-3} \\
\cdot & & & & & \cdot & q_n & q_{n-1}
\end{vmatrix}
$$

Case II: This is the case where $n = 2m + 1$ is odd. The proof proceeds along similar lines. $\square$

9.2 Exercises

1. Let the real polynomial $q(z) = \sum_{i=0}^n q_i z^i$ have zeros $\mu_1, \ldots, \mu_n$. Prove **Orlando's formula**, that is, that for the Hurwitz determinants defined in Theorem 9.1.10, we have

$$\det \mathbf{H}_{n-1} = (-1)^{\frac{n(n-1)}{2}} q_n^{n-1} \prod_{i<k} (\mu_i + \mu_k).$$

2. (a) Show that a real polynomial p is a Hurwitz polynomial if and only if the zeros and poles of

$$f(z) = \frac{p(z) - p(-z)}{p(z) + p(-z)}$$

are simple, located on the imaginary axis, and they mutually separate each other.

(b) Show that a real polynomial $p(z) = \sum_{i=0}^{n} p_i z^i$ has all of its zeros in the open unit disk if and only if $|p_n| > |p_0|$, and the zeros and poles of

$$f(z) = \frac{p(z) - p^\sharp(z)}{p(z) + p^\sharp(z)}$$

are simple, located on the unit circle, and they mutually separate each other. Here $p^\sharp(z) = z^n p(z^{-1})$ is the reciprocal polynomial to p.

3. Given a polynomial $f \in \mathbf{R}[x]$ of degree n, we define

$$S(f) = f(x)f''(x) - f'(x)^2.$$

Show that f has n distinct real roots if and only if

$$S(f) < 0, S(f') < 0, \ldots, S(f^{(n-2)}) < 0.$$

Show that $f(x) = x^3 + 3ux + 2v$ has three distinct real roots if and only if $u < 0$ and $v^2 + u^3 < 0$.

9.3 Notes and Remarks

The approach to problems of root location via the use of quadratic forms goes back to Jacobi. The greatest impetus to this line of research was given by Hermite [1856].

There are other approaches to the stability analysis of polynomials. Some of them are based on complex analysis, and in particular on the principle of argument. A very general approach to stability was developed by Liapunov [1893]. For the case of a linear system $\dot{x} = Ax$, this leads to the celebrated Liapunov matrix equation $AX + XA^* = -Q$. Surprisingly, it took half a century to clarify the connection between Lyapunov stability theory and the algebraic stability criteria. For such a derivation, in the spirit of this book, see Willems and Fuhrmann [1992].

10
Elements of System Theory

10.1 Introduction

This chapter is devoted to a short introduction to algebraic system theory. We shall focus on the main conceptual underpinnings of the theory, more specifically on the themes of external and internal representations of systems and the associated realization theory. We feel that these topics are to be considered as an essential part of linear algebra. In fact, the notions of reachability and observability, introduced by Kalman (see Kalman [1968] and the references within), fill a gap that the notion of cyclicity leaves open. Also, they have such a strong intuitive appeal that it will be rather perverse not to use them and search instead for sterile, but "pure," substitute terms.

We saw the central role that polynomials played in the structure theory of linear transformations. The same role is played by rational functions in the context of algebraic system theory. In fact, realization theory is, for rational functions, what the shift operator and the companion matrices were for polynomials.

From the purely mathematical point of view, realization theory has, in the algebraic context, as its main theme a special type of representation for rational functions.

Note that, given a quadruple of matrices (A, b, c, d), which are of sizes $n \times n, n \times 1, 1 \times n$ and 1×1, respectively, the function g defined by

$$g(z) = d + c(zI - A)^{-1}b \qquad (10.1)$$

is a scalar, proper rational function. The realization problem is corresponding inverse problem. Namely, given a scalar, proper rational function $g(z)$, we want to find a quadruple of matrices (A, b, c, d), for which Eq. (10.1) holds.

In this sense, the realization problem is an extension of a simpler problem that we solved earlier. That problem was the construction of an operator, or a matrix, that had a given polynomial as its characteristic polynomial. In the solution to that problem, shift operators, and their matrix representations in terms of companion matrices, played a central role. Therefore, it is natural to expect that the same objects will play a similar role in the solution of the realization problem, and this in fact is the case.

10.2 Systems and Their Representations

Generally we associate the word *system* with dynamics, that is, with the way a system evolves with time. Time itself can be modeled in various ways, most commonly as continuous or discrete, as the case may be. Contrary to some approaches to the study of systems, we will focus on the proverbial black box approach, given by the following diagram:

Here Σ denotes the system, u the input or control signal, and y the output or observation signal. The way the output signal depends on the input signal is called the **input/output relation**. Such a description is termed an **external representation**. An **internal representation** of a system is a model, usually given in terms of difference or differential equations, that explains or is compatible with the external representation. Unless further assumptions are made on the properties of the input/output relations (linearity, continuity), there is not much of interest that can be said. Because of the context of linear algebra and the elementary nature of our approach, with the tendency to emphasize linear algebraic properties, it is natural for us to restrict ourselves to linear time-invariant finite-dimensional systems. We proceed to introduce these systems.

Definition 10.2.1

1. *A discrete-time finite-dimensional linear time-invariant system is a triple $\{U, X, Y\}$ of finite-dimensional vector spaces over a field F and a quadruple of linear transformations $A \in L(X, X)$, $B \in L(U, X)$, $C \in L(X, Y)$, and $D \in L(U, Y)$, with the system equations given by*

$$\begin{cases} x_{n+1} &= Ax_n + Bu_n \\ y_n &= Cx_n + Du_n. \end{cases} \quad (10.2)$$

2. *A continuous-time finite-dimensional linear time-invariant system is a triple $\{U, X, Y\}$ of finite-dimensional vector spaces over the real or complex field and a quadruple of linear transformations $A \in L(X, X)$, $B \in L(U, X)$, $C \in L(X, Y)$, and $D \in L(U, Y)$, with the system equations given by*

$$\begin{cases} \dot{x} & = & Ax + Bu \\ y & = & Cx + Du. \end{cases} \tag{10.3}$$

The spaces U, X, Y are called the **input space**, **state space**, *and* **output space**, *respectively. Usually we identify these spaces with F^m, F^n, F^p, respectively, and then the transformations A, B, C, D are given by matrices. In both case we will use the notation*

$$\left(\begin{array}{c|c} A & B \\ \hline C & D \end{array} \right)$$

to describe the system. Such representations are called **state space realizations**.

Since the development of discrete-time linear systems does not depend on analysis, we will concentrate our attention on this class.

Let us start from the state equation $x_{n+1} = Ax_n + Bu_n$. Substituting in this $x_n = Ax_{n-1} + Bu_{n-1}$ we get $x_{n+1} = A^2 x_{n-1} + ABu_{n-1} + Bu_n$, and, proceeding by induction, we get

$$x_{n+1} = A^k x_{n-k+1} + A^{k-1} Bu_{n-k+1} + \cdots Bu_n. \tag{10.4}$$

Our standing assumption is that before a finite time all signals were zero, that is, in the remote past the system was at rest. Thus, for some n_0, $u_n = 0, x_n = 0$ for $n < n_0$. With this Eq. (10.4) reduces to

$$x_{n+1} = \sum_{j=0}^{\infty} A^j Bu_{n-j},$$

and, in particular,

$$x_0 = \sum_{j=0}^{\infty} A^j Bu_{-j-1}.$$

Thus, for $n \geq 0$, Eq. (10.4) also could be written as

$$\begin{aligned} x_{n+1} & = \sum_{j=0}^{n} A^j Bu_{n-j} + \sum_{j=n+1}^{\infty} A^j Bu_{n-j} \\ & = A^{n+1} \sum_{j=0}^{\infty} A^j Bu_{-j-1} + \sum_{j=0}^{n} A^j Bu_{n-j} \\ & = A^{n+1} x_0 + \sum_{j=0}^{n} A^j Bu_{n-j}. \end{aligned}$$

This is the state evolution equation based on initial conditions at time zero.

From the preceeding, the input/output relations become

$$y_{n+1} = \sum_{j=0}^{\infty} CA^j Bu_{n-j}. \tag{10.5}$$

We write $y = \tilde{f}(u)$ and call $\tilde{f}$ the input/output map of the system.

Suppose we look now at sequences of input and output signals shifted by one time unit. Let us write $\theta_n = u_{n-1}$ and denote the corresponding states and outputs by ξ_n and η_n, respectively. Then

$$\xi_n = \sum_{j=0}^{\infty} A^j B\theta_{n-j-1} = \sum_{j=0}^{\infty} A^j Bu_{n-j-2} = x_{n-1},$$

and

$$\eta_n = C\xi_n = Cx_{n-1} = y_{n-1}.$$

Let us now introduce the shift operator σ acting on time signals by $\sigma(u_j) = u_{j-1}$. Then the input/output map satisfies

$$\tilde{f}(\sigma(u)) = \sigma(y) = \sigma(\tilde{f}(u)). \tag{10.6}$$

If the signal, which was zero in the remote past, goes over to the truncated Laurent series $\sum_{j=-\infty}^{\infty} u_{-j}z^j$, we conclude that $\sigma(u)$ is mapped into

$$\sum_{j=-\infty}^{\infty} u_{-j-1}z^j = \sum_{j=-\infty}^{\infty} u_{-j-1}z^{j+1} = z \cdot \sum_{j=-\infty}^{\infty} u_{-j}z^j,$$

that is, in $U((z^{-1}))$ the shift σ acts as multiplication by z. Since $\tilde{f}$ is a linear map, we get by induction, for an arbitrary polynomial p,

$$\tilde{f}(p \cdot u) = p \cdot \tilde{f}(u).$$

This means that, with the polynomial module structure induced by σ, the input/output map $\tilde{f}$ is an $F[z]$-module homomorphism.

We find it convenient to associate with the infinite sequence $\{u_j\}_{-\infty}^{n_0}$ the truncated Laurent series $\sum_{j \geq n_0} u_j z^{-j}$. Thus, positive powers of z are associated with past signals, whereas negative powers are associated with future signals. With this convention also applied to the state and output sequences, the input/output relation (10.5) now can be written as

$$y(z) = G(z)u(z), \tag{10.7}$$

where $y(z) = \sum_{j \geq n_0} y_j z^{-j}$ and

$$G(z) = D + \sum_{i=1}^{\infty} \frac{CA^{i-1}B}{z^i} = D + C(zI - A)^{-1}B. \tag{10.8}$$

The function $G(z)$ defined in Eq. (10.8) is called the **transfer function** of the system. From this we can conclude the following:

Proposition 10.2.1 *Let*

$$\Sigma = \left(\frac{A \mid B}{C \mid D} \right)$$

be a finite-dimensional linear time-invariant system. Then its **transfer function** $G(z) = D + C(zI - A)^{-1}B$ *is proper rational.*

Proof: Follows from the equality

$$(zI - A)^{-1} = \frac{\text{adj}\,(zI - A)}{\det(zI - A)}. \qquad \square$$

From the system equations (10.2) it is clear that, given that the system is at rest in the remote past, there will be no output from the system prior to an input being sent into the system. In terms of signal spaces, the space of future inputs is mapped into the space of future output signals. This is the concept of **causality**, and it is built into the internal representation. We formalize this as follows:

Definition 10.2.2 *Let U and Y be finite-dimensional linear spaces over the field F.*

1. *An* **input/output map** *is an $F[z]$-module homomorphism $\tilde{f} : U((z^{-1}))$ $\longrightarrow Y((z^{-1}))$.*

2. *An input/output map $\tilde{f}$ is* **causal** *if $\tilde{f}(U[[z^{-1}]]) \subset Y[[z^{-1}]]$ and* **strictly causal** *if $\tilde{f}(U[[z^{-1}]]) \subset z^{-1}Y[[z^{-1}]]$.*

We have the following characterization:

Proposition 10.2.2

1. *$\tilde{f}(U((z^{-1}))) \subset Y((z^{-1}))$ is an $F[z]$-module homomorphism if and only if there exists a $G \in L(U,Y)((z^{-1}))$ for which $\tilde{f}(u) = G \cdot u$.*

2. *$\tilde{f}$ is causal if and only if $G \in L(U,Y)[[z^{-1}]]$ and strictly causal if and only if $G \in z^{-1}L(U,Y)[[z^{-1}]]$.*

Proof:

1. Assume that $\tilde{f} : U((z^{-1})) \longrightarrow Y((z^{-1}))$ is defined by $\tilde{f}(u) = G \cdot u$ for some $G \in L(U,Y)((z^{-1}))$. Then

$$z \cdot \tilde{f}(u) = z(Gu) = G(zu) = \tilde{f}(z \cdot u),$$

that is, $\tilde{f}$ is an $F[z]$-module homomorphism.

Conversely, assume that $\tilde{f}$ is an $F[z]$-module homomorphism. Let $e_1, \ldots, e_m$ be a basis in U. Set $g_i = \tilde{f}(e_i)$. Let G have columns g_i. Then, using the fact that $\tilde{f}$ is a homomorphism, we compute

$$\tilde{f}(u) = \tilde{f}\sum_{i=1}^{m} u_i z^i = \sum_{i=1}^{m} z^i \tilde{f}(u_i) = \sum_{i=1}^{m} z^i G u_i = Gu.$$

2. Assume that G has the expansion $G(z) = \sum_{i=-n}^{\infty} G_i/z^i$ and $u(z) = \sum_{i=0}^{\infty} u_i/z^i$. The polynomial part, not including the constant term, of Gu is given by $\sum_{k=0}^{n} \sum_{i=0}^{k} G_{-n+k-i} u_i z^k$. This vanishes if and only if $\sum_{i=0}^{k} G_{-n+k-i} u_i = 0$ for all choices of $u_0, \ldots, u_{n-1}$. This in turn is equivalent to $G_{-n} = \cdots = G_{-1} = 0$.

Strict causality is handled similarly. $\qquad\qquad\qquad\qquad\qquad\qquad\square$

Let us digress a bit on the notion of state. Heuristically, the state of the system is the minimum amount of information we need to know on the system at present so that its future outputs can be computed, given that we have access to future inputs. In a sense, the present state of the system is the combined effect of all past inputs. But this type of information is overly redundant. In particular, many past inputs may lead to the same state. Since the input/output map is linear, we can easily eliminate the future inputs from the definition. Therefore, given an input/output map $\tilde{f}$, it is natural to introduce the following notion of equivalence of past inputs. We say that, with $u, v \in U[z]$, $u \simeq_f v$ if $\pi_- \tilde{f} u = \pi_- \tilde{f} v$. This means that the past input sequences u and v have the same future outputs, and thus they will be indistinguishable based on future observations. The moral of this is that, in order to elucidate the notion of state, it is convenient to introduce an auxiliary input/output map.

Definition 10.2.3 *Let $\tilde{f}(U((z^{-1}))) \subset Y((z^{-1}))$ be an input/output map. Then the **restricted input/output map** $f : U[z] \longrightarrow z^{-1}Y[[z^{-1}]]$ is defined by $f(u) = \pi_- \tilde{f}(u)$.*

If $\tilde{f}$ has the transfer function G, then the restricted input/output map is given by $f(u) = \pi_- Gu$, that is, it is the Hankel operator defined by G. The relation between the input/output map and the restricted input/output map is best described by the following commutative diagram:

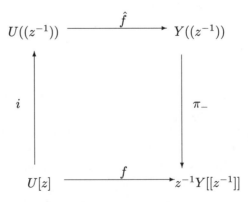

Here, $i : U[z] \longrightarrow U((z^{-1}))$ is the natural embedding.

10.3   Realization Theory

We now turn to the realization problem. Given a causal input/output map $\hat{f}$ with proper transfer function G, we want to find a state space realization of it, that is, we want to find linear transformations A, B, C, D such that

$$G(z) = \left(\begin{array}{c|c} A & B \\ \hline C & D \end{array} \right) = D + C(zI - A)^{-1}B.$$

Since $G(z) = \sum_{i=0}^{\infty} G_i/z^i$, we must have for the realization

$$\begin{cases} \quad D &= G_0 \\ \\ CA^{i-1}B &= G_i, \;\; i \geq 1. \end{cases}$$

The coefficients G_i are called the **Markov parameters** of the system.

To this end we go back to a state space system

$$\left(\begin{array}{c|c} A & B \\ \hline C & D \end{array} \right)$$

We define the **reachability map** $\mathcal{R} : U[z] \longrightarrow X$ by

$$\mathcal{R} \sum_{i=0}^{n} u_i z^i = \sum_{i=0}^{n} A^i B u_i, \tag{10.9}$$

and the **observability map** $\mathcal{O} : X \longrightarrow z^{-1}Y[[z^{-1}]]$ by

$$\mathcal{O}x = \sum_{i=1}^{\infty} \frac{CA^{i-1}x}{z^i}. \tag{10.10}$$

Proposition 10.3.1 *Given the state space system*

$$\left(\begin{array}{c|c} A & B \\ \hline C & D \end{array} \right)$$

with state space X. Let X carry the $F[z]$-module structure induced by A as in Eq. (4.7). Then:

1. *The reachability and observability maps are $F[z]$-module homomorphisms.*

2. *If f is the restricted input/output map of the system and $G(z) = D + C(zI - A)^{-1}B$ the transfer function, then*

$$f = H_G = \mathcal{O}\mathcal{R}. \tag{10.11}$$

Proof:

1. We compute

$$
\mathcal{R}z\sum_{i=0}^{n}u_i z^i = \mathcal{R}\sum_{i=0}^{n}u_i z^{i+1} = \sum_{i=0}^{n}A^{i+1}Bu_i
$$

$$
= A\sum_{i=0}^{n}A^i Bu_i = A\mathcal{R}\sum_{i=0}^{n}u_i z^i. \tag{10.12}
$$

Also,

$$
\mathcal{O}Ax = \sum_{i=1}^{\infty}\frac{CA^{i-1}A\xi}{z^i} = \sum_{i=1}^{\infty}\frac{CA^i\xi}{z^i}
$$

$$
= S_-\sum_{i=1}^{\infty}\frac{CA^{i-1}\xi}{z^i} = S_-\mathcal{O}\xi. \tag{10.13}
$$

2. Let $\xi \in U$ and consider it as a constant polynomial in $U[z]$. Then we have

$$
\mathcal{O}\mathcal{R}\xi = \mathcal{O}(B\xi) = \sum_{i=1}^{\infty}\frac{CA^{i-1}Bx}{z^i} = G(z)\xi.
$$

Since both $\mathcal{O}$ and $\mathcal{R}$ are $F[z]$-module homomorphisms, it follows that

$$
\mathcal{O}\mathcal{R}(z^i\xi) = \mathcal{O}(A^i B\xi) = \sum_{j=1}^{\infty}\frac{CA^{j-1}(A^i B\xi)}{z^j}
$$

$$
= \sum_{j=1}^{\infty}\frac{CA^{i+j-1}B\xi}{z^j} = \pi_- z^i\sum_{j=1}^{\infty}\frac{CA^{j-1}B\xi}{z^j}
$$

$$
= \pi_- z^i G(z)\xi = \pi_- G(z)z^i\xi.
$$

By linearity we get, for $u \in U[z]$, that Eq. (10.11) holds. $\quad\square$

So far nothing has been assumed that would imply further properties of the reachability and observability maps. In particular, recalling the canonical factorizations of maps discussed in Chapter 1, we are interested in the case where $\mathcal{R}$ is surjective and $\mathcal{O}$ is injective. This leads to the following definition:

Definition 10.3.1 *Given the state space system Σ with transfer function*

$$
G = \left(\begin{array}{c|c} A & B \\ \hline C & D \end{array}\right)
$$

with state space X. Then:

1. *The system Σ is called* **reachable** *if the reachability map $\mathcal{R}$ is surjective.*

2. *The system Σ is called* **observable** *if the observability map $\mathcal{O}$ is injective.*

3. *The system is called* **canonical** *if it is both reachable and observable.*

Proposition 10.3.2 *Given the state space system Σ with transfer function*

$$G = \left(\begin{array}{c|c} A & B \\ \hline C & D \end{array} \right)$$

with the n-dimensional state space X. Then:

1. *The following statements are equivalent:*

   (a) *The system Σ is reachable.*

   (b) *The zero state can be steered to an arbitrary state $\xi \in X$ in a finite number of steps.*

   (c) *We have*
   $$\operatorname{rank}(B, AB, \ldots, A^{n-1}B) = n. \tag{10.14}$$

   (d)
   $$\bigcap_{i=0}^{n-1} \operatorname{Ker} B^*(A^*)^i = \{0\}. \tag{10.15}$$

   (e)
   $$\bigcap_{i=0}^{\infty} \operatorname{Ker} B^*(A^*)^i = \{0\}. \tag{10.16}$$

2. *The following statements are equivalent:*

   (a) *The system Σ is observable.*

   (b) *The only state with all future outputs zero is the zero state.*

   (c)
   $$\bigcap_{i=0}^{\infty} \operatorname{Ker} CA^i = \{0\}. \tag{10.17}$$

   (d)
   $$\bigcap_{i=0}^{n-1} \operatorname{Ker} CA^i = \{0\}. \tag{10.18}$$

   (e) *We have*
   $$\operatorname{rank}(C^*, A^*C^*, \ldots, (A^*)^{n-1}C^*) = n. \tag{10.19}$$

Proof:

1. $(a) \Rightarrow (b)$ Given a state $\xi \in X$, there exists a $v(z) = \sum_{i=0}^{k-1} v_i z^i$ such that $\xi = \mathcal{R}v = \sum_{i=0}^{k-1} A^i B v_i$. Setting $u_i = v_{k-1-i}$ and using this control sequence, the solution to the state equation, with zero initial condition, is given by $x_k = \sum_{i=0}^{k-1} A^i B u_{k-1-i} = \sum_{i=0}^{k-1} A^i B v_i = \xi$.

$(b) \Rightarrow (c)$ Since every state $\xi \in X$ has a representation $\xi = \sum_{i=0}^{k-1} A^i B u_i$ for some k, it follows by an application of the Cayley–Hamilton theorem that we can assume without loss of generality that $k = n$. Thus, the map $(B, AB, \ldots, A^{n-1}B) : U^n \longrightarrow X$ defined by $(u_0, \ldots, u_n) \mapsto \sum_{i=0}^{n-1} A^i B u_i$ is surjective. Hence, $\text{rank}\,(B, AB, \ldots, A^{n-1}B) = n$.

$(c) \Rightarrow (d)$ The adjoint to the map $(B, AB, \ldots, A^{n-1}B)$ is

$$\begin{pmatrix} B^* \\ B^* A^* \\ \cdot \\ \cdot \\ B^*(A^*)^{n-1} \end{pmatrix} : X^* \longrightarrow (U^*)^n.$$

Applying Theorem 4.4.2, we have

$$\text{Ker} \begin{pmatrix} B^* \\ B^* A^* \\ \cdot \\ \cdot \\ B^*(A^*)^{n-1} \end{pmatrix} = \bigcap_{i=0}^{n-1} \text{Ker}\, B^*(A^*)^i = \{0\}.$$

$(d) \Rightarrow (e)$ This follows from the inclusion $\cap_{i=0}^{\infty} \text{Ker}\, B^*(A^*)^i \subset \cap_{i=0}^{n-1} \text{Ker}\, B^*(A^*)^i$.

$(e) \Rightarrow (a)$ Let $\phi \in (\text{Im}\,\mathcal{R})^{\perp} = \text{Ker}\,\mathcal{R}^* = \cap_{i=0}^{\infty} \text{Ker}\, B^*(A^*)^i = \{0\}$. Then $\phi = 0$ and the system is reachable.

2. This can be proved similarly. Alternatively, it follows from the first part by duality considerations. $\qquad \square$

Proposition 10.3.1 implied that any realization of an input/output map leads to a factorization of the restricted input/output map. We proceed to prove the converse.

Theorem 10.3.1 *Let $\tilde{f} : U((z^{-1})) \longrightarrow Y((z^{-1}))$ be an input/output map. Then to any realization of f corresponds a unique factorization $f = hg$, with $h : X \longrightarrow Y((z^{-1}))$ and $g : U((z^{-1})) \longrightarrow X$ $F[z]$-module homomorphisms.*

Conversely, given any factorization $f = hg$ into a product of $F[z]$-module homomorphisms, there exists a unique associated realization. The factorization is canonical, that is, g is surjective and h injective if and only if the realization is canonical.

Proof: We saw that a realization leads to a factorization $f = \mathcal{OR}$, with $\mathcal{R}, \mathcal{O}$ the reachability and observability maps, respectively.

Conversely, assume that we are given a factorization

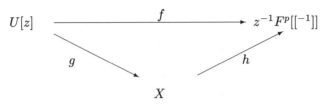

with X an $F[z]$-module and g, h both $F[z]$-module homomorphisms. We define a triple of maps $A : X \longrightarrow X$, $B : U \longrightarrow X$, and $C : X \longrightarrow Y$ by

$$\begin{cases} Ax & = & z \cdot x \\ Bu & = & g(i(u)) \\ Cx & = & \pi_1 h(x). \end{cases} \qquad (10.20)$$

Here, $i : U \longrightarrow U[z]$ is the embedding of U in $U[z]$, which identifies a vector with the corresponding constant polynomial vector, and $\pi_1 : z^{-1} Y[[z^{-1}]]$ is the map that reads the coefficient of z^{-1} in the expansion of an element of $z^{-1} Y[[z^{-1}]]$. It is immediate to check that these equations constitute a realization of f with g and h the reachability and observability maps, respectively. In particular, the realization is canonical if and only if the factorization is canonical. $\qquad \square$

There are natural concepts of isomorphism for both factorizations as well as realizations.

Definition 10.3.2

1. *Let $f = h_i g_i, i = 1, 2$ be two factorizations of the restricted input/output map f through the $F[z]$-modules X_i, respectively. We say that the two factorizations are **isomorphic** if there exists an $F[z]$-module isomorphism ϕ for which the following diagram is commutative:*

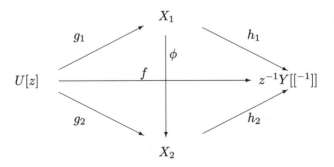

2. *We say that two realizations*

$$\left(\begin{array}{c|c} A_1 & B_1 \\ \hline C_1 & D_1 \end{array}\right) \quad \text{and} \quad \left(\begin{array}{c|c} A_2 & B_2 \\ \hline C_2 & D_2 \end{array}\right)$$

are **isomorphic** *if there exists an isomorphism* $Z : X_1 \longrightarrow X_2$ *such that the following diagram is commutative:*

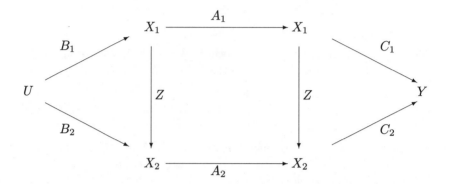

Theorem 10.3.2 (State space isomorphism)

1. *Given two canonical factorizations of a restricted input/output map, then they are isomorphic. Moreover, the isomorphism is unique.*

2. *Assume that*

$$G = \left(\begin{array}{c|c} A_1 & B_1 \\ \hline C_1 & D_1 \end{array}\right) = \left(\begin{array}{c|c} A_2 & B_2 \\ \hline C_2 & D_2 \end{array}\right)$$

and the two realizations are canonical. Then $D_1 = D_2$ *and the two realizations are isomorphic and, moreover, the isomorphism is unique.*

Proof:

1. Let $f = h_1 g_1 = h_2 g_2$ be two canonical factorizations through the modules X_1, X_2, respectively. We define $\phi : X_1 \longrightarrow X_2$ by $\phi(g_1(u)) = g_2(u)$. First we show that ϕ is well defined. We note that the injectivity of h_1, h_2 implies that $\operatorname{Ker} g_i = \operatorname{Ker} f$. So $g_1(u_1) = g_1(u_2)$ implies that $u_1 - u_2 \in \operatorname{Ker} g_1 = \operatorname{Ker} f = \operatorname{Ker} g_2$ and hence also $g_2(u_1) = g_2(u_2)$.

   Next we compute

   $$\phi(z \cdot g_1(u)) = \phi(g_1(z \cdot u)) = g_2(z \cdot u) = z \cdot g_2(u) = z \cdot \phi(g_1(u)).$$

   This shows that ϕ is an $F[z]$-homomorphism.

It remains to show that $h_2\phi = h_1$. Now, given $x \in X_1$, we have $x = g_1(u)$ and $h_1(x) = h_1 g_1(u) = f(u)$. On the other hand, $h_2\phi(x) = h_2\phi(g_1(u)) = h_2 g_2(u) = f(u)$ and the equality $h_2\phi = h_1$ is proved.

To prove uniqueness, assume that ϕ_1, ϕ_2 are two isomorphisms from X_1 to X_2 which make the isomorphism diagram commutative. Then $f = h_2 g_2 = h_2\phi_1 g_1 = h_2\phi_2 g_1$, which implies that $h_2(\phi_2 - \phi_1)g_1 = 0$. Since g_1 is surjective and h_2 injective, the equality $\phi_2 = \phi_1$ follows.

2. Let $\mathcal{R}_1, \mathcal{R}_2$ be the reachability maps of the two realizations, respectively, and $\mathcal{O}_1, \mathcal{O}_2$ the respective observability maps. All four maps are $F[z]$-homomorphisms and, by our assumptions, the reachability maps are surjective while the observability maps are injective. Thus the factorizations $f = \mathcal{O}_i \mathcal{R}_i$ are canonical factorizations. By the first part, there exists a unique $F[z]$-isomorphism $Z : X_1 \longrightarrow X_2$ such that $Z\mathcal{R}_1 = \mathcal{R}_2$ and $\mathcal{O}_2 Z = \mathcal{O}_1$. That Z is a module isomorphism implies that Z is an invertible map satisfying $ZA_1 = A_2 Z$. Restricting $Z\mathcal{R}_1 = \mathcal{R}_2$ to constant polynomials implies that $ZB_1 = B_2$. Finally, looking at coefficients of z^{-1} in the equality $\mathcal{O}_2 Z = \mathcal{O}_1$ implies that $C_2 Z = C_1$. These equalities, taken together, imply the commutativity of the diagram. Uniqueness follows from part 1. $\qquad\square$

We have used the concept of canonical factorizations, but so far we have not demonstrated their existence. This is easy and is taken up next.

Proposition 10.3.3 *Let $f : U[z] \longrightarrow z^{-1}Y[[z^{-1}]]$ be an $F[z]$-homomorphism, and $f = H_G$ the Hankel operator of the transfer function. Then canonical factorizations exist. Specifically, the following factorizations are canonical:*

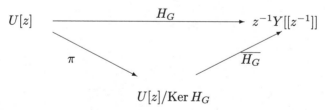

Here, π is the canonical projection and $\overline{H_G}$ the map induced by H_G on the quotient space $U[z]/\mathrm{Ker}\,H_G$.

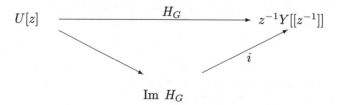

Here, i is the embedding of $\operatorname{Im} H_G$ in $z^{-1}Y[[z^{-1}]]$ and H is the Hankel operator considered as a map from $U[z]$ onto $\operatorname{Im} H_G$.

Proof: We saw in Chapter 8 that the Hankel operator is an $F[z]$-homomorphism. Thus, we can apply Exercise 3 in Chapter 1 to conclude that $H_G|_{U[z]/\operatorname{Ker} H_G}$ is an injective $F[z]$-homomorphism.   □

From now on we will treat only the case of single-input/single-output systems. These are the systems with scalar-valued transfer functions. If we further assume that the systems have finite-dimensional realizations, then by Kronecker's theorem this is equivalent to the transfer function being rational. In this case, both $\operatorname{Ker} H_g$ and $\operatorname{Im} H_g$ are submodules of $F[z]$ and $z^{-1}F[[^{-1}]]$, respectively. Moreover, as linear spaces, $\operatorname{Ker} H_g$ has finite codimension and $\operatorname{Im} H_g$ is finite-dimensional. We can take advantage of the results on representation of submodules contained in Theorems 1.3.5 and 5.3.1 to go from the abstract formulation of realization theory to very concrete state space representations. Since $\operatorname{Im} H_g$ is a finite-dimensional S_--invariant subspace, it is necessarily of the form X^q for some uniquely determined monic polynomial q. This leads to the coprime factorization $g = pq^{-1} = q^{-1}p$. A similar argument applies to $\operatorname{Ker} H_g$, which is necessarily of the form $qF[z]$, and this leads to the same coprime factorization.

However, rather than start with a coprime factorization, our starting point will be any factorization $g = pq^{-1} = q^{-1}p$ with no coprimeness assumptions. We note that $\pi_1(f) = [f, 1]$ with the form defined by Eq. (5.27), in particular, $\pi_1(f) = 0$ for any polynomial f.

Theorem 10.3.3 *Let g be a strictly proper rational function and assume that $g = pq^{-1} = q^{-1}p$ with q monic. Then the following are realizations of g:*

1. *Assume that $g = q^{-1}p$. In the state space X_q define*

$$\left\{ \begin{array}{rcl} Af & = & S_q f \\ B\xi & = & p \cdot \xi \\ Cf & = & <f, 1>. \end{array} \right. \qquad (10.21)$$

Then

$$g = \left(\begin{array}{c|c} A & B \\ \hline C & 0 \end{array} \right).$$

This realization is observable. The reachability map $\mathcal{R} : F[z] \longrightarrow X_q$ is given by

$$\mathcal{R}u = \pi_q(pu). \qquad (10.22)$$

The reachable subspace is $\operatorname{Im} \mathcal{R} = q_1 X_{q_2}$, where $q = q_1 q_2$ and q_1 is the greatest common divisor of p and q. Hence, the realization is reachable if and only if p and q are coprime.

2. *Assume that $g = pq^{-1}$. In the state space X_q define*

$$\begin{cases} Af & = & S_q f \\ B\xi & = & \xi \\ Cf & = & <f, p>. \end{cases} \qquad (10.23)$$

Then

$$g = \left(\begin{array}{c|c} A & B \\ \hline C & 0 \end{array} \right).$$

This realization is reachable, and the reachability map $\mathcal{R} : F[z] \longrightarrow X_q$ is given by

$$\mathcal{R}u = \pi_q u, \qquad u \in F[z]. \qquad (10.24)$$

The unobservable subspace is given by $\operatorname{Ker} \mathcal{O} = q_1 X_{q_2}$, where $q = q_1 q_2$ and q_2 is the greatest common divisor of p and q. The system is observable if and only if p and q are coprime.

3. *Assume that $g = q^{-1}p$. In the state space X^q define*

$$\begin{cases} Ah & = & S^q h \\ B\xi & = & g \cdot \xi \\ Ch & = & [h, 1]. \end{cases} \qquad (10.25)$$

Then

$$g = \left(\begin{array}{c|c} A & B \\ \hline C & 0 \end{array} \right).$$

This realization is observable. It is reachable if and only if p and q are coprime.

4. *Assume that $g = pq^{-1}$. In the state space X^q define*

$$\begin{cases} Ah & = & S^q h \\ B\xi & = & q^{-1} \cdot \xi \\ Ch & = & [h, p]. \end{cases} \qquad (10.26)$$

Then

$$g = \left(\begin{array}{c|c} A & B \\ \hline C & 0 \end{array} \right).$$

This realization is reachable. It is observable if and only if p and q are coprime.

*We refer to these realizations as **shift realizations**.*

5. *If p and q are coprime, then all four realizations are isomorphic. In particular, the following commutative diagram shows the isomorphism of realizations (10.21) and (10.23):*

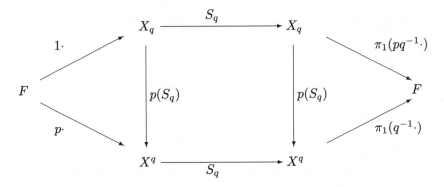

Proof:

1. Let $\xi \in F$. We compute

$$
\begin{aligned}
CA^{i-1}B\xi &= \, < S_q^{i-1}p\xi, 1 >= [q^{-1}q\pi_- q^{-1}z^{i-1}p\xi, 1] \\
&= [z^{i-1}g\xi, 1] = [g, z^{i-1}]\xi = g_i\xi.
\end{aligned}
$$

Hence, Eq. (10.21) is a realization of g.

To show the observability of this realization, assume that $f \in \cap_{i=0}^{\infty}\mathrm{Ker}\,CA^i$. Then, for $i \geq 0$,

$$
0 =< S_q^i f, 1 >= [q^{-1}q\pi_- q^{-1}z^i f, 1] = [q^{-1}f, z^i].
$$

This shows that $q^{-1}f$ is necessarily a polynomial. However, $f \in X_q$ implies that $q^{-1}f$ is strictly proper. So $q^{-1}f = 0$ and hence also $f = 0$.

We proceed to compute the reachability map $\mathcal{R}$. We have, for $u(z) = \sum_i u_i z^i$,

$$
\mathcal{R}u = \mathcal{R}\sum_i u_i z^i = \sum_i S_q^i p u_i = \pi_q \sum_i p u_i z^i = \pi_q(pu).
$$

Clearly, as $\mathcal{R}$ is an $F[z]$-module homomorphism, $\mathrm{Im}\,\mathcal{R}$ is a submodule of X_q and hence of the form $\mathrm{Im}\,\mathcal{R} = q_1 X_{q_2}$ for a factorization $q = q_1 q_2$ into monic factors. Since $p \in \mathrm{Im}\,\mathcal{R}$, it follows that $p = q_1 p_1$, and therefore q_1 is a common factor of p and q.

Now let q' be any other common factor of p and q. Then $q = q'q''$ and $p = q'p'$. For any polynomial u we have, by Lemma 1.3.3,

$$
\mathcal{R}u = \pi_q(pu) = q'\pi_{q''}(p'u) \in q'X_{q''}.
$$

So we have obtained the inclusion $q_1 X_{q_2} \subset q'X_{q''}$, which, by Proposition 5.1.3, implies that $q'|q_1$, that is, q_1 is the greatest common divisor of p and q.

Clearly, $\mathrm{Im}\,\mathcal{R} = q_1 X_{q_2} = X_q$ if and only if $q_1 = 1$, that is, p and q are coprime.

2. Let $\xi \in F$. We compute

$$CA^{i-1}B\xi = <S_q^{i-1}\xi, p> = [q^{-1}q\pi_-q^{-1}z^{i-1}\xi, p]$$
$$= [gz^{i-1}, 1]\xi = g_i\xi.$$

Hence, Eq. (10.23) is a realization of g.

We compute the reachability map of this realization:

$$\mathcal{R}u = \mathcal{R}\sum_i u_i z^i = \sum_i S_q^i u_i = \sum_i \pi_q z^i u_i$$
$$= \pi_q \sum_i z^i u_i = \pi_q u.$$

This implies the surjectivity of $\mathcal{R}$ and hence the reachability of this realization.

Now the unobservable subspace is a submodule of X_q and hence of the form $q_1 X_{q_2}$ for a factorization $q = q_1 q_2$. So for every $f \in q_1 X_{q_2}$ we have

$$0 = CA^i f = <S_q^i f, p> = [q^{-1}q\pi_-q^{-1}z^i f, p] = [q^{-1}z^i f, p]$$
$$= [q_2^{-1}q_1^{-1}z^i q_1 f_2, p] = [pq_2^{-1}f_2, z^i].$$

In particular, choosing $f_1 = 1$, we conclude that pq_2^{-1} is a polynomial, say p_2. So $p = p_2 q_2$, and q_2 is a common factor of p and q. Now let q'' be any other common factor of p and q, and set $q = q'q''$, $p = p'q''$. For $f = q'f' \in q'X_{q''}$ we have

$$CA^i f = <S_q^i q' f', p> = [q^{-1}q\pi_-q^{-1}z^i q' f', p]$$
$$= [p(q'')^{-1}z^i f', 1] = [p'f', z^i] = 0.$$

So we get the inclusion $q'X_{q''} \subset q_1 X_{q_2}$ and hence $q''|q_2$, which shows that q_2 is indeed the greatest common factor of p and q.

Clearly, this realization is observable if and only if Ker $\mathcal{O} = q_1 X_{q_2} = \{0\}$, which is the case only if q_2 is constant and hence are p and q are coprime.

3. Using the isomorphism (5.21) of the polynomial and rational models, we get the isomorphisms of the realizations (10.21) and (10.25).

4. By the same method, we get the isomorphisms of the realizations (10.23) and (10.26). These isomorphisms are independent of coprimeness assumptions.

5. By parts 3 and 4 it suffices to show that the realizations (10.21) and (10.23) are isomorphic. By the coprimeness of p and q and Theorem 5.1.3, the transformation $p(S_q)$ is invertible and certainly satisfies $p(S_q)S_q = S_q p(S_q)$, that is, it is a module homomorphism. Since also $p(S_q)1 = \pi_q(p \cdot 1) = p$ and $< p(S_q)f, 1 > = < f, p >$, it follows that the diagram is commutative. $\qquad\square$

Based on the shift realizations, by a suitable choice of bases, whether in X_q or X^q, we can get concrete realizations given in terms of matrices.

Theorem 10.3.4 *Let g be a strictly proper rational function, and assume that $g = pq^{-1} = q^{-1}p$ with q monic. Assume that $q(z) = z^n + q_{n-1}z^{n-1} + \cdots + q_0$, $p(z) = p_{n-1}z^{n-1} + \cdots + p_0$ and $g(z) = \sum_{i=1}^{\infty} g_i/z^i$. Then the following are realizations of g:*

1. **Controller realization:**

$$
A = \begin{pmatrix} 0 & 1 & & & \\ & & \ddots & & \\ & & & \ddots & 1 \\ -q_0 & \cdots & \cdots & & -q_{n-1} \end{pmatrix},
$$

$$
B = \begin{pmatrix} 0 \\ \vdots \\ 0 \\ 1 \end{pmatrix}, \qquad C = \begin{pmatrix} p_0 & \cdots & \cdots & p_{n-1} \end{pmatrix}.
$$

2. **Controllability realization:**

$$
A = \begin{pmatrix} 0 & \cdots & \cdots & -q_0 \\ 1 & & & \vdots \\ & \ddots & & \vdots \\ & & 1 & -q_{n-1} \end{pmatrix},
$$

$$
B = \begin{pmatrix} 1 \\ 0 \\ \vdots \\ \\ 0 \end{pmatrix}, \qquad C = \begin{pmatrix} g_1 & \cdots & \cdots & g_n \end{pmatrix}.
$$

3. **Observability realization:**

$$
A = \begin{pmatrix} 0 & 1 & & & \\ & & \ddots & & \\ & & & \ddots & 1 \\ -q_0 & \cdots & \cdots & & -q_{n-1} \end{pmatrix},
$$

$$
B = \begin{pmatrix} g_1 \\ \vdots \\ \\ \vdots \\ g_n \end{pmatrix}, \qquad C = \begin{pmatrix} 1 & 0 & \cdots & 0 \end{pmatrix}.
$$

4. **Observer realization:**

$$A \ = \ \begin{pmatrix} 0 & \cdot & \cdot & \cdot & -q_0 \\ 1 & & & & \cdot \\ & \cdot & & & \cdot \\ & & \cdot & & \cdot \\ & & & 1 & -q_{n-1} \end{pmatrix},$$

$$B \ = \ \begin{pmatrix} p_0 \\ \cdot \\ \cdot \\ \cdot \\ p_{n-1} \end{pmatrix}, \qquad C = \begin{pmatrix} 0 & \cdot & \cdot & 0 & 1 \end{pmatrix}.$$

5. If p and q are coprime, then the previous four realizations are all canonical and hence isomorphic. The isomorphisms are given by means of the following diagram:

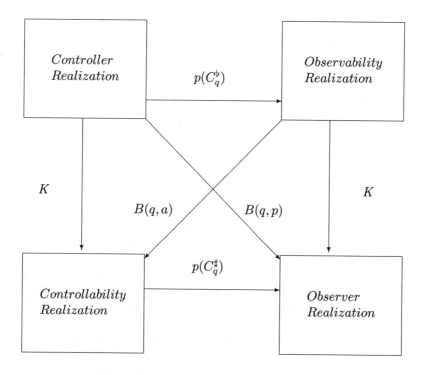

Here, $B(q,p)$ and $B(q,a)$ are Bezoutians with the polynomial a arising from the solution to the Bezout equation $ap + bq = 1$. K is the Hankel matrix

$$K = [I]_{co}^{st} = \begin{pmatrix} q_1 & \cdot & \cdot & \cdot & q_{n-1} & 1 \\ \cdot & & & \cdot & 1 & \\ \cdot & & \cdot & & & \\ \cdot & & & & & \\ q_{n-1} & 1 & & & & \\ 1 & & & & & \end{pmatrix}.$$

Proof:

1. We use the representation $g = pq^{-1}$ and take the matrix representation of the shift realization (10.23) with respect to the control basis $\{e_1, \ldots, e_n\}$. In fact, we have $A = [S_q]_{co}^{co} = C_q^{\flat}$, and, as $e_n(z) = 1$, B also has the required form. Finally, $Cf = < p, f > = [\widetilde{p}]^{st}[f]^{co}$ and the obvious fact that $[\widetilde{p}]^{st} = (\begin{array}{cccc} p_0 & \cdot & \cdot & \cdot & p_{n-1} \end{array})$.

2. We use the representation $g = pq^{-1}$ and take the matrix representation of the shift realization (10.23) with respect to the standard basis $\{1, z, \ldots, z^{n-1}\}$. In this case, $A = [S_q]_{st}^{st} = C_q^{\sharp}$. The form of B is immediate. To conclude, we use Eq. (8.51), that is, $p(z) = \sum_{i=1}^{n} g_i e_i(z)$, to get $[\widetilde{p}]^{co} = (\begin{array}{cccc} g_1 & \cdot & \cdot & \cdot & g_n \end{array})$.

3. We use the representation $g = q^{-1}p$ and take the matrix representation of the shift realization (10.21) with respect to the control basis. Here again, $A = [S_q]_{co}^{co} = C_q^{\flat}$ and B again is obtained by applying Eq. (8.51). The matrix C is obtained as $[\widetilde{1}]^{st}$.

4. We use the representation $g = q^{-1}p$ and take the matrix representation of the shift realization (10.21) with respect to the standard basis. The derivation follows as before.

5. We note that the controller and controllability realizations are different matrix representations of realization (10.21) taken with respect to the control and standard bases. Thus, the isomorphism is given by $K = [I]_{co}^{st}$, which has the required Hankel form. A similar observation holds for the isomorphism of the observability and observer realizations.

   On the other hand, the controller and observability realizations are matrix representations of realizations (10.21) and (10.23), both taken with respect to the control basis. By Theorem 10.3.3, under the assumption of coprimeness for p and q, these realizations are isomorphic, and the isomorphism is given by the map $p(S_q)$. Taking the matrix representation of this map with respect to the control basis, we get $[p(S_q)]_{co}^{co} = p([S_q]_{co}^{co}) = p(C_q^{\flat})$. A similar derivation holds for the isomorphism of the controllability and observer realizations.

   The controller and observer realizations are matrix representations of (10.21) and (10.23) taken with respect to the control and standard bases, respectively. Hence, based on the diagram in Theorem 10.3.3 and using Theorem 8.4.2, the isomorphism is given by $[p(S_q)]^{st}_{co} = B(q, p)$.

Finally, since by Theorem 5.1.3 the inverse of $p(S_q)$ is $a(S_q)$, the polynomial a coming from a solutionto the Bezout equation, the same reasoning as before shows that the isomorphism between the observability and controllability realizations is given by $[a(S_q)]^{st}_{co} = B(q, a)$.

□

Note that the controller and observer realizations are dual to each other, and the same is true for the controllability and observability realizations.

The continued fraction representation (8.55) of a rational function can be translated immediately into a canonical form realization that incorporates the atoms.

Theorem 10.3.5 *Let the strictly proper transfer function $g = p/q$ have the sequence of atoms $\{\beta_i, a_{i+1}(z)\}$, and assume that $\alpha_1, \ldots, \alpha_r$ are the degrees of the atoms and*

$$a_k(z) = z^{\alpha_k} + \sum_{i=0}^{\alpha_k - 1} a_i^{(k)} z^i. \tag{10.27}$$

Then g has a realization of the form (A, b, c), of the form

$$A = \begin{pmatrix} A_{11} & A_{12} & & & \\ A_{21} & A_{22} & \cdot & & \\ & \cdot & \cdot & \cdot & \\ & & \cdot & \cdot & A_{r-1\,r} \\ & & & A_{r\,r-1} & A_{rr} \end{pmatrix}, \tag{10.28}$$

where

$$A_{ii} = \begin{pmatrix} 0 & \cdot & \cdot & \cdot & -a_0^{(i)} \\ 1 & & & & \cdot \\ & \cdot & & & \cdot \\ & & \cdot & & \cdot \\ & & & 1 & -a_{\alpha_i-1}^{(i)} \end{pmatrix}, \quad i = 1, \ldots r, \tag{10.29}$$

$$A_{i+1\,i} = \begin{pmatrix} 0 & \cdot & \cdot & 0 & 1 \\ & \cdot & & & 0 \\ \cdot & & & & \cdot \\ \cdot & & & & \cdot \\ 0 & \cdot & \cdot & \cdot & 0 \end{pmatrix}, \tag{10.30}$$

$A_{i\,i+1} = \beta_{i-1}A_{i+1\,i},$

$$b = \begin{pmatrix} 1 \\ 0 \\ \cdot \\ \cdot \\ \cdot \\ 0 \end{pmatrix}, \qquad c = \begin{pmatrix} 0 & \cdot & \cdot & \beta_0 & 0 & \cdot & \cdot & 0 \end{pmatrix}, \qquad (10.31)$$

the β_0 being in the α_1 position.

Proof: We use the shift realization (10.23) of g, which, by the coprimeness of p and q, is canonical. Taking its matrix representation with respect to the orthogonal basis $B_{or} = \{1, z, \ldots, z^{n_1-1}, Q_1, zQ_1, \ldots, z^{n_2-1}Q_1, \ldots, Q_{r-1}, \ldots, z^{n_{r-1}-1}Q_{r-1}\}$, with Q_i the Lanczos polynomials of the first kind, proves the theorem. In the computation of the matrix representation, we lean heavily on the recursion formula (8.58) for the Lanczos polynomials.
□

10.4 Stabilization

The purpose of modeling, and realization theory is part of that process, is to gain a better understanding of the system in order to improve its performance by controlling it.

The most fundamental control problem is that of stabilization. We say that a system is **stable** if, given any initial condition, the system tends to its rest point with increasing time. If the system is given by a difference equation $x_{n+1} = Ax_n$, the stability condition is that all eigenvalues of A lie in the open unit disk. If we have a continuous time system given by $\dot{x} = Ax$, the stability condition is that all eigenvalues of A lie in the open left half-plane.

Since all of the information needed for the computation of the future behavior of the system is given in the system equations and the present state, that is, the initial condition, of the system, we expect that a control law should be a function of the state, that is, a map, linear in the context in which we work, $F : X \longrightarrow U$. In the case of a system given by Eqs. (10.2), we have

$$\begin{cases} x_{n+1} &= Ax_n + Bu_n \\ u_n &= Fx_n + v_n, \end{cases} \qquad (10.32)$$

and the closed loop equation becomes

$$x_{n+1} = (A + BF)x_n + Bv_n.$$

In terms of flow diagrams, we have the following feedback configuration:

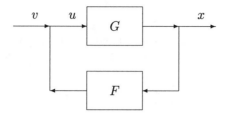

There are two extreme idealizations inherent in the assumptions of this scheme. The first is the assumption that we have a precise model of our system, which is hardly ever the case. The other idealization is the assumption that we have access to full state information. This difficulty is more easily overcome, via the utilization of observers, that is, auxiliary systems that reconstruct, asymptotically, the state, or, more generally, by the use of dynamic controllers. The first problem is more serious, and one way to cope with it is to develop the theory of robust control. This, however, is outside the scope of this book.

It is still instructive to study the original problem of stabilizing a given linear system with full state information. The following theorem on pole shifting states that, using state feedback, we have full freedom in assigning the characteristic polynomial of the closed-loop system, provided the pair (A, B) is reachable. Here we handle only the single-input case.

Theorem 10.4.1 *Let the pair (A, b) be reachable with A an $n \times n$ matrix and b an n vector. Then, for each monic polynomial s of degree n, there exists a control law $k : F^n \longrightarrow F$ for which the characteristic polynomial of $A - bk$ is s.*

Proof: The characteristic polynomial is independent of the basis chosen. We choose to work in the control basis corresponding to $q(z) = det(zI - A) = z^n + q_{n-1}z^{n-1} + \cdots + q_0$. With respect to this basis, the pair (A, b) has the matrix representation

$$
A = \begin{pmatrix} 0 & 1 & & & \\ & & \cdot & & \cdot \\ & & & \cdot & \\ & & & & 1 \\ -q_0 & \cdot & \cdot & \cdot & -q_{n-1} \end{pmatrix}, \qquad b = \begin{pmatrix} 0 \\ \cdot \\ \cdot \\ \cdot \\ 1 \end{pmatrix}.
$$

Let $s(z) = z^n + s_{n-1}z^{n-1} + \cdots + s_0$. Then $k = \begin{pmatrix} q_0 - s_0 & \cdots & q_{n-1} - s_{n-1} \end{pmatrix}$ is the required control law. This theorem explains the reference to the control basis. $\qquad \square$

We now pass on to the analysis of stabilization when we have no direct access to full state information. We assume that the system is given by the equations

$$\begin{cases} x_{n+1} &=& Ax_n + Bu_n \\ y_n &=& Cx_n. \end{cases} \tag{10.33}$$

We construct an auxiliary system, called an **observer**, that utilizes all of the available information, namely, the control signal and the observation signal:

$$\begin{cases} \xi_{n+1} &=& A\xi_n + Bu_n - H(y_n - C\xi_n) \\ \eta_n &=& \xi_n. \end{cases} \tag{10.34}$$

We define the error signal by

$$e_n = x_n - \xi_n.$$

The object of the construction of an observer is to choose the matrix H so that the error e_n tends to zero. This means that with increasing time the output of the observer gives an approximation of the current state of the system. Subtracting the two system equations, we get

$$x_{n+1} - \xi_{n+1} = A(x_n - \xi_n) + H(Cx_n - C\xi_n)$$

or

$$e_{n+1} = (A + HC)e_n.$$

Thus, we see that the observer construction problem is exactly dual to that of stabilization by state feedback.

Theorem 10.4.2 *An observable, linear, single-output system has an observer or, equivalently, an asymptotic state estimator.*

Proof: We apply Theorem 10.4.1. This shows that the characteristic polynomial of $A + HC$ can be chosen arbitrarily. $\square$

Of course, state estimation and the state feedback law can be combined into a single framework. Thus we use the observer to get an asymptotic approximation for the state, and we feed this back into the system as if it were the true state. The next theorem shows that, as far as stabilization is concerned, this is sufficient.

Theorem 10.4.3 *Given a canonical single-input/single-output system*

$$\left(\begin{array}{c|c} A & B \\ \hline C & 0 \end{array} \right).$$

Choose linear maps H, F so that $A + BF$ and $A + HC$ are both stable matrices. Then:

  1. The closed loop system given by

$$\begin{cases} x_{n+1} &=& Ax_n + Bu_n \\ y_n &=& Cx_n \\ \xi_{n+1} &=& A\xi_n + Bu_n - H(y_n - C\xi_n) \\ u_n &=& -F\xi_n + v_n \end{cases} \tag{10.35}$$

  is stable.

2. *The transfer function of the closed-loop system, that is, the transfer function from v to y, is $G_c(z) = C(zI - A + BF)^{-1}B$.*

Proof:

1. We can rewrite the closed-loop system equations as

$$\begin{cases} \begin{pmatrix} x_{n+1} \\ \xi_{n+1} \end{pmatrix} = \begin{pmatrix} A & -BF \\ HC & A - HC - BF \end{pmatrix} \begin{pmatrix} x_n \\ \xi_n \end{pmatrix} \\ \qquad y_n = \begin{pmatrix} C & 0 \end{pmatrix} \begin{pmatrix} x_n \\ \xi_n \end{pmatrix}, \end{cases} \qquad (10.36)$$

and it remains to compute the characteristic polynomial of the new state matrix. We use the similarity

$$\begin{pmatrix} I & 0 \\ -I & I \end{pmatrix} \begin{pmatrix} A & -BF \\ HC & A - HC - BF \end{pmatrix} \begin{pmatrix} I & 0 \\ I & I \end{pmatrix}$$

$$= \begin{pmatrix} A - BF & -BF \\ 0 & A - HC \end{pmatrix}.$$

Therefore, the characteristic polynomial of

$$\begin{pmatrix} A & -BF \\ HC & A - HC - BF \end{pmatrix}$$

is the product of the characteristic polynomials of $A - BF$ and $A - HC$. This also shows that the controller and observer for the original system can be designed independently.

2. We compute the transfer function of the closed-loop system, which is given by

$$G_c(z) = \left(\begin{array}{cc|c} A & -BF & B \\ HC & A - HC - BF & B \\ \hline C & 0 & 0 \end{array} \right).$$

Utilizing the previous similarity, we get

$$G_c(z) = \left(\begin{array}{cc|c} A - BF & -BF & B \\ 0 & A - HC & 0 \\ \hline C & 0 & 0 \end{array} \right) = C(zI - A + BF)^{-1}B.$$

Note that the closed-loop transfer function does not depend on the specifics of the observer. This is the case as the transfer function desribes the steady state of the system. The observer does effect the transients in the system. $\qquad \square$

Let us review the previous observer–controller construction from a transfer function point of view. Let $G(z) = C(zI - A)^{-1}B$ have the polynomial coprime factorization p/q. Going back to the closed-loop system equations, we can write $\xi_{n+1} = (A - HC)\xi_n + Hy_n + Bu_n$. Hence the transfer function from (u, y) to the control signal is given by $(F(zI - A + HC)^{-1}B \quad F(zI - A + HC)^{-1}H)$, and the overall transfer function is determined from

$$\begin{cases} y &= Gu \\ u &= v - H_u u - H_y y, \end{cases} \tag{10.37}$$

where

$$\begin{cases} H_u &= F(zI - A + HC)^{-1}B \\ H_y &= F(zI - A + HC)^{-1}H. \end{cases}$$

The flow chart of this configuration is given by:

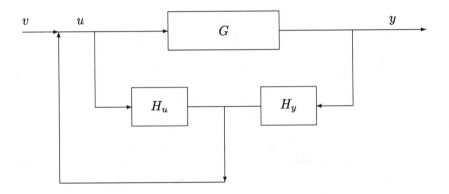

Now note that, given that we chose H to stabilize $A - HC$, the two transfer functions H_u, H_y have the following representations:

$$H_u = \frac{p_u}{q_c}, \qquad H_y = \frac{p_y}{q_c},$$

where $q_c(z) = \det(zI - A + HC)$ is a stable polynomial with $\deg q_c = \deg q$. From Eq. (10.37) we get $(I + H_u)u = v - H_y y$ or $u = (I + H_u)^{-1}v - (I + H_u)^{-1}H_y y$. This implies that $y = Gu = G(I + H_u)^{-1}v - G(I + H_u)^{-1}H_y y$. So the closed-loop transfer function is given by

$$G_c = (I + G(I + H_u)^{-1}H_y)^{-1}G(I + H_u)^{-1}.$$

Since we are dealing here with single-input/single-output systems, all transfer functions are scalar-valued and have polynomial coprime factorizations. When these are substituted into the previous expression, we obtain

$$G_c = \cfrac{1}{1 + \cfrac{p}{q} \cdot \cfrac{1}{1 + \cfrac{p_u}{q_c}} \cdot \cfrac{p_y}{q_c}} \cdot \frac{p}{q} \cdot \cfrac{1}{1 + \cfrac{p_u}{q_c}}$$

$$= \cfrac{1}{1 + \cfrac{p}{q} \cdot \cfrac{p_y}{q_c + p_u}} \cdot \frac{p_y}{q_c} \cdot \frac{p}{q} \cdot \frac{q_c}{q_c + p_u} = \frac{pq_c}{q(q_c + p_u) + pp_y}.$$

The denominator polynomial has degree $2 \deg q$ and has to be chosen to be stable. So let t be an arbitrary stable polynomial of degree $2 \deg q$. Since p and q are coprime, the equation $ap + bq = t$ is solvable, with an infinite number of solutions. We can specify a unique solution by imposing the extra requirement that $\deg a < \deg q$. Since $\deg ap < 2 \deg q - 1$, we must have $\deg(bq) = \deg t$. So $\deg b = \deg q$, and we can write $b = q_c + p_u$ and $a = p_y$. Moreover, p_u, p_y have degrees smaller than $\deg q$.

A very special case is obtained if we choose a stable polynomial s of degree equal to the degree of q, and let $t = s \cdot q_c$. The closed-loop transfer function in this case reduces to p/s, which is the case in the observer–controller configuration.

We now go back to the equation

$$q(q_c + p_u) + pp_y = sq_c.$$

and divide by the right-hand side to obtain

$$\frac{q}{s} \cdot \frac{q_c + p_u}{q_c} + \frac{p}{s} \cdot \frac{p_y}{q_c} = 1. \tag{10.38}$$

Now all four rational functions appearing in this equation are proper and stable, that is, they belong to $\mathbf{RH}_+^\infty$. Moreover, Eq. (10.38) is just a Bezout identity in $\mathbf{RH}_+^\infty$, and $g = (p/s)/(q/s)$ is a coprime factorization of G over the ring $\mathbf{RH}_+^\infty$.

This analysis is important as we now can reverse our reasoning. We see that for stabilization it was sufficient to solve a Bezout equation over $\mathbf{RH}_+^\infty$, which is based on a coprime factorization of G over the same ring. We can take this as a starting point of a more general approach.

10.5 The Youla–Kucera Parametrization

We now go back to the stabilization problem. We would like to design a compensator, that is, an auxiliary system, that takes as its input the output of the original system, and its output is taken as an additional control signal in such a way that the closed-loop system is stable. In terms of flowcharts, we are back to the following configuration:

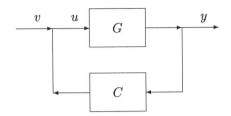

with the sole difference that we no longer assume the state to be accessible.
It is a simple exercise to check that the closed-loop transfer function is
given by $G_c = (I - GC)^{-1}G = G(I - CG)^{-1}$.

Now a natural condition for stabilization would be that G_c should be
stable. Indeed, from an external point of view, this seems to be the right
condition. However, this condition does not guarrantee that internal signals
will stay bounded. To see what the issue is, we note that there are two
natural entry point for disturbances, one is the observation errors and the
other is that of errors in the control signals. Thus, we are lead to the
following standard control configuration:

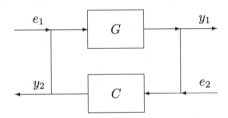

and the internal stability requirement is that the four transfer functions
from the error variables u_1, u_2 to the internal variables e_1, e_2 should be
stable. Since the system equations are

$$\begin{cases} e_2 &= u_2 + Ge_1 \\ e_1 &= u_1 + Ce_2, \end{cases}$$

we get

$$\begin{pmatrix} e_1 \\ e_2 \end{pmatrix} = \begin{pmatrix} (I - CG)^{-1} & (I - CG)^{-1}C \\ (I - GC)^{-1}G & (I - GC)^{-1} \end{pmatrix}.$$

We denote the 2×2 block transfer function by $H(G, C)$. This leads to the
following.

Definition 10.5.1 *Given a linear system with transfer function G and
a compensator C, we say that the pair (G, C) is* **internally stable** *if
$H(G, C) \in \mathbf{RH}^\infty_+$.*

Thus, internal stability is a stronger requirement than just closed-loop
stability, which only requires $(I - GC)^{-1}G$ to be stable.

The next result gives a criterion for internal stability via a coprimeness condition over the ring $\mathbf{RH}_+^\infty$. We use the fact that any rational transfer function has a coprime factorization over the ring $\mathbf{RH}_+^\infty$.

Proposition 10.5.1 *Given the rational transfer function G and the compensator C, let $G = N/M$ and $C = U/V$ be coprime factorizations over* $\mathbf{RH}_+^\infty$. *Then (G, C) is internally stable if and only if $\Delta = MV - NU$ is invertible in* $\mathbf{RH}_+^\infty$.

Proof: Assume that $\Delta^{-1} \in \mathbf{RH}_+^\infty$. Then, clearly,

$$H(G, C) = \Delta^{-1} \begin{pmatrix} MV & MU \\ NV & MV \end{pmatrix} \in \mathbf{RH}_+^\infty.$$

Conversely, assume that $H(G, C) \in \mathbf{RH}_+^\infty$. Since

$$(I - GC)^{-1}GC = (I - GC)^{-1}GC(I - (I - GC)) \in \mathbf{RH}_+^\infty,$$

it follows that $\Delta^{-1}NU \in \mathbf{RH}_+^\infty$. This implies that

$$\begin{pmatrix} M \\ N \end{pmatrix} \Delta^{-1} \begin{pmatrix} V & U \end{pmatrix} \in \mathbf{RH}_+^\infty.$$

Since coprimeness in $\mathbf{RH}_+^\infty$ is equivalent to the Bezout identities, it follows that $\Delta^{-1} \in \mathbf{RH}_+^\infty$. $\qquad\square$

Corollary 10.5.1 *Let $G = N/M$ be a coprime factorization over* $\mathbf{RH}_+^\infty$. *Then there exists a stabilizing compensator C if and only if $C = U/V$ and $MV - NU = I$.*

Proof: If there exist U, V such that $MV - NU = I$, then $C = U/V$ is easily checked to be a stabilizing controller.

Conversely, if $C = U_1/V_1$ is a stabilizing controller, then, by Proposition 10.5.1, $\Delta = MV_1 - NU_1$ is invertible in $\mathbf{RH}_+^\infty$. Defining $U = U_1\Delta^{-1}, V = V_1\Delta^{-1}$. Thus $C = U/V$ and $MV - NU = I$. $\qquad\square$

The following result, which goes by the name of the **Youla–Kucera parametrization**, is the cornerstone of modern stabilization theory.

Theorem 10.5.1 *Let $G = N/M$ be a coprime factorization. Let U, V be any solution of the Bezout equation $MV - NU = I$. Then the set of all stabilizing controllers is given by*

$$\left\{ C = \frac{U + MQ}{V + NQ} \,\middle|\, Q \in \mathbf{RH}_+^\infty, V + NQ \neq 0 \right\}.$$

Proof: Suppose that $C = (U+MQ)/(V+NQ)$ with $Q \in \mathbf{RH}_+^\infty$. We check that $M(V+NQ) - N(U+MQ) = MV - NU = I$. Thus, by Corollary 10.5.1, C is a stabilizing compensator.

Conversely, assume that C stabilizes G. We know, by Corollary 10.5.1 that $C = U_1/V_1$ with $MV_1 - NU_1 = I$. By subtraction we get $M(V_1 - V) = N(U_1 - U)$. Since M, N are coprime, M divides $U_1 - U$, and so there exists a $Q \in \mathbf{RH}_+^\infty$ such that $U_1 = U + MQ$. This immediately implies that $V_1 = V + NQ$. □

10.6 Exercises

1. (The Hautus test). Consider the pair (A, b) with A a real $n \times n$ matrix and b an n vector.

   (a) Show that (A, b) is reachable if and only if rank $\begin{pmatrix} zI - A & b \end{pmatrix} = n$ for all $z \in \mathbf{C}$.

   (b) Show that (A, b) is stabilizable by state feedback if and only if rank $\begin{pmatrix} zI - A & b \end{pmatrix} = n$ for all z in the closed right half-plane.

2. Let $\phi_n(z), \psi_n(z)$ be the monic orthogonal polynomials of the first and second kind, respectively, associated with the positive sequence $\{c_n\}$, which were defined in exercise 8.6b. Show that

$$\frac{1}{2} \frac{\psi_n}{\phi_n} = \left(\begin{array}{c|c} A_n & B_n \\ \hline C_n & D_n \end{array} \right),$$

where the matrices A_n, B_n, C_n, D_n are given by

$$A_n = \begin{pmatrix} \gamma_1 & \gamma_2(1-\gamma_1^2) & \gamma_3\prod_{i=1}^{2}(1-\gamma_i^2) & \cdot & \cdot & \gamma_n\prod_{i=1}^{n-1}(1-\gamma_i^2) \\ 1 & -\gamma_2\gamma_1 & \gamma_2(1-\gamma_1^2) & \cdot & \cdot & -\gamma_n\gamma_1\prod_{i=2}^{n-1}(1-\gamma_i^2) \\ & 1 & -\gamma_3\gamma_2 & & \cdot & \\ & & \cdot & \cdot & & \\ & & & \cdot & 1 & -\gamma_n\gamma_{n-2}(1-\gamma_{n-1}^2) \\ & & & & 1 & -\gamma_n\gamma_{n-1} \end{pmatrix}$$

$$B_n = \begin{pmatrix} 1 \\ 0 \\ \cdot \\ \cdot \\ \cdot \\ 0 \end{pmatrix},$$

$$C_n = \begin{pmatrix} \gamma_1 & \gamma_2(1-\gamma_1^2) & \cdots & \gamma_n\prod_{i=1}^{n-1}(1-\gamma_i^2) \end{pmatrix}$$

$$D_n = \frac{1}{2}.$$

3. Show that $\det A_n = (-1)^{n-1}\gamma_n$.

4. Show that the Toeplitz matrix can be diagonalized via the following congruence transformation:

$$
\begin{pmatrix}
1 & & & & \\
\phi_{1,0} & 1 & & & \\
& & \ddots & & \\
\phi_{n,0} & & & \phi_{n,n-1} & 1
\end{pmatrix}
\begin{pmatrix}
c_0 & \cdot & \cdot & \cdot & \cdot & c_n \\
\cdot & \cdot & \cdot & \cdot & \cdot & \cdot \\
\cdot & \cdot & \cdot & \cdot & \cdot & \cdot \\
\cdot & \cdot & \cdot & \cdot & \cdot & \cdot \\
\cdot & \cdot & \cdot & \cdot & \cdot & \cdot \\
c_n & \cdot & \cdot & \cdot & \cdot & c_0
\end{pmatrix}
$$

$$
\times
\begin{pmatrix}
1 & \phi_{1,0} & & \phi_{n,0} \\
& 1 & & \phi_{n,1} \\
& & \ddots & \\
& & & \phi_{n,n-1} \\
& & & 1
\end{pmatrix}
$$

$$
= c_0
\begin{pmatrix}
h_0 & & & \\
& h_1 & & \\
& & \ddots & \\
& & & \ddots \\
& & & h_n
\end{pmatrix}
$$

$$
= c_0
\begin{pmatrix}
1 & & & \\
& (1-\gamma_1^2) & & \\
& & \ddots & \\
& & & (1-\gamma_{n-1}^2)\cdots(1-\gamma_1^2)
\end{pmatrix}.
$$

5. Show that

$$\det T_n = c_0(1-\gamma_1^2)^{n-1}(1-\gamma_2^2)^{n-2}\cdots(1-\gamma_{n-1}^2). \qquad (10.39)$$

6. Define a sequence of kernel functions by

$$K_n(z,w) = \sum_{j=0}^{n} \frac{\phi_j(z)\overline{\phi_j(w)}}{h_j}.$$

Show that the kernels $K_n(z,w)$ have the following properties.

(a) We have

$$K_n(z,w) = \overline{K_n(w,z)}.$$

(b) The matrix $K_n = (k_{ij})$, defined through $K_n(z, w) = \sum_{i=0}^{n} \sum_{j=0}^{n} K_{ij} z^i \overline{w}^j$, is Hermitian, that is, $k_{ij} = \overline{k_{ji}}$.

(c) The kernels K_n are reproducing kernels, that is, for each $f \in X_{\phi_{n+1}}$ we have $< f, K(\cdot, w) >_C = f(w)$.

(d) The reproducing kernels have the following determinantal representation:

$$K_n(z, w) = -\frac{1}{\det T_n} \begin{vmatrix} c_0 & . & . & . & c_n & 1 \\ & & . & . & . & \overline{w} \\ & & . & . & . & . \\ & & . & . & . & . \\ & & . & . & . & . \\ c_n & . & . & . & c_0 & \overline{w}^n \\ 1 & z & . & . & z^n & 0 \end{vmatrix}.$$

(e) The following representation for the reproducing kernels holds:

$$K_n(z, w) = \sum_{j=0}^{n} \frac{\phi_j(z)\overline{\phi_j(w)}}{h_j} = \frac{\phi_n^\sharp(z)\overline{\phi_n^\sharp(w)} - z\overline{w}\phi_n(z)\overline{\phi_n(w)}}{h_n(1 - z\overline{w})}.$$

This representation of the reproducing kernels $K_n(z, w)$ is called the **Christoffel–Darboux formula**.

7. Show that all of the zeros of the orthogonal polynomials ϕ_n are located in the open unit disc.

8. Prove the following Gohberg–Semencul formula:

$$K_n = \frac{1}{h_n}[\phi_n^\sharp(S)\phi_n^\sharp(\tilde{S}) - S\phi_n(\tilde{S})\phi_n(S)\tilde{S}]$$

$$= \left[\begin{pmatrix} 1 & & & \\ \phi_{n,n-1} & . & & \\ . & . & . & \\ . & . & . & . \\ \phi_{n,1} & . & . & \phi_{n,n-1} & 1 \end{pmatrix} \begin{pmatrix} 1 & \phi_{n,n-1} & . & . & \phi_{n,1} \\ & . & & . & . \\ & & . & . & . \\ & & & . & \phi_{n,n-1} \\ & & & & 1 \end{pmatrix} \right.$$

$$\left. - \begin{pmatrix} 0 & & & \\ \phi_{n,0} & . & & \\ . & . & . & \\ . & . & . & . \\ \phi_{n,n-1} & . & . & \phi_{n,0} & 0 \end{pmatrix} \begin{pmatrix} 0 & \phi_{n,0} & . & . & \phi_{n,n-1} \\ & . & & . & . \\ & & . & . & . \\ & & & . & \phi_{n,0} \\ & & & & 0 \end{pmatrix} \right].$$

9. With

$$K_n(z, w) = \sum_{j=0}^{n} \frac{\phi_j(z)\overline{\phi_j(w)}}{h_j} = \sum_{i=0}^{n} \sum_{j=0}^{n} K_{ij} z^i \overline{w}^j,$$

show that $K_n = T_n^{-1}$.

10.7   Notes and Remarks

The study of systems, linear and nonlinear, is the work of many people; however, the contributions of Kalman to this topic were and still remain central. His insight into the connection of an applied engineering subject to abstract module theory is of fundamental importance. This insight led to the wide usage of polynomial techniques in linear system theory. An early exposition of this is given in Kalman, Falb, and Arbib [1969]. Another approach to systems problems, using polynomial techniques, was developed by Rosenbrock [1970], who also proved the ultimate result on pole placement by state feedback. The connection between the abstract module approach and Rosenbrock's method of polynomial system matrices was clarified in Fuhrmann [1976].

11

Hankel Norm Approximation

11.1 Introduction

We pick as our theme for the last chapter in this book the theory of Hankel norm approximation problems. This theory has come to be known generally as the AAK theory, in honor of the pioneering and very influential contribution made by Adamyan, Arov, and Krein [1968a, 1968b, 1971, 1978].

The reason for the choice of this topic is twofold. First and foremost, it is an extremely interesting and rich circle of ideas, and it allows us the consideration of several distinct problems within a unified framework. The second reason is just as important. We shall use the problems under discussion as a vehicle for the development of many results that are counterparts in this new setting of results obtained previously in polynomial terms. This development can be considered as the construction of a bridge that connects algebra and analysis.

The natural setting for the development of the AAK theory is that of Hardy spaces, whose study is part of complex and functional analysis. However, to keep our exposition algebraic, we shall assume the rationality of functions and the finite dimensionality of model spaces. Moreover, to stay in line with the exposition in previous chapters, we shall treat only the case of scalar-valued transfer functions. There is another choice we are making, and that is to study Hankel operators on Hardy spaces corresponding to the open left and right half-planes, rather than the spaces corresponding to the unit disk and its exterior. Thus in effect we are studying continuous-time systems, although we shall not delve into this in great detail. The advantage of this choice is greater symmetry. The price we pay for this is that, to a certain extent, shift operators have to be replaced by translation semigroups; but even this can be corrected. Once we get the identification of the Hankel operator ranges with rational models, shift operators reenter the picture.

Of course, with the choices we have made, the results are not the most general. However, we gain the advantage of simplicity. As a result, this material can be explained to the undergraduate without any reference to the more advanced areas of mathematics, such as functional analysis and operator theory. In fact, the material presented in this chapter covers, albeit in simplified form, many of the most important results in modern operator theory and system theory. Thus it also can be taken as an introduction to these topics.

The chapter is structured as follows. In Section 11.2 we collect basic information on Hankel operators, invariant subspaces, and their representation via Beurling's theorem. Next we introduce model intertwining operators. We do this using the frequency domain representation of the right translation semigroup. We study the basic properties of intertwining maps and in particular their invertibility properties. The important point here is the connection of invertibility to the solvability of an H_+^∞ Bezout equation. We follow this by defining Hankel operators. For the case of a rational, antistable function, we give specific Beurling-type representations for the cokernel and the image of the corresponding Hankel operator. Of importance is the connection between the Hankel operators and the intertwining maps. This connection, coupled with the invertibility properties of intertwining maps, are the key to duality theory.

In Section 11.3 we do a detailed analysis of Schmidt pairs of a Hankel operator with a scalar, rational symbol. Some important lemmas are derived in our setting from an algebraic point of view. These lemmas lead to a polynomial formulation of the singular-value/singular-vector equation of the Hankel operator. This equation, which we refer to as the Fundamental Polynomial Equation (FPE), is easily reduced, using the theory of polynomial models, to a standard eigenvalue problem. Using nothing more than the polynomial division algorithm, the subspace of all singular vectors corresponding to a given singular value is parametrized via the minimal degree solution of the FPE. We obtain a connection between the minimal degree solution and the multiplicity of the singular value.

The FPE can be transformed using a simple algebraic manipulation to a form that leads immediately to lower bound estimates on the the number of antistable zeroes of p_k, the minimal degree solution corresponding to the kth Hankel singular value. This lower bound is shown to actually coincide with the degree of the minimal degree solution for the special case of the smallest singular value. Thus this polynomial turns out to be antistable. Another algebraic manipulation of the FPE leads to a Bezout equation over H_+^∞. This provides the key to duality.

Section 11.4 has duality theory as its main theme. Using the previously obtained Bezout equation, we invert the intertwining map corresponding to the initial Hankel operator. The inverse intertwining map is related to a new Hankel operator that has inverse singular values to those of the original one. Moreover, we can compute the Schmidt pairs corresponding to this Hankel operator in terms of the original Schmidt pairs. The estimates on the number of antistable zeroes of the minimum degree solutions of the FPE that were obtained for the original Hankel operator Schmidt vectors now are applied to the new Hankel operator Schmidt vectors. Thus, we obtain a second set of inequalities. The two sets of inequalities, taken together, lead to precise information on the number of antistable zeroes of the minimal degree solutions corresponding to all singular values. Utilizing this information leads to the solution of the Nehari problem as well as that of the general Hankel norm approximation problem.

Section 11.5 is a brief introduction to the Nevanlinna–Pick interpolation, that is, interpolation by functions that minimize the H^∞-norm.

Finally, in Section 11.6, we study the singular values and singular vectors of both the Nehari complement and the best Hankel norm approximant. In both cases, the singular values are a subset of the set of singular values of the original Hankel operator, and the corresponding singular vectors are obtained by orthogonal projections on the respective model spaces.

11.2   Preliminaries

Analytic Hankel operators are generally defined in the time domain, and via the Fourier transform their frequency domain representation is obtained. We will skip this part and introduce Hankel operators directly as frequency domain objects. Our choice is to develop the theory of continuous-time systems. This means that the relevant frequency domain spaces are the Hardy spaces of the left and right half-planes. Thus we will study Hankel operators defined on half-plane Hardy spaces rather than on those of the unit disk as was done by Adamjan, Arov, and Krein [1971]. In this we follow the choice of Glover [1984]. This choice seems to be a very convenient one as all results on duality simplify significantly, due to the greater symmetry between the two half-planes in comparison to the unit disk and its exterior.

Rational Hardy Spaces

Our principal interest in the following will be centered around the study of Hankel operators with rational symbols. In the algebraic approach taken in Chapter 8, the Hankel operators were defined using the direct sum decomposition $F((z^{-1})) = F[z] \oplus z^{-1}F[[z^{-1}]]$ or, alternatively, $F(z) = F[z] \oplus F_-(z)$. The way to think of this direct sum decomposition is that it is a decomposition based on the location of singularities of rational functions. Functions in $F_-(z)$ have only finite singularities and behave nicely at ∞, whereas polynomials behave nicely at all points but have a singularity at ∞.

We shall now consider analogous spaces. However, as we are passing to an area bordering on analysis, we restrict the field to be the real field. Most of the results go through immediately if the underlying field is taken to be the field of complex numbers.

The two domains in which we shall consider singularities are the open left and right half-planes. This choice is dictated by considerations of symmetry that somewhat simplify the development of the theory. In particular, many results concerning duality are made more easily accessible.

Our setting will be that of Hardy spaces. Thus H_+^2 is the Hilbert space of all analytic functions in the open right half-plane with

$$||f||^2 = \sup_{x>0} \frac{1}{\pi} \int_{-\infty}^{\infty} |f(x+iy)|^2 dy.$$

The space H_-^2 is similarly defined in the open left half-plane. It is a theorem of Fatou (see Hoffman [1962]) that guarantees the existence of boundary values of $H_\pm^2$-functions on the imaginary axis. Thus the spaces $H_\pm^2$ can be considered as closed subspaces of $L^2(i\mathbf{R})$ and the space of Lebesgue square integrable functions on the imaginary axis. It follows from the Fourier–Plancherel and Paley–Wiener theorems that

$$L^2(i\mathbf{R}) = H_+^2 \oplus H_-^2,$$

with H_+^2 and H_-^2 the Fourier–Plancherel transforms of $L^2(0, \infty)$ and $L^2(-\infty, 0)$, respectively. Also, H_+^∞ and H_-^∞ will denote the spaces of bounded analytic functions on the open right and left half-planes, respectively. These spaces can be considered as subspaces of $L^\infty(i\mathbf{R})$, the space of Lebesgue measurable and essentially bounded functions on the imaginary axis. We will define $f^*(z) = \overline{f(-\overline{z})}$.

This describes the general analytic setting. Since we want to keep the exposition as elementary, that is, algebraic, as possible, we shall not work with these spaces but rather with the subsets of these spaces consisting of their rational elements. We use the letter R in front of the various spaces to denote **real rational**. The following definition introduces the spaces of interest.

Definition 11.2.1 We denote by $\mathbf{RL}^\infty, \mathbf{RH}_+^\infty, \mathbf{RH}_-^\infty, \mathbf{RL}^2, \mathbf{RH}_+^2, \mathbf{RH}_-^2$ the sets of all real rational functions that are contained in $L^\infty, H_+^\infty, H_-^\infty, L^2, H_+^2, H_-^2$, respectively.

The following proposition, the proof of which is obvious, gives a characterization of the elements of these spaces.

Proposition 11.2.1 *In terms of polynomials, the spaces of Definition 11.2.1 have the following characterizations:*

$$\mathbf{RL}^\infty = \left\{ \frac{p}{q} \,|\, p, q \in R[z], \deg p \le \deg q, q(\zeta) \ne 0 \text{ for } \zeta \in i\mathbf{R} \right\} \quad (11.1)$$

$$\mathbf{RH}^\infty_+ = \left\{ \frac{p}{q} \in \mathbf{RL}^\infty | q \text{ stable} \right\} \quad (11.2)$$

$$\mathbf{RH}^\infty_- = \left\{ \frac{p}{q} \in \mathbf{RL}^\infty | q \text{ antistable} \right\} \quad (11.3)$$

$$\mathbf{RL}^2 = \left\{ \frac{p}{q} \in \mathbf{RL}^\infty | \deg p < \deg q \right\} \quad (11.4)$$

$$\mathbf{RH}^2_+ = \left\{ \frac{p}{q} \in \mathbf{RL}^2 | q \text{ stable} \right\} \quad (11.5)$$

$$\mathbf{RH}^2_- = \left\{ \frac{p}{q} \in \mathbf{RL}^2 | q \text{ antistable} \right\}. \quad (11.6)$$

We note that all of the spaces that are introduced here are obviously infinite-dimensional.

Proposition 11.2.2 *We have the following additive decompositions:*

$$\mathbf{RL}^\infty = \mathbf{RH}^\infty_+ + \mathbf{RH}^\infty_- \quad (11.7)$$

$$\mathbf{RL}^2 = \mathbf{RH}^2_+ \oplus \mathbf{RH}^2_-. \quad (11.8)$$

Proof: Follows by applying partial fraction decomposition. The sum in Eq. (11.7) is not a direct sum, as the constants appear in both summands. The fact that the direct sum in Eq. (11.8) is orthogonal follows from the most elementary properties of complex contour integration and the partial fraction decomposition. The details are omitted. $\qquad \square$

We shall denote by P_+ and P_- the orthogonal projections of $\mathbf{RL}^2$ on $\mathbf{RH}^2_+$ and $\mathbf{RH}^2_-$, respectively.

Theorem 11.2.1

1. $\mathbf{RL}^\infty$ *is a ring with identity under the usual operation of addition and multiplication of rational functions.*

2. $\mathbf{RL}^2$ *is is a linear space but also a module over* $\mathbf{RL}^\infty$ *as well as over* $\mathbf{RH}^\infty_+$ *and* $\mathbf{RH}^\infty_-$.

3. *With the* $\mathbf{RH}^\infty_+$ *module structure on* $\mathbf{RL}^2$, $\mathbf{RH}^2_+$ *is a submodule.*

4. $\mathbf{RH}_-^2$ *has an* $\mathbf{RH}_+^\infty$ *module structure given by*

$$\psi \cdot h = P_-(\psi h), \qquad h \in \mathbf{RH}_-^2.$$

We wish to point out that, as $z \notin \mathbf{RL}^\infty$, multiplication by the z operator is not defined in these spaces. This forces us to some minor departures from the algebraic versions of some statements.

With all of these spaces at hand we can proceed to introduce several classes of operators.

Definition 11.2.2 *Let* $\psi \in \mathbf{RL}^\infty$. *Then:*

1. *The operator* $L_\psi : \mathbf{RL}^2 \longrightarrow \mathbf{RL}^2$ *defined by*

$$L_\psi f = \psi f$$

is called the **Laurent operator** *with symbol* ψ.

2. *The operator* $T_\psi : \mathbf{RH}_+^2 \longrightarrow \mathbf{RH}_+^2$ *defined by*

$$T_\psi f = P_+ \psi f$$

is called the **Toeplitz operator** *with symbol* ψ.

3. *The operator* $H_\psi : \mathbf{RH}_+^2 \longrightarrow \mathbf{RH}_-^2$ *defined by*

$$H_\psi f = P_- \psi f$$

is called the **Hankel operator** *with symbol* ψ. *Similarly, the operator* $\hat{H}_\psi : \mathbf{RH}_-^2 \longrightarrow \mathbf{RH}_+^2$,

$$\hat{H}_\psi f = P_+ \psi f,$$

is called the **involuted Hankel operator** *with symbol* ψ.

Proposition 11.2.3 *Relative to the inner product in* $\mathbf{RL}^2$ *and the induced inner products in* $\mathbf{RH}_\pm^2$, *we have for* $\psi \in \mathbf{RL}^\infty$

$$L_\psi^* = L_{\psi^*},$$

$$T_\psi^* = T_{\psi^*},$$

and

$$H_\psi^* = \hat{H}_{\psi^*}.$$

Here, $\psi^*(z) = \overline{\psi(-\bar{z})}$. Note that on the imaginary axis we have $\psi^*(it) = \overline{\psi(it)}$.

Proof: By elementary computations.   □

Proposition 11.2.4 *Given a real stable polynomial q, then:*

1. *The map $P_q : \mathbf{RH}_+^2 \longrightarrow \mathbf{RH}_+^2$ defined by*

$$P_q f = \frac{q^*}{q} P_- \frac{q}{q^*} f, \qquad f \in \mathbf{RH}_+^2$$

   is an orthogonal projection.

2. *We have*

$$\operatorname{Ker} P_q = \frac{q^*}{q} \mathbf{RH}_+^2 \tag{11.9}$$

   and

$$\operatorname{Im} P_q = X^q. \tag{11.10}$$

3. *We have the orthogonal direct sum decomposition*

$$\mathbf{RH}_+^2 = X^q \oplus \frac{q^*}{q} \mathbf{RH}_+^2, \tag{11.11}$$

   as well as

$$(X^q)^\perp = \frac{q^*}{q} \mathbf{RH}_+^2.$$

4. *We have*

$$\dim X^q = \dim \left\{ \frac{q^*}{q} \mathbf{RH}_+^2 \right\}^\perp = \deg q.$$

Proof:

1. We compute

$$P_q^2 f = \frac{q^*}{q} P_- \frac{q}{q^*} \frac{q^*}{q} P_- \frac{q}{q^*} f = \frac{q^*}{q} P_-^2 \frac{q}{q^*} f = \frac{q^*}{q} P_- \frac{q}{q^*} f = P_q f.$$

   So P_q is indeed a projection.

   For $f, g \in \mathbf{RH}_+^2$ we have

$$\begin{aligned}
(P_q f, g) &= (\frac{q^*}{q} P_- \frac{q}{q^*} f, g) = (P_- \frac{q}{q^*} f, \frac{q}{q^*} g) \\
&= (\frac{q}{q^*} f, P_- \frac{q}{q^*} g) = (f, \frac{q^*}{q} P_- \frac{q}{q^*} g) = (f, P_q g).
\end{aligned}$$

   This shows that P_q is an orthogonal projection.

2. Assume that $f \in (q^*/q)\mathbf{RH}_+^2$. Then $f = (q^*/q)g$ and

$$P_q f = \frac{q^*}{q} P_- \frac{q}{q^*} \frac{q^*}{q} g = \frac{q^*}{q} P_- g = 0.$$

Thus $(q^*/q)\mathbf{RH}_+^2 \subset \operatorname{Ker} P_q$.

Conversely, let $f \in \operatorname{Ker} P_q$. Then $(q^*/q)P_-(q/q^*)f = 0$. This shows that $(q/q^*)f \in \mathbf{RH}_+^2$ and hence $f \in (q^*/q)\mathbf{RH}_+^2$ or $\operatorname{Ker} P_q \subset (q^*/q)$ $\times \mathbf{RH}_+^2$. Thus equality (11.9) follows.

Assume that $f = p/q \in X^q$. Then

$$P_q f = \frac{q^*}{q} P_- \frac{q}{q^*} \frac{p}{q} = \frac{q^*}{q} P_- \frac{p}{q^*} = \frac{p}{q} = f.$$

So $X^q \subset \operatorname{Im} P_q$.

Conversely, suppose that $f \in \operatorname{Im} P_q$, that is, $f = (q^*/q)h$ with $h = P_-(q/q^*)f_1 \in \mathbf{RH}_+^2$. This representation shows that $f \in X^q$. Thus $\operatorname{Im} P_q \subset X^q$ and equality follows.

3. Follows from $I = P_q + (I - P_q)$.

4. Follows from Proposition 5.1.1.   □

Invariant Subspaces

Before the introduction of Hankel operators we digress a bit on invariant subspaces of $\mathbf{RH}_+^2$. Since we are using the half-planes for our definition of the $\mathbf{RH}_+^2$ spaces, we do not have the shift operators coveniently at our disposal. This forces us to a slight departure from the usual convention.

Definition 11.2.3

1. A subspace $M \subset \mathbf{RH}_+^2$ is called an **invariant subspace** if, for each $\psi \in \mathbf{RH}_+^\infty$, we have

$$T_\psi M \subset M.$$

2. A subspace $M \subset \mathbf{RH}_+^2$ is called a **backward-invariant subspace** if, for each $\psi \in \mathbf{RH}_+^\infty$, we have

$$T_\psi^* M \subset M.$$

Since $\mathbf{RH}_+^2$ is a subspace of $\mathbf{R}_-(z)$, the space of strictly proper rational functions, we have in it two notions of backward invariance. One is the analytic one given by Definition 11.2.3 and the other one is the algebraic one, namely, invariance with respect to the shift S_- of Eq. (5.20) restricted to $\mathbf{R}_-(z)$. The following proposition shows that in this case analysis and algebra meet, and the two notions of invariance coincide.

Proposition 11.2.5 *Let $M \subset \mathbf{RH}_+^2$ be a subspace. Then M is a backward-invariant subspace if and only if M is S_--invariant.*

Proof: Assume first that M is S_--invariant. Let $f = p/q \in M$ with p, q coprime and q stable. We show first that $X^q \in M$. By the coprimeness of p and q, there exist real polynomials a, b that solve the Bezout equation $ap + bq = 1$. Then

$$a(S_-)f = \pi_- a \cdot \frac{p}{q} = \pi_- \cdot \frac{ap + bq}{q} = \pi_- \cdot \frac{1}{q} = \frac{1}{q}.$$

We conclude that $1/q \in M$ and hence $z^i/q \in M$ for $i = 1, \ldots, \deg q - 1$. Thus we have $X^q \subset M$. For f as before and $\psi = r/s \in \mathbf{RH}_+^\infty$, we compute now, taking partial fractions and using the fact that q and s^* are coprime,

$$\frac{r^*}{s^*} \frac{p}{q} = \frac{v}{s^*} + \frac{u}{q}$$

and hence

$$P_+ \psi^* f = P_+ \frac{r^*}{s^*} \frac{p}{q} = \frac{u}{q} \in X^q \subset M.$$

This shows that M is backward-invariant.

Conversely, let $M \subset \mathbf{RH}_+^2$ be backward-invariant, and let $f = p/q \in M$. Clearly, there exists a scalar α such that $zp/q = \alpha + t/q$, and hence $S_- f = t/q$. To show algebraic invariance, we have to show the existence of a proper rational stable function $\psi = r/s$ for which $P_+(r^*/s^*)(p/q) = (t/q)$. This of course is equivalent to the partial fraction decomposition

$$\frac{r^*}{s^*} \frac{p}{q} = \frac{u}{q} + \frac{v}{s^*}.$$

Let us now choose s to be an arbitrary stable polynomial, satisftying $\deg s = \deg q - 1$. Let $e^* = \pi_q z s^*$, that is, there exists a constant γ such that $e^* = zs^* - \gamma q$ with $\deg e < \deg q$. We now compute

$$\begin{aligned}
\frac{e^*}{s^*} \frac{p}{q} &= \frac{zs^* - \gamma q}{s^*} \frac{p}{q} = \frac{zp}{q} - \frac{\gamma p}{s^*} \\
&= \alpha + \frac{t}{q} - \frac{\gamma p}{s^*} = \frac{t}{q} + \frac{\alpha s^* - \gamma p}{s^*} \\
&= \frac{u}{q} + \frac{v}{s^*}.
\end{aligned}$$

and hence

$$P_+ \frac{e^*}{s^*} \frac{p}{q} = \frac{u}{q},$$

and the proof is complete. □

Clearly, orthogonal complements of invariant subspaces are backward-invariant subspaces. However, in inner product spaces, we have proper subspaces whose orthogonal complement is trivial. The next result is of this type.

Proposition 11.2.6 *Let $M \subset \mathbf{RH}_+^2$ be an infinite-dimensional, backward-invariant subspace. Then $M^\perp = \{0\}$.*

Proof: Let $f \in M^\perp$. Since f is rational, it has an irreducible representation $f = e/d$, with $\deg e < \deg d = \delta$. Since M is infinite-dimensional, we can find δ linearly independent functions, $g_1, \ldots, g_\delta$ in M. Let $M_1 \subset M$ be the smallest backward-invariant subspace of $\mathbf{RH}_+^2$ containing all of the g_i. We claim that M_1 is finite-dimensional. For this it suffices to show that, if $g = r/s$ and $\phi \in \mathbf{RH}_+^\infty$, then $T_\phi^* = t/s \in X^s$. Assume therefore that $\phi = a/b$. Then by partial fraction decomposition

$$\phi^* g = \frac{a^*}{b^*} \cdot \frac{r}{s} = \frac{\alpha}{s} + \frac{\beta}{b^*},$$

and hence $T_\phi^* g = P_+ \phi^* g = \alpha/s \in X^s$. Now, if $g_i = r_i/s_i$, then clearly $M_1 \subset X^{s_1 \cdots s_\delta}$, and so M_1 is finite-dimensional. By Proposition 5.3.1, there exists a polynomial q for which $M_1 = X^q$. Since $g_1, \ldots, g_\delta \in M_1$, it follows that $\delta \leq \dim M_1 = \deg q$. Now, by Proposition 11.2.4, $M_1^\perp = (q^*/q)\mathbf{RH}_+^2$. So for $f = e/d$ we have the representation $f = (q^*/q) f_1$ with $f_1 \in \mathbf{RH}_+^2$. There can be no pole-zero cancellation between the zeros of q^* and those of the denominator of f_1. This clearly implies that q^* divides e. But $\deg e < \delta \leq \deg q$. Necessarily $e = 0$, and of course also $f = 0$. □

Definition 11.2.4

1. *A function $m \in \mathbf{RL}^\infty$ is called an* **all-pass** *function if it satisfies $m^* m = 1$ on the imaginary axis.*

2. *A function $m \in \mathbf{RH}_+^\infty$ is called* **inner** *if it is also an all-pass function.*

Rational inner functions have a simple characterization.

Proposition 11.2.7 *$m \in \mathbf{RH}_+^\infty$ is an inner function if and only if it has a representation*

$$m(z) = \pm \frac{p^*}{p} \tag{11.12}$$

for some real, stable polynomial p.

Proof: Assume that p is a real, stable polynomial. Then m defined by Eq. (11.12) is clearly in $\mathbf{RH}_+^\infty$. Moreover, on the imaginary axis, we have

$$m^* m = \overline{\left(\frac{p^*}{p}\right)} \left(\frac{p^*}{p}\right) = 1.$$

Conversely, assume that $m \in \mathbf{RH}_+^\infty$ is inner. Let $m = r/p$ with p a real, stable polynomial. We may assume without loss of generality that r and p

are coprime. Since m is inner, we get $(r/p)(r^*/p^*) = 1$, or $rr^* = pp^*$. Using our coprimeness assumption, we have $r|p^*|r$. This implies that $r = \alpha p^*$, with $|\alpha| = 1$. As r is a real polynomial, we must have $\alpha = \pm 1$.   □

We proceed to prove the basic tool for all that follows, namely, the representation result for invariant subspaces. Actually, Beurling [1949] proved the theorem for invariant subspaces of H^2. We give an algebraic version of his theorem, which is better suited to our needs.

Theorem 11.2.2 (Beurling) $M \subset \mathbf{RH}_+^2$ *is a nonzero invariant subspace if and only if*

$$M = m\mathbf{RH}_+^2,$$

where m is a rational inner function.

Proof: If m is inner, then clearly $M = m\mathbf{RH}_+^2$ is an invariant subspace.

To prove the converse, assume that $M \subset \mathbf{RH}_+^2$ is a nonzero invariant subspace. By Proposition 11.2.6 it follows that $M^\perp$ is necessarily finite-dimensional. Since $M^\perp$ is backward-invariant, it is also S_--invariant. We invoke now Proposition 5.3.1 to conclude that $M^\perp = X^q$ for some polynomial q. Using Proposition 11.2.4, it follows that $M = (q^*/q)\mathbf{RH}_+^2$.   □

There is a natural division relation for inner functions. If $m = m_1 m_2$ with m, m_1, m_2 inner functions in $\mathbf{RH}_+^\infty$, then we say that m_1 divides m. The greatest common inner divisor of m_1, m_2 is a common inner divisor that is divided by any other common inner divisor. The least common inner multiple is defined similarly.

As was the case with polynomials, covered by Proposition 1.3.4, this division relation has a geometric interpretation.

Proposition 11.2.8 *Let $m, m_i \in \mathbf{RH}_+^\infty$, $i = 1, \ldots, s$, be inner functions. Then:*

1. *We have $m\mathbf{RH}_+^2 \subset m_1\mathbf{RH}_+^2$ if and only if m_1 divides m.*

2. *We have*

$$\sum_{i=1}^s m_i\mathbf{RH}_+^2 = m\mathbf{RH}_+^2,$$

   where m is the greatest common inner divisor of all m_i, $i = 1, \ldots, s$.

3. *We have*

$$\bigcap_{i=1}^s m_i\mathbf{RH}_+^2 = m\mathbf{RH}_+^2,$$

   where m is the least common inner multiple of all m_i.

Proof: The proof follows along the same lines as that of Proposition 1.3.4. We omit the details. □

Two functions $f_1, f_2 \in \mathbf{RH}_+^\infty$ are called **coprime** if their greatest common inner factor is the 1. Clearly, this is equivalent to the existence of a $\delta > 0$ for which $|f_1(z)| + |f_2(z)| \geq \delta$.

Proposition 11.2.9 *Let $f_1, f_2 \in \mathbf{RH}_+^\infty$. Then f_1, f_2 are coprime if and only if there exist functions $a_1, a_2 \in \mathbf{RH}_+^\infty$ such that the Bezout identity holds:*

$$a_1(z)f_1(z) + a_2(z)f_2(z) = 1. \tag{11.13}$$

Proof: Clearly, if Eq. (11.13) holds, then f_1, f_2 cannot have a nontrivial common inner factor.

Conversely, assume that f_1, f_2 are coprime. Then $f_1\mathbf{RH}_+^2, f_2\mathbf{RH}_+^2$ are both invariant subspaces of $\mathbf{RH}_+^2$ and so is their sum. Therefore, there exists a rational inner function m such that

$$f_1\mathbf{RH}_+^2 + f_2\mathbf{RH}_+^2 = m\mathbf{RH}_+^2. \tag{11.14}$$

This implies that m is a common inner factor of the f_i. Hence necessarily we have $m = 1$. Let us choose $\alpha > 0$. Then $1/(z + \alpha) \in \mathbf{RH}_+^2$. Equation (11.14), with $m = 1$, shows that there exist strictly proper functions $b_1, b_2 \in \mathbf{RH}_+^2$ such that $b_1 f_1 + b_2 f_2 = 1/(z + \alpha)$. Defining $a_i = (z + \alpha)b_i$, we have $a_i \in \mathbf{RH}_+^\infty$, and they satisfy Eq. (11.13). □

Model Operators and Intertwining Maps

The $\mathbf{RH}_+^\infty$ module structure on $\mathbf{RH}_+^2$ induces a similar module structure on backward-invariant subspaces.

Given an inner function $m \in \mathbf{RH}_+^\infty$, we consider the backward-invariant subspace

$$H(m) = \{m\mathbf{RH}_+^2\}^\perp = \mathbf{RH}_+^2 \ominus m\mathbf{RH}_+^2.$$

Of course, since m is rational, we may assume without loss of generality that $m = q^*/q$ for some stable polynomial q. Therefore, in this case $H(m) = X^q$. The algebra $\mathbf{RH}_+^\infty$ or, equivalently, the algebra of analytic Toeplitz operators, induces an algebra of bounded operators in $\{m\mathbf{RH}_+^2\}^\perp$. Thus, for $\theta \in \mathbf{RH}_+^\infty$ the maps $T_\theta : H(m) \longrightarrow H(m)$ are defined by

$$T_\theta f = P_{H(m)}\theta f, \quad \text{for } f \in H(m). \tag{11.15}$$

Clearly, if $\theta \in \mathbf{RH}_+^\infty$, we have $\|T_\theta\| \leq \|\theta\|_\infty$.

The next theorem sums up the duality properties of intertwining operators.

Theorem 11.2.3 *Let $\theta, m \in \mathbf{RH}_+^\infty$ with m an inner function, and let T_θ be defined by*

$$T_\theta f := P_{H(m)}\theta f, \quad \text{for } f \in H(m).$$

Then:

1. *Its adjoint T_θ^* is given by*

$$T_\theta^* f = P_+ \theta^* f, \quad \text{for } f \in H(m).$$

2. *The operator $\tau_m : H(m) \longrightarrow H(m)$ defined by*

$$\tau_m f := m f^*$$

   is unitary.

3. *The operators T_{θ^*} and T_θ^* are unitarily equivalent. More specifically, we have*

$$T_\theta \tau_m = \tau_m T_\theta^*.$$

Proof:

1. Let $f, g \in H(m)$. Then

$$
\begin{aligned}
(T_\theta f, g) &= (P_{H(m)}\theta f, g) = (m P_- m^* \theta f, g) \\
&= (P_- m^* \theta f, m^* g) = (m^* \theta f, P_- m^* g) \\
&= (m^* \theta f, m^* g) = (\theta f, g) = (f, \theta^* g) \\
&= (P_+ f, \theta^* g) = (f, P_+ \theta^* g) = (f, T_\theta^* g).
\end{aligned}
$$

   Here we used the fact that $g \in H(m)$ if and only if $m^* g \in \mathbf{RH}_-^2$.

2. Clearly, the map τ_m, as a map in $\mathbf{RL}^2$, is unitary. From the orthogonal direct sum decomposition,

$$\mathbf{RL}^2 = \mathbf{RH}_-^2 \oplus H(m) \oplus m\mathbf{RH}_+^2,$$

   it follows by conjugation that

$$\mathbf{RL}^2 = m^* \mathbf{RH}_-^2 \oplus \{\mathbf{RH}_-^2 \ominus m^* \mathbf{RH}_-^2\} \oplus \mathbf{RH}_+^2.$$

   Hence $m\{\mathbf{RH}_-^2 \ominus m^* \mathbf{RH}_-^2\} = H(m)$.

3. We compute

$$T_\theta \tau_m f = T_\theta m f^* = P_{H(m)} \theta m f^* = m P_- m^* \theta m f^* = m P_- \theta f^*.$$

   Now

$$\tau_m T_\theta^* = \tau_m (P_+ \theta^* f) = m(P_+ \theta^* f)^* = m P_- \theta f^*. \qquad \square$$

We proceed to study the invertibility properties of the maps T_θ. This will be instrumental in the analysis of Hankel operators restricted to their cokernels.

Theorem 11.2.4 *Let $\theta, m \in \mathbf{RH}_+^\infty$ with m an inner function. The following statements are equivalent:*

1. *The operator T_θ defined in Eq. (11.15) is invertible.*

2. *There exists a $\delta > 0$ such that*

$$|\theta(z)| + |m(z)| \geq \delta, \quad for\ all\ z\ with\ \mathrm{Re}\ z > 0. \tag{11.16}$$

3. *There exist $\xi, \eta \in \mathbf{RH}_+^\infty$ that solve the Bezout equation*

$$\xi\theta + \eta m = 1. \tag{11.17}$$

In this case we have
$$T_\theta^{-1} = T_\xi.$$

Proof: $1 \Rightarrow 2$ We prove this by contradiction. If no such δ exists, then necessarily there exists, a point α in the open right half-plane such that $\theta(\alpha) = m(\alpha) = 0$. Since both θ and m are real, we also have $\theta(\overline{\alpha}) = m(\overline{\alpha}) = 0$. By Proposition 11.2.4, all real linear combinations of $1/(z+\alpha), 1/(z+\overline{\alpha})$ are in $H(m)$. Still we compute

$$T_\theta^* \frac{1}{z+\alpha} = P_+\theta^* \frac{1}{z+\alpha} = P_+ \left[\frac{\theta^*(z) - \overline{\theta(\alpha)}}{z+\overline{\alpha}} + \frac{\overline{\theta(\alpha)}}{z+\overline{\alpha}} \right] = \frac{\overline{\theta(\alpha)}}{z+\overline{\alpha}}.$$

This holds as θ^* is in $\mathbf{RH}_-^\infty$ and the numerator vanishes at $z = -\alpha$. Since $\theta(\alpha) = 0$, we get that $1/(z+\overline{\alpha}) \in \mathrm{Ker}\ T_\theta^*$. The same reasoning applies to $1/(z+\alpha)$. This shows that $\mathrm{Ker}\ T_\theta^*$ contains all real linear combinations of $1/(z+\alpha), 1/(z+\overline{\alpha})$, and hence it cannot be invertible.

$2 \Rightarrow 3$ That coprimeness implies the solvability of the Bezout equation was proved in Proposition 11.2.9.

$3 \Rightarrow 1$ Assume that there exist $\xi, \eta \in \mathbf{RH}_+^\infty$ which solve the Bezout equation (11.17). We show that $T_\theta^{-1} = T_\xi$. Let $f \in H(m)$. Then

$$T_\xi T_\theta f = P_{H(m)}\xi P_{H(m)}\theta f = P_{H(m)}\xi\theta f = P_{H(m)}(1 - m\eta)f = f.$$

Here we used the fact that $\xi \mathrm{Ker}\ P_{H(m)} = \xi m\mathbf{RH}_+^2 \subset m\mathbf{RH}_+^2 = \mathrm{Ker}\ P_{H(m)}$. $\square$

Hankel Operators

Our next topic is the detailed study of Hankel operators, which were introduced in Defininition 11.2.2. We shall study their kernel and image, and their relation to Beurling's theorem. We shall describe the connection between Hankel operators and intertwining maps.

Definition 11.2.5 *Given a function* $\phi \in L^{\infty}(i\mathbf{R})$, *the* Hankel *operator* $H_{\phi} : H_+^2 \longrightarrow H_-^2$ *is defined by*

$$H_{\phi}f = P_-(\phi f), \qquad for \quad f \in H_+^2. \tag{11.18}$$

The adjoint operator $(H_{\phi})^* : H_-^2 \longrightarrow H_+^2$ *is given by*

$$(H_{\phi})^* f = P_+(\phi^* f), \qquad for \quad f \in H_-^2.$$

Here, $\phi^*(z) = \overline{\phi(-\bar{z})}$.

Thus assume that $\phi = (n/d) \in H^{\infty}_{-}$ and $n \wedge d = 1$. So our assumption is that d is antistable. In spite of the slight ambiguity, we will write $n = \deg d$. It will always be clear from the context what n means. This leads to

$$\phi = \frac{n}{d} = \frac{n}{d^*} \frac{d^*}{d}.$$

Thus $\phi = m^*\eta = m^{-1}\eta$ with $\eta = n/d^*, m = d/d^*$ is a coprime factorization over $\mathbf{RH}_+^{\infty}$. This particular coprime factorization, where the denominator is an inner function, is called the **DSS factorization** [Douglas, Shapiro, and Shields, 1971].

The next theorem discusses the functional equation of Hankel operators.

Theorem 11.2.5

1. *For every* $\psi \in \mathbf{RL}^{\infty}$ *the Hankel operator* H_{ϕ} *satisfies the* **Hankel functional equation**

$$P_-\psi H_{\phi}f = H_{\phi}\psi f, \qquad f \in H_+^2.$$

2. $\operatorname{Ker} H_{\phi}$ *is an invariant subspace, that is, for* $f \in \operatorname{Ker} H_{\phi}$ *and* $\psi \in \mathbf{RH}_+^{\infty}$ *we have* $\psi f \in \operatorname{Ker} H_{\phi}$.

Proof:

1. We compute

$$\begin{aligned} P_-\psi H_{\phi}f &= P_-\psi P_-\phi f = P_-\psi\phi f \\ &= P_-\phi\psi f = H_{\phi}\psi f. \end{aligned}$$

2. Follows from the Hankel functional equation.   □

The following theorem shows that the Hankel functional equation characterizes the class of Hankel operators.

Theorem 11.2.6 *Let* $H : \mathbf{RH}_+^2 \longrightarrow \mathbf{RH}_-^2$ *satisfy the Hankel functional equation. Then there exists a function* $\psi \in \mathbf{RH}^{\infty}_{-}$ *for which*

$$H = H_{\psi}.$$

Proof: Choose $\alpha > 0$. Then $1/(z + \alpha) \in \mathbf{RH}_+^\infty$. Moreover, we clearly have

$$\left\{ \frac{\theta}{z + \alpha} | \theta \in \mathbf{RH}_+^\infty \right\} = \mathbf{RH}_+^2.$$

If we had $H = H_\psi$ with $\psi \in \mathbf{RH}_-^\infty$, then

$$H \frac{1}{z + \alpha} = P_- \frac{\psi(z)}{z + \alpha} = P_- \frac{\psi(z) - \psi(-\alpha) + \psi(-\alpha)}{z + \alpha}$$

$$= \frac{\psi(z) - \psi(-\alpha)}{z + \alpha} = \phi.$$

Therefore we define

$$\psi(z) = \phi(z)(z + \alpha) + \psi(-\alpha).$$

Obviously, $\psi \in \mathbf{RH}_-^\infty$. We clearly have

$$P_- \psi \frac{1}{z + \alpha} = P_- \frac{\phi(z)(z + \alpha) + \psi(-\alpha)}{z + \alpha}$$

$$= \phi(z).$$

Therefore, for any $\theta \in \mathbf{RH}_+^\infty$,

$$H \frac{\theta}{z + \alpha} = P_- \theta H \frac{1}{z + \alpha} = P_- \theta P_- \psi \frac{1}{z + \alpha}$$

$$= P_- \theta \psi \frac{1}{z + \alpha} = P_- \psi \frac{\theta}{z + \alpha}.$$

So $Hf = P_- \psi f = H_\psi f$ for every $f \in \mathbf{RH}_+^2$. This shows that $H = H_\psi$. $\square$

It follows from Beurling's theorem that $\operatorname{Ker} H_\phi = m\mathbf{RH}_+^2$ for some inner function $m \in \mathbf{RH}_+^\infty$. Since we are dealing with the rational case, the next theorem can make this more specific, and characterizes the kernel and image of a Hankel operator; it also clarifies the connection between them and polynomial and rational models.

Theorem 11.2.7 *Let $\phi = (n/d) \in \mathbf{RH}_-^\infty$ and $n \wedge d = 1$. Then:*

1.

$$\operatorname{Ker} H_\phi = \frac{d}{d^*} \mathbf{RH}_+^2.$$

2.

$$\{\operatorname{Ker} H_\phi\}^\perp = \{ \frac{d}{d^*} \mathbf{RH}_+^2 \}^\perp = X^{d^*}.$$

3.

$$\operatorname{Im} H_\phi = \mathbf{RH}_-^2 \ominus \frac{d^*}{d} \mathbf{RH}_-^2 = X^d.$$

4.

$$\dim \operatorname{Im} H_\phi = \deg d.$$

Proof:

1. $\{\operatorname{Ker} H_\phi\}^\perp$ contains only rational functions. Let

$$f = \frac{p}{q} \in \{\frac{d}{d^*}\mathbf{RH}_+^2\}^\perp.$$

Then $(d^*p/dq) \in \mathbf{RH}_-^2$. So $q \mid d^*p$. But, as $p \wedge q = 1$, it follows that $q \mid d^*$, that is, $d^* = qr$. Hence $f = (rp/d^*) \in X^{d^*}$.

Conversely, let $(p/d^*) \in X^{d^*}$, then $(p/d^*) = (p/d)(d/d^*)$ or (d^*/d) $\times (p/d^*) \in \mathbf{RH}_-^2$. Therefore, we conclude that $(p/d^*) \in \{(d/d^*)\mathbf{RH}_+^2\}^\perp$.

2 to 4. These statements follow from Proposition 11.2.4. □

As a corollary we can state the following theorem.

Theorem 11.2.8 (Kronecker) *Let $\phi \in \mathbf{RL}^\infty$. Then* rank $H_\phi = k$ *if and only if ϕ has k poles in the right half-plane.*

Proof: Taking a partial fraction decomposition into stable and antistable parts, we may assume without loss of generality that ϕ is antistable. □

The previous theorem, although of an elementary nature, is central to all further development as it provides the direct link between the infinite-dimensional object, namely, the Hankel operator, and the well-developed theory of polynomial and rational models. This link will be exploited continuously.

We have an $\mathbf{RH}_+^\infty$-module structure on any finite-dimensional backward-invariant subspace $H(m)$. The $\mathbf{RH}_+^\infty$-module homomorphisms are the maps $X : H(m) \longrightarrow H(m)$ that satisfy $XT_\theta = T_\theta X$ for every $\theta \in \mathbf{RH}_+^\infty$. At the same time, the abstract Hankel operators, that is, the operators $H : \mathbf{RH}_+^2 \longrightarrow \mathbf{RH}_+^2$, that satisfy the Hankel functional equation are also $\mathbf{RH}_+^\infty$-module homomorphisms. The next result relates these two classes of module homomorphisms.

Theorem 11.2.9 *A map $H : \mathbf{RH}_+^2 \longrightarrow \mathbf{RH}_+^2$ is a Hankel operator with a nontrivial kernel $m\mathbf{RH}_+^2$, with m rational inner, if and only if we have*

$$H = m^* X P_{H(m)}, \tag{11.19}$$

where $X : H(m) \longrightarrow H(m)$ is an intertwining map, that is, it satisfies

$$XT_\psi = T_\psi X$$

for every $\psi \in \mathbf{RH}_+^\infty$.

Proof: Assume that $X : H(m) \longrightarrow H(m)$ is an intertwining map. We define $H : \mathbf{RH}_+^2 \longrightarrow \mathbf{RH}_-^2$ by Eq. (11.19). Now let $\psi \in \mathbf{RH}_+^\infty$ and $f \in \mathbf{RH}_+^2$. Then

$$
\begin{aligned}
H\psi f &= m^* X P_{H(m)} \psi f = m^* X P_{H(m)} \psi P_{H(m)} f \\
&= m^* X T_\psi P_{H(m)} f = m^* T_\psi X P_{H(m)} f = m^* P_{H(m)} \psi X P_{H(m)} f \\
&= m^* P_{H(m)} \psi X P_{H(m)} f = m^* m P_- m^* \psi X P_{H(m)} f \\
&= P_- \psi m^* X P_{H(m)} f = P_- \psi H f.
\end{aligned}
$$

So H satisfies the Hankel functional equation.

Conversely, if H is a Hankel operator with kernel $m\mathbf{RH}_+^2$, then we define $X : H(m) \longrightarrow H(m)$ by $X = mH|_{H(m)}$. For $f \in H(m)$ and $\psi \in \mathbf{RH}_+^\infty$ we compute,

$$
\begin{aligned}
X T_\psi f &= mH P_{H(m)} \psi f = mH\psi f = mP_- \psi H f \\
&= mP_- m^* m\psi H f = P_{H(m)} \psi mH f = T_\psi X f.
\end{aligned}
$$

So X is intertwining. Obviously, Eq. (11.19) holds. $\square$

As a corollary we obtain the all-important characterization of intertwining maps.

Theorem 11.2.10 *Let m be a rational inner function. Then $X : H(m) \longrightarrow H(m)$ is an intertwining map if and only if for some $\psi \in \mathbf{RH}_+^\infty$ we have*

$$
X = T_\psi.
$$

Proof: If $X = T_\psi$ for some $\psi \in \mathbf{RH}_+^\infty$, we clearly have $T_\theta T_\psi = T_\psi T_\theta$ for every $\theta \in \mathbf{RH}_+^\infty$.

Conversely, assume that X is an intertwining map. Define the Hankel operator $H : \mathbf{RH}_+^2 \longrightarrow \mathbf{RH}_-^2$ by $H = m^* X P_{H(m)}$. By Theorem 11.2.6, H is a Hankel operator and there exists a $\phi \in \mathbf{RH}_-^\infty$ for which $H = H_\phi$. Since obviously $\mathrm{Ker}\, H \supset m\mathbf{RH}_+^2$, it follows that $P_- \phi m f = 0$ for all $f \in \mathbf{RH}_+^2$. This shows that $\psi = m\phi \in \mathbf{RH}_+^\infty$. In terms of H, the operator X is given by $X = mH_\phi|_{H(m)}$, that is, for $f \in H(m)$,

$$
\begin{aligned}
Xf &= mH_\phi f = mP_- \phi f = mP_- m^* m\phi f \\
&= P_{H(m)}(m\phi)f = T_\psi f.
\end{aligned}
$$
    $\square$

Hankel operators in general and those with rational symbols in particular are never invertible. Still we may want to invert the Hankel operator as a map from its cokernel, that is, the orthogonal complement of its kernel, to its image. We saw that such a restriction of a Hankel operator is of considerable interest because of its connection to intertwining maps of model operators. Theorem 11.2.4 gave a full characterization of invertibility properties of intertwining maps. These now can be applied to the inversion of the restricted Hankel operators, which will turn out to be of great importance in the development of duality theory.

11.3   Schmidt Pairs of Hankel Operators

We saw in Section 7.5 that singular values of operators are closely related to the problem of best approximation by operators of finite rank. That this basic method could be applied to the approximation of Hankel operators by Hankel operators of lower ranks through the detailed analysis of singular values and the corresponding Schmidt pairs is a fundamental contribution of Adamjan, Arov, and Krein [1971].

We recall that, given a linear operator A on an inner product space, μ is a singular value of A if there exists a nonzero vector f such that $A^*Af = \mu^2 f$. Rather than solve the previous equation, we let $g = \frac{1}{\mu}Af$ and go over to the equivalent system

$$\begin{cases} Af & = \mu g \\ A^*g & = \mu f, \end{cases}$$

that is, μ is a singular value of both A and A^*.

The analysis of Schmidt pairs of Hankel operators goes back to Adamjan, Arov, and Krein. Here, for the rational case, we present an algebraic derivation of some of their results.

Assume without loss of generality that $\phi = (n/d) \in \mathbf{RH}^\infty_-$ is strictly proper with n and d coprime polynomials. Necessarily, d is antistable. We proceed to compute the singular vectors of the Hankel operator H_ϕ. In view of Eq. (7.14), we have to solve

$$\begin{cases} H_\phi f = \mu g \\ H_\phi^* g = \mu f \end{cases}$$

or, equivalently,

$$\begin{cases} P_- \dfrac{n}{d} \dfrac{p}{d^*} = \mu \dfrac{\hat{p}}{d} \\ P_+ \dfrac{n^*}{d^*} \dfrac{\hat{p}}{d} = \mu \dfrac{p}{d^*}. \end{cases}$$

This means that there exist polynomials π and ξ such that

$$\begin{cases} \dfrac{n}{d} \dfrac{p}{d^*} = \mu \dfrac{\hat{p}}{d} + \dfrac{\pi}{d^*} \\ \dfrac{n^*}{d^*} \dfrac{\hat{p}}{d} = \mu \dfrac{p}{d^*} + \dfrac{\xi}{d}. \end{cases}$$

These equations can be rewritten as polynomial equations,

$$np = \mu d^* \hat{p} + d\pi \tag{11.20}$$

$$n^* \hat{p} = \mu dp + d^* \xi. \tag{11.21}$$

Equation (11.20), considered as an equation modulo the polynomial d, is not an eigenvalue equation as there are too many unknowns. More specifically, we have to find the coefficients of both p and $\hat{p}$. To overcome this difficulty, we study in more detail the structure of Schmidt pairs of Hankel operators.

Lemma 11.3.1 *Let $\{p/d^*, \hat{p}/d\}$ and $\{q/d^*, \hat{q}/d\}$ be two Schmidt pairs of the Hankel operator $H_{n/d}$, corresponding to the same singular value μ. Then*

$$\frac{p}{\hat{p}} = \frac{q}{\hat{q}},$$

that is, this ratio is independent of the Schmidt pair.

Proof: The polynomials p, $\hat{p}$ correspond to one Schmidt pair, and let the polynomials q, $\hat{q}$ correspond to another Schmidt pair, that is,

$$nq = \mu d^* \hat{q} + d\rho \tag{11.22}$$

$$n^* \hat{q} = \mu dq + d^* \eta. \tag{11.23}$$

Now, from Eqs. (11.20) and (11.23) we get

$$0 = \mu d(p\hat{q} - q\hat{p}) + d^*(\xi\hat{q} - \eta\hat{p}).$$

Since d and d^* are coprime, it follows that $d^* \mid p\hat{q} - q\hat{p}$. On the other hand, from Eqs. (11.20) and (11.22), we get

$$0 = \mu d^*(\hat{p}q - \hat{q}p) + d(\pi q - \rho p),$$

and hence that $d \mid \hat{p}q - \hat{q}p$. Now both d and d^* divide $\hat{p}q - \hat{q}p$, and, as $\deg(\hat{p}q - \hat{q}p) < \deg d + \deg d^*$, it follows that

$$\hat{p}q - \hat{q}p = 0.$$

Equivalently,

$$\frac{p}{\hat{p}} = \frac{q}{\hat{q}}$$

that is, $p/\hat{p}$ is independent of the particular Schmidt pair associated to the singular value μ. □

Lemma 11.3.2 *Let $\{p/d^*, \hat{p}/d\}$ be a Schmidt pair associated with the singular value μ. Then $p/\hat{p}$ is unimodular or all pass.*

Proof: Going back to Eq. (11.21) and the dual of Eq. (11.20), we have

$$n^* \hat{p} = \mu dp + d^* \xi$$

$$n^* p^* = \mu d(\hat{p})^* + d^* \pi^*.$$

It follows that

$$0 = \mu d(pp^* - \hat{p}(\hat{p})^*) + d^*(\xi p^* - \pi^*\hat{p})$$

and hence $d^* \mid (pp^* - \hat{p}(\hat{p})^*)$. By symmetry, $d \mid (pp^* - \hat{p}(\hat{p})^*)$ also, and so necessarily $pp^* - \hat{p}(\hat{p})^* = 0$. This can be rewritten as $(p/\hat{p})[p^*/(\hat{p})^*] = 1$, that is, $p/\hat{p}$ is all pass.   □

We will say that a pair of polynomials $(p, \hat{p})$, with $\deg p, \deg \hat{p} < \deg d$, is a **solution pair** if there exist polynomials π and ξ such that Eqs. (11.20) and (11.21) are satisfied.

The next lemma characterizes all solution pairs.

Lemma 11.3.3 Let μ be a singular value of the Hankel operator $H_{n/d}$. Then there exists a unique, up to a constant factor, solution pair $(p, \hat{p})$ of minimal degree. The set of all solutions pairs is given by

$$\{(q, \hat{q}) \mid q = pa, \ \hat{q} = \hat{p}a \ , \deg a < \deg q - \deg p\}.$$

Proof: Clearly, if μ is a singular value of the Hankel operator, then a nonzero solution pair $(p, \hat{p})$ of minimal degree exists. Let $(q, \hat{q})$ be any other solution, paired with $\deg q, \ \deg \hat{q} < \deg d$. By the division rule for polynomials, $q = ap + r$ with $\deg r < \deg p$. Similarly, $\hat{q} = \hat{a}\hat{p} + \hat{r}$ with $\deg \hat{r} < \deg \hat{p}$. From Eq. (11.20) we get

$$n(ap) = \mu d^*(a\hat{p}) + d(a\pi), \tag{11.24}$$

whereas Eq. (11.22) yields

$$n(ap + r) = \mu d^*(\hat{a}\hat{p} + \hat{r}) + d(\tau). \tag{11.25}$$

By subtraction we obtain

$$nr = \mu d^*((\hat{a} - a)\hat{p} + \hat{r}) + d(\tau - a\pi). \tag{11.26}$$

Similarly, from Eq. (11.23) we get

$$n^*(\hat{a}\hat{p} + \hat{r}) = \mu d(ap + r) + d^*\xi, \tag{11.27}$$

whereas Eq. (11.21) yields

$$n^*(a\hat{p}) = \mu d(ap) + d^*(a\xi). \tag{11.28}$$

Subtracting the two gives

$$n^*((\hat{a} - a)\hat{p} + \hat{r}) = \mu dr + d^*(\eta - a\xi). \tag{11.29}$$

Equations (11.26) and (11.29) imply that

$$\left\{ \frac{r}{d^*}, \frac{(\hat{a} - a)\hat{p} + \hat{r})}{d} \right\}$$

is a μ Schmidt pair. Since necessarily $\deg r = \deg(\hat{a} - a)\hat{p} + \hat{r})$, we get $\hat{a} = a$. Finally, since we assumed $(p, \hat{p})$ to be of minimal degree, we must have $r = \hat{r} = 0$.

Conversely, if a is any polynomial satisfying $\deg a < \deg d - \deg p$, then, from Eqs. (11.20) and (11.21), it follows by multiplication that $(pa, \hat{p}a)$ is also a solution pair. □

Lemma 11.3.4 *Let p, q be coprime polynomials with real coefficients such that p/q is all-pass. Then $q = \pm p^*$.*

Proof: Since p/q is all-pass, it follows that $p/qp^*/q^* = 1$, or $pp^* = qq^*$. As the polynomials p and q are coprime, it follows that $p \mid q^*$ and hence $q^* = \pm p$. □

In the general case we have the following.

Lemma 11.3.5 *Let p, q be polynomials with real coefficients such that $p \wedge q = 1$ and p/q is all-pass. Then, with $r = p \wedge \hat{p}$, we have $p = rs$ and $\hat{p} = \pm rs^*$.*

Proof: Write $p = rs$, $\hat{p} = r\hat{s}$. Then $s \wedge \hat{s} = 1$ and $s/\hat{s}$ is all-pass. The result follows by applying the previous lemma. □

The next theorem is of central importance due to the fact that it reduces the analysis to one polynomial. Thus we get an equation that is easily reduced to an eigenvalue problem.

Theorem 11.3.1 *Let μ be a singular value of H_ϕ, and let $(p, \hat{p})$ be a nonzero, minimal degree solution pair of Eqs. (11.20) and (11.21). Then p is a solution of*

$$np = \lambda d^* p^* + d\pi, \tag{11.30}$$

with λ real and $|\lambda| = \mu$.

Proof: Let $(p, \hat{p})$ be a nonzero, minimal degree solution pair of Eqs. (11.20) and (11.21). By taking their adjoints we can easily see that $(\hat{p}^*, p^*)$ is also a nonzero, minimal degree solution pair. By uniqueness of such a solution, that is, by Lemma 11.3.3, we have

$$\hat{p}^* = \epsilon p. \tag{11.31}$$

Since $\hat{p}/p$ is all-pass and both polynomials are real, we have $\epsilon = \pm 1$. Let us put $\lambda = \epsilon\mu$. Then Eq. (11.31) can be rewritten as $\hat{p} = \epsilon p^*$, and so Eq. (11.30) follows from Eq. (11.20). □

We will refer to Eq. (11.30) as the **fundamental polynomial equation**. It will be the basis for all future derivations.

Corollary 11.3.1 *Let μ_i be a singular value of H_ϕ, and let p_i be the minimal degree solution of the fundamental polynomial equation, that is,*

$$np_i = \lambda_i d^* p_i^* + d\pi_i.$$

Then:

1.
$$\deg p_i = \deg p_i^* = \deg \pi_i.$$

2. *Putting $p_i(z) = \sum_{j=0}^{n-1} p_{i,j} z^j$ and $\pi_i(z) = \sum_{j=0}^{n-1} \pi_{i,j} z^j$, we have the equality*

$$\pi_{i,n-1} = \lambda_i p_{i,n-1}. \tag{11.32}$$

Corollary 11.3.2 *Let p be a minimal degree solution of Eq. (11.30). Then:*

1. *The set of all singular vectors of the Hankel operator $H_{n/d}$, corresponding to the singular value μ, is given by*

$$\mathrm{Ker}\,(H_{n/d}^* H_{n/d} - \mu^2 I) = \left\{ \frac{pa}{d^*} \mid a \in R[z],\ \deg a < \deg d - \deg p \right\}.$$

2. *The multiplicity of $\mu = \|H_\phi\|$ as a singular value of H_ϕ is equal to $m = \deg d - \deg p$, where p is the minimum degree solution of Eq. (11.30).*

3. *There exists a constant c such that $c + (n/d)$ is a constant multiple of an antistable all-pass function if and only if $\mu_1 = \cdots = \mu_n$.*

Proof: We will prove part 3 only. Assume that all singular values are equal to μ. Thus the multiplicity of μ is $\deg d$. Hence the minimal degree solution p of Eq. (11.30) is a constant and so is π. Putting $c = (\pi/p)$, then Eq. (11.30) can be rewritten as

$$\frac{n}{d} + c = \lambda \frac{d^* p^*}{dp},$$

and this is a multiple of an antistable all-pass function.

Conversely, assume, without loss of generality, that $(n/d) + c$ is antistable all-pass. Then the induced Hankel operator is isometric and all of its singular values are equal to 1. □

The following simple proposition is important in the study of zeroes of singular vectors.

Proposition 11.3.1 *Let μ_k be a singular value of H_ϕ, and let p_k be the minimal degree solution of*

$$np_k = \lambda_k d^* p_k^* + d\pi_k.$$

Then:

1. *The polynomials p_k and p_k^* are coprime.*

2. *The polynomial p_k has no imaginary axis zeroes.*

Proof:

1. Let $e = p_k \wedge p_k^*$. Without loss of generality we may assume that $e = e^*$. The polynomial e has no imaginary axis zeroes, for that would imply that e and π_k have a nontrivial common divisor. Thus the fundamental polynomial equation could be divided by a suitable polynomial factor. This is in contradiction to the assumption that p_k is a minimal degree solution.

2. This clearly follows from the first part.   □

The fundamental polynomial equation is easily reduced, using polynomial models, to either a generalized eigenvalue equation or to a regular eigenvalue equation. Starting from Eq. (11.30), we apply the standard functional calculus, and the fact that $d(S_d) = 0$, that is, the Cayley–Hamilton theorem, to obtain

$$n(S_d)p_i = \lambda_i d^*(S_d)p_i^*. \tag{11.33}$$

Now d, d^* are coprime as d is antistable and d^* is stable. Thus, by Theorem 5.1.3, $d^*(S_d)$ is invertible. In fact, the inverse of $d^*(S_d)$ is easily computed through the solution of the Bezout equation $a(z)d(z) + b(z)d^*(z) = 1$, with $\deg a$, $\deg b < \deg d$. In this case, the polynomials a and b are uniquely determined, which by virtue of symmetry forces the equality $a = b^*$. Hence

$$b^*(z)d(z) + b(z)d^*(z) = 1. \tag{11.34}$$

From this we get $b(S_d)d^*(S_d) = I$ or $d^*(S_d)^{-1} = b(S_d)$.

Because of the symmetry in the Bezout equation (11.34), we expect that some reduction in the computational complexity should be possible. This indeed turns out to be the case.

Given an arbitrary polynomial f, we let

$$\begin{cases} f_+(z^2) &= \dfrac{f(z) + f^*(z)}{2} \\[2mm] f_-(z^2) &= \dfrac{f(z) - f^*(z)}{2z}. \end{cases}$$

The Bezout equation can be rewritten as

$$(b_+(z^2) - zb_-(z^2))(d_+(z^2) + zd_-(z^2))$$
$$+(b_+(z^2) + zb_-(z^2))(d_+(z^2) - zd_-(z^2)) = 1$$

or

$$2(b_+(z^2)d_+(z^2) - z^2 b_-(z^2)d_-(z^2)) = 1.$$

We can of course solve the lower degree Bezout equation

$$2(b_+(z)d_+(z) - zb_-(z)d_-(z)) = 1.$$

This is possible as, by the assumption that d is antistable, d_+ and zd_- are coprime. Putting $b(z) = b_+(z^2) + zb_-(z^2)$, we get a solution to the Bezout equation (11.34).

Going back to Eq. (11.33), we have

$$n(S_d)b(S_d)p_i = \lambda_i p_i^*. \tag{11.35}$$

To simplify, we let $r = \pi_d(bn) = (bn) \, mod(d)$. Then Eq. (11.35) is equivalent to

$$r(S_d)p_i = \lambda_i p_i^*. \tag{11.36}$$

If $K : X_d \longrightarrow X_d$ is given by $Kp = p^*$, then Eq. (11.36) is equivalent to the generalized eigenvalue equation

$$r(S_d)p_i = \lambda_i K p_i.$$

Since K is obviously invertible and $K^{-1} = K$, the last equation transforms into the regular eigenvalue equation

$$K r(S_d)p_i = \lambda_i p_i.$$

To get a matrix equation one can take the matrix representation with respect to any choice of basis in X_d.

We now begin the study of the zero location of the numerator polynomials of singular vectors. This is of course the same as the study of the zeroes of minimal degree solutions of Eq. (11.30). The following proposition provides a lower bound on the number of zeroes the minimal degree solutions of Eq. (11.30) can have in the open left half-plane. However, the lower bound is sharp in one special case. This is enough to lead us eventually to a full characterization.

Proposition 11.3.2 Let $\phi = (n/d) \in \mathbf{RH}_-^\infty$. Let μ_k be a singular value of H_ϕ satisfying

$$\mu_1 \geq \cdots \geq \mu_{k-1} > \mu_k = \cdots = \mu_{k+\nu-1} > \mu_{k+\nu} \geq \cdots \geq \mu_n,$$

that is, μ_k is a singular value of multiplicity ν. Let p_k be the minimum degree solution of Eq. (11.30) corresponding to μ_k. Then the number of antistable zeroes of p_k is $\geq k - 1$.

If μ_n is the smallest singular value of H_ϕ and is of multiplicity ν, that is,

$$\mu_1 \geq \cdots \geq \mu_{n-\nu} > \mu_{n-\nu+1} = \cdots = \mu_n,$$

and $p_{n-\nu+1}$ is the corresponding minimum degree solution of Eq. (11.30), then all of the zeroes of $p_{n-\nu+1}$ are antistable.

Proof: From Eq. (11.30), that is, $np_k = \lambda_k d^* p_k^* + d\pi_k$, we get, dividing by dp_k,

$$\frac{n}{d} - \frac{\pi_k}{p_k} = \lambda_k \frac{d^* p_k^*}{dp_k},$$

which implies of course that

$$\|H_{n/d} - H_{\pi_k/p_k}\| \leq \|\frac{n}{d} - \frac{\pi_k}{p_k}\|_\infty = \mu_k \|\frac{d^* p_k^*}{dp_k}\|_\infty = \mu_k.$$

This means, by the definition of singular values, that rank $H_{\pi_k/p_k} \geq k - 1$. But this implies, by Kronecker's theorem, that the number of antistables poles of π_k/p_k, which is the same as the number of antistable zeroes of p_k, is $\geq k - 1$.

If μ_n is the smallest singular value and has multiplicity ν, and $p_{n-\nu+1}$ is the minimal degree solution of Eq. (11.30), then it has degree $n - \nu$. But by the previous part it must have at least $n - \nu$ antistable zeroes. So this implies that all of the zeroes of $p_{n-\nu+1}$ are antistable.   □

The previous result is extremely important from our point of view. It shifts the focus from the largest singular value, the starting point in all derivations so far, to the smallest singular value. Certainly the derivation is elementary, inasmuch as we use only the definition of singular values and Kronecker's theorem. The great advantage is that at this stage we can solve an important Bezout equation which is the key to duality theory.

We now have at hand all that is needed to obtain the optimal Hankel norm approximant corresponding to the smallest singular value. We shall delay this analysis to a later stage and develop duality theory first.

From Eq. (11.30) we obtain, dividing by $\lambda_n d^* p_n^*$, the Bezout equation

$$\frac{n}{d^*}\left(\frac{1}{\lambda_n}\frac{p_n}{p_n^*}\right) - \frac{d}{d^*}\left(\frac{1}{\lambda_n}\frac{\pi_n}{p_n^*}\right) = 1. \qquad (11.37)$$

Since the polynomials p_n and d are antistable, all four functions appearing in the Bezout equation are in $\mathbf{RH}_+^\infty$. We shall discuss next the implications of this.

11.4   Duality and Hankel Norm Approximation

In this section we develop a duality theory in the context of Hankel norm approximation problems. There are three operations applied to a given, antistable, transfer function: namely, inversion of the restricted Hankel operator, taking the adjoint map, and, finally, one-sided multiplication by unitary operators. The last two operations do not change the singular values, whereas the first operation inverts them.

We will say that two inner product space operators $T : H_1 \longrightarrow H_2$ and $T' : H_3 \longrightarrow H_4$ are **equivalent** if there exist unitary operators $U : H_1 \longrightarrow H_3$ and $V : H_2 \longrightarrow H_4$ such that $VT = T'U$. Clearly, this is an equivalence relation.

Lemma 11.4.1 *Let* $T : H_1 \longrightarrow H_2$ *and* $T' : H_3 \longrightarrow H_4$ *be equivalent. Then* T *and* T' *have the same singular values.*

Proof: Let $T^*Tx = \mu^2 x$. Since $VT = T'U$, it follows that

$$U^*T'^*T'Ux = T^*V^*VTx = T^*Tx = \mu^2 x,$$

or $T'^*T'(Ux) = \mu^2(Ux)$.   □

The following proposition borders on the trivial and no proof need be given. However, when applied to Hankel operators, it has far-reaching implications. In fact, it provides a key to duality theory and leads eventually to the proof of the central results.

Proposition 11.4.1 *Let* T *be an invertible linear transformation. Then, if* x *is a singular vector of the operator* T *corresponding to the singular value* μ, *that is,* $T^*Tx = \mu^2 x$, *then,*

$$T^{-1}(T^{-1})^*x = \mu^{-2}x,$$

that is, x *is also a singular vector for* $(T^{-1})^*$ *corresponding to the singular value* μ^{-1}.

In view of this proposition, it is of interest to compute $[(H_\phi|H(m))^{-1}]^*$. Before proceeding with this, we compute the inverse of a related operator. This is a special case of Theorem 11.2.4. Note that, since $\|T_\theta^{-1}\| = \mu_n^{-1}$, there exists, by Theorem 11.2.4, a $\xi \in H_+^\infty$ such that $T_\theta^{-1} = T_\xi$ and $\|\xi\|_\infty = \mu_n^{-1}$. The next theorem provides this ξ.

Theorem 11.4.1 *Let* $\phi = (n/d) \in \mathbf{RH}_-^\infty$. *Then* $\theta = (n/d^*) \in \mathbf{RH}_+^\infty$. *The operator* T_θ *defined by Eq. (11.15) is invertible, and its inverse given by* $T_{(1/\lambda_n)(p_n/p_n^*)}$, *where* λ_n *is the last signed singular value of* H_ϕ *and* p_n *is the minimal degree solution of*

$$np_n = \lambda_n d^* p_n^* + d\pi_n.$$

Proof: From the previous equation we obtain the Bezout equation

$$\frac{n}{d^*}\left(\frac{1}{\lambda_n}\frac{p_n}{p_n^*}\right) - \frac{d}{d^*}\left(\frac{\pi_n}{\lambda_n p_n^*}\right) = 1.$$

By Theorem 11.3.2, the polynomial p_n is antistable so $(p_n/p_n^*) \in \mathbf{RH}_+^\infty$. This, by Theorem 11.2.4, implies the result.   □

We know (see Corollary 10.5.1) that stabilizing controllers are related to solutions of Bezout equations over $\mathbf{RH}_+^\infty$. Thus we expect Eq. (11.37) to lead to a stabilizing controller. The next corollary is a result of this type.

Corollary 11.4.1 *Let $\phi = (n/d) \in \mathbf{RH}^\infty$. The controller $k = (p_n/\pi_n)$ stabilizes ϕ. If the multiplicity of μ_n is m, there exists a stabilizing controller of degree $n - m$.*

Proof: Since p_n is antistable, we get from Eq. (11.30) that $np_n - d\pi_n = \lambda_n d^* p_n^*$ is stable. We compute

$$\frac{\phi}{1 - k\phi} = \frac{\dfrac{n}{d}}{1 - \dfrac{p_n}{\pi_n}\dfrac{n}{d}} = \frac{-n\pi_n}{np_n - d\pi_n} = \frac{-n\pi_n}{\lambda_n d^* p_n^*} \in \mathbf{RH}^\infty_+. \qquad \square$$

Theorem 11.4.2 *Let $\phi = (n/d) \in \mathbf{RH}^\infty_-$. Let $H : X^{d^*} \longrightarrow X^d$ be defined by $H = H_\phi|X^{d^*}$. Then:*

1. $H_\phi^{-1} : X^d \longrightarrow X^{d^*}$ is given by

$$H_\phi^{-1}h = \frac{1}{\lambda_n}\frac{d}{d^*}P_-\frac{p_n}{p_n^*}h. \tag{11.38}$$

2. $(H_\phi^{-1})^* : X^{d^*} \longrightarrow X^d$ is given by

$$(H_\phi^{-1})^* f = \frac{1}{\lambda_n}\frac{d^*}{d}P_+\frac{p_n^*}{p_n}f. \tag{11.39}$$

Proof:

1. Let $m = d/d^*$ and let T be the map given by $T = mH_{n/d}$. Thus we have the following commutative diagram:

$$
\begin{array}{ccc}
X^{d^*} & \xrightarrow{\ H_\phi\ } & X^d \\
 & T \searrow & \downarrow \ m \\
 & & X^{d^*}
\end{array}
$$

Now

$$Tf = \frac{d}{d^*}P_-\frac{n}{d}f = \frac{d}{d^*}P_-\frac{d^*}{d}\frac{n}{d^*}f = P_{H(d/d^*)}\frac{n}{d^*}f = P_{X^{d^*}}\frac{n}{d^*}f,$$

that is, $T = T_\theta$, where $\theta = n/d^*$. Now, from $T_\theta = mH_{n/d}$, we have, by Theorem 11.4.1, $T_\theta^{-1} = T_{(1/\lambda_n)(p_n/p_n^*)}$. So, for $h \in X^d$,

$$
\begin{aligned}
H_{n/d}^{-1}h &= \frac{1}{\lambda_n}P_{H(d/d^*)}\frac{p_n}{p_n^*}\frac{d}{d^*}h = \frac{1}{\lambda_n}\frac{d}{d^*}P_-\frac{d^*}{d}\frac{p_n}{p_n^*}\frac{d}{d^*}h \\
&= \frac{1}{\lambda_n}\frac{d}{d^*}P_-\frac{p_n}{p_n^*}h.
\end{aligned}
$$

2. The previous equation also can be written as

$$H_{n/d}^{-1}h = T_{(1/\lambda_n)(p_n/p_n^*)}mh.$$

Therefore, using Theorem 11.2.3, we have, for $f \in X^{d^*}$,

$$(H_\phi^{-1})^*f = m^*(T_{(1/\lambda_n)(p_n/p_n^*)})^*f = \frac{d^*}{d}P_+\frac{1}{\lambda_n}\frac{p_n^*}{p_n}f = \frac{1}{\lambda_n}\frac{d^*}{d}P_+\frac{p_n}{p_n^*}f.$$

$\square$

Corollary 11.4.2 *There exist polynomials α_i, of degree $\leq n-2$, such that*

$$\lambda_i p_n^* p_i - \lambda_n p_n p_i^* = \lambda_i d^*\alpha_i, \quad i = 1,\dots,n-1. \tag{11.40}$$

This holds also formally for $i = n$ with $\alpha_n = 0$.

Proof: Since

$$H_{\frac{n}{d}}\frac{p_i}{d^*} = \lambda_i\frac{p_i^*}{d},$$

it follows that

$$(H_{\frac{n}{d}}^{-1})^*\frac{p_i}{d^*} = \lambda_i^{-1}\frac{p_i^*}{d}.$$

So, using Eq. (11.39), we have

$$\frac{1}{\lambda_n}\frac{d^*}{d}P_+\frac{p_n^*}{p_n}\frac{p_i}{d^*} = \frac{1}{\lambda_i}\frac{p_i^*}{d},$$

that is,

$$\frac{\lambda_n}{\lambda_i}\frac{p_i^*}{d^*} = P_+\frac{p_n^*}{p_n}\frac{p_i}{d^*}.$$

This implies, using a partial fraction decomposition, the existence of polynomials α_i, $i = 1,\dots,n$ such that $\deg\alpha_i < \deg p_n = n-1$, and

$$\frac{p_n^*}{p_n}\frac{p_i}{d^*} = \frac{\lambda_n}{\lambda_i}\frac{p_i^*}{d^*} + \frac{\alpha_i}{p_n},$$

that is, Eq. (11.40) follows. $\square$

We saw, in Theorem 11.4.2, that for the Hankel operator H_ϕ the map $(H_\phi^{-1})^*$ is not a Hankel map. However, there is an equivalent Hankel map. We sum this up in the following.

Theorem 11.4.3 *Let $\phi = n/d \in \mathbf{RH}_-^\infty$. Let $H : X^{d^*} \longrightarrow X^d$ be defined by $H = H_\phi|X^{d^*}$. Then:*

1. *The operator $(H_\phi^{-1})^*$ is equivalent to the Hankel operator $H_{(1/\lambda_n)(d^*p_n/dp_n^*)}$.*

2. *The Hankel operator $H_{(1/\lambda_n)(d^*p_n/dp_n^*)}$ has singular values $\mu_1^{-1} < \cdots < \mu_n^{-1}$, and its Schmidt pairs are $\{p_i^*/d^*,\ p_i/d\}$.*

Proof:

1. We saw that

$$(H_{n/d}^{-1})^* = \frac{d^*}{d} T_{(1/\lambda_n)(p_n/p_n^*)}^*.$$

Since multiplication by d^*/d is a unitary map of X^{d^*} onto X^d, the operator $(H_{n/d}^{-1})^*$ has, by Lemma 11.4.1, the same singular values as $T_{(1/\lambda_n)(p_n/p_n^*)}^*$. These are the same as those of the adjoint operator $T_{(1/\lambda_n)(p_n/p_n^*)}$. However, the last operator is equivalent to the Hankel operator $H_{(1/\lambda_n)(d^* p_n/dp_n^*)}$. Indeed, we compute

$$\frac{d^*}{d} T_{(1/\lambda_n)(p_n/p_n^*)} f = \frac{d^*}{d} P_{H(d/d^*)} \frac{1}{\lambda_n} \frac{p_n}{p_n^*} f = \frac{d^*}{d} \frac{d}{d^*} P_- \frac{d^*}{d} \frac{1}{\lambda_n} \frac{p_n}{p_n^*} f$$

$$= H_{(1/\lambda_n)(d^* p_n/dp_n^*)} f.$$

2. Next we show that this Hankel operator has singular values $\mu_1^{-1} < \cdots < \mu_n^{-1}$ and its Schmidt pairs are $\{p_i^*/d^*, \ p_i/d\}$. Indeed,

$$H_{(d^* p_n)/(dp_n^*)} \frac{p_i^*}{d^*} = P_- \frac{d^* p_n}{dp_n^*} \frac{p_i^*}{d^*} = P_- \frac{p_n p_i^*}{dp_n^*}.$$

Now, from Eq. (11.40) we get $p_n p_i^* = (\lambda_i/\lambda_n) p_n^* p_i - (\lambda_i/\lambda_n) d^* \alpha_i$ or, taking the dual of that equation, $p_n p_i^* = (\lambda_n/\lambda_i) p_n^* p_i + d\alpha_i^*$. So

$$\frac{p_n p_i^*}{dp_n^*} = \frac{\lambda_n}{\lambda_i} \frac{p_n^* p_i}{dp_n^*} + \frac{d\alpha_i^*}{dp_n^*} = \frac{\lambda_n}{\lambda_i} \frac{p_i}{d} + \frac{\alpha_i^*}{p_n^*}.$$

Hence

$$P_- \frac{p_n p_i^*}{dp_n^*} = \frac{\lambda_n}{\lambda_i} \frac{p_i}{d}.$$

Therefore,

$$\frac{1}{\lambda_n} H_{(d^* p_n)/(dp_n^*)} \frac{p_i^*}{d^*} = \frac{1}{\lambda_i} \frac{p_i}{d}. \qquad \square$$

The duality results obtained before allow us now to complete our study of the zero structure of minimal degree solutions of the fundamental polynomial equation (11.30). This in turn leads to an elementary proof of the central theorem in the AAK theory.

Theorem 11.4.4 (Adamjan, Arov, and Krein) *Let* $\phi = (n/d) \in$ **RH**$_-^\infty$.

1. *Let* μ_k *be a singular value of* H_ϕ *satisfying*

$$\mu_1 \geq \cdots \geq \mu_{k-1} > \mu_k = \cdots = \mu_{k+\nu-1} > \mu_{k+\nu} \geq \cdots \geq \mu_n,$$

that is, μ_k *is a singular value of multiplicity* ν. *Let* p_k *be the minimum degree solution of Eq. (11.30) corresponding to* μ_k. *Then the number of antistable zeroes of* p_k *is exactly* $k - 1$.

2. If μ_1 is the largest singular value of H_ϕ and is of multiplicity ν, that
is,

$$\mu_1 = \cdots = \mu_\nu > \mu_{\nu+1} \geq \cdots \geq \mu_n,$$

and p_1 is the corresponding minimum degree solution of Eq. (11.30),
then all of the zeroes of p_1 are stable; this is equivalent to saying that
p_1/d is outer.

Proof:

1. We saw, in the proof of Proposition 11.3.2, that the number of antistable zeroes of p_k is $\geq k - 1$. Now, by Theorem 11.4.3, p_k^* is the
   minimum degree solution of the fundamental equation corresponding to the transfer function $(1/\lambda_n)(d^* p_n/dp_n^*)$ and the singular value
   $\mu_{k+\nu-1}^{-1} = \cdots = \mu_k^{-1}$. Clearly we have

   $$\mu_n^{-1} \geq \cdots \geq \mu_{k+\nu}^{-1} > \mu_{k+\nu-1}^{-1} = \cdots = \mu_k^{-1} > \mu_{k-1}^{-1} \geq \cdots \geq \mu_1^{-1}.$$

   In particular, applying Proposition 11.3.2, the number of antisatble
   zeroes of p_k^* is $\geq n - k - \nu + 1$. Since the degree of p_k^* is $n - \nu$, it
   follows that the number of stable zeroes of p_k^* is $\leq k - 1$. However,
   this is the same as saying that the number of antistable zeroes of p_k
   is $\leq k-1$. Combining the two inequalities, it follows that the number
   of antistable zeroes of p_k is exactly $k - 1$.

2. The first part implies that the minimum degree solution of Eq. (11.30)
   has only stable zeroes. $\square$

We now come to apply some results of the previous section to the case of
a Hankel norm approximation. We use here the characterization, obtained
in Section 7.5, of singular values $\mu_1 \geq \mu_2 \geq \cdots$ of a linear transformation
$A : V_1 \longrightarrow V_2$ as approximation numbers, namely,

$$\mu_k = \inf\{\|A - A_k\| \mid \operatorname{rank} A_k \leq k - 1\}.$$

We shall denote by $\mathbf{RH}_{[k-1]}^\infty$ the set of all rational functions in $\mathbf{RL}^\infty$
that have at most $k - 1$ antistable poles.

Theorem 11.4.5 (Adamjan, Arov, and Krein) *Let* $\phi = (n/d) \in \mathbf{RH}^\infty$
be a scalar, strictly proper, transfer function with n *and* d *coprime polynomials and* d *a monic of degree* n. *Assume that*

$$\mu_1 \geq \cdots \geq \mu_{k-1} > \mu_k = \cdots = \mu_{k+\nu-1} > \mu_{k+\nu} \geq \cdots \geq \mu_n > 0$$

are the singular values of H_ϕ. *Then*

$$\mu_k = \inf\{\|H_\phi - A\| \mid \operatorname{rank} A \leq k - 1\}$$

$$= \inf\{\|H_\phi - H_\psi\| \mid \operatorname{rank} H_\psi \leq k - 1\}$$

$$= \inf\left\{\|\phi - \psi\|_\infty \mid \psi \in \mathbf{RH}_{[k-1]}^\infty\right\}.$$

Moreover, the infimum is attained on a unique function $\psi_k = \phi - (H_\phi f_k/f_k) = \phi - \mu(g/f)$, where $\{f_k, g_k\}$ is an arbitrary Schmidt pair of H_ϕ that corresponds to μ_k.

Proof: Given $\psi \in \mathbf{RH}^\infty_{[k-1]}$, we have, by Kronecker's theorem, that rank $H_\psi = k - 1$. Therefore, we clearly have

$$\mu_k = \inf\{\|H_\phi - A\| \,|\, \text{rank } A \le k - 1\}$$

$$\le \inf\{\|H_\phi - H_\psi\| \,|\, \text{rank } H_\psi \le k - 1\}$$

$$\le \inf\{\|\phi - \psi\|_\infty \,|\, \psi \in \mathbf{RH}^\infty_{[k-1]}\},$$

so the proof will be complete if we can exhibit a function $\psi_k \in \mathbf{RH}^\infty_{[k-1]}$ for which the equality $\mu_k = \|\phi - \psi\|_\infty$ holds. To this end, let p_k be the minimal degree solution of Eq. (11.30), and define $\psi_k = \pi_k/p_k$. From the equation

$$np_k = \lambda_k d^* p_k^* + d\pi_k$$

we get, dividing by dp_k, that

$$\frac{n}{d} - \frac{\pi_k}{p_k} = \lambda_k \frac{d^* p_k^*}{dp_k}.$$

This is of course equivalent to

$$\psi_k = \frac{\pi_k}{p_k} = \frac{n}{d} - \lambda_k \frac{d^* p_k^*}{dp_k} = \phi - \frac{H_\phi f_k}{f_k},$$

as for $f_k = p_k^*/d$ we have $H_\phi f_k = \lambda_k(p_k/d^*)$ and, by Lemma 11.3.1, the ratio $H_\phi f_k/f_k$ is independent of the particular Schmidt pair. So

$$\|\phi - \psi\|_\infty = \|\frac{n}{d} - \frac{\pi_k}{p_k}\|_\infty = \|\lambda_k \frac{d^* p_k^*}{dp_k}\|_\infty = \mu_k.$$

Moreover, $(\pi_k/p_k) \in \mathbf{RH}^\infty_{[k-1]}$, as p_k has exactly $k - 1$ antistable zeroes. □

Corollary 11.4.3 *The polynomials π_k and p_k have no common antistable zeroes.*

Proof: Follows from the fact that rank $H_{\pi_k/p_k} \ge k - 1$. □

We now are ready to give a simple proof of Nehari's theorem in our rational context.

Theorem 11.4.6 (Nehari) *Given a rational function $\phi = (n/d) \in H_-^\infty$ and $n \wedge d = 1$. Then*

$$\sigma_1 = ||H_\phi|| = \inf ||\phi - q||_\infty \quad q \in H_+^\infty,$$

and this infimum is attained on a unique function $q = \phi - \sigma_1(g/f)$, where $\{f, g\}$ is an arbitrary σ_1-Schmidt pair of H_ϕ.

Proof: Let $\sigma_1 = ||H_\phi||$. It follows from Eq. (11.18) and the fact that, for $q \in H_+^\infty$ we have $H_q = 0$, that

$$\sigma_1 = ||H_\phi|| = ||H_\phi - H_q|| = ||H_{\phi-q}|| \leq ||\phi - q||_\infty,$$

and so $\sigma_1 \leq \inf_{q \in H_+^\infty} ||\phi - q||_\infty$.

To complete the proof we will show that there exists a $q \in H_+^\infty$ for which equality holds.

We saw, in Theorem 11.4.4, that for $\sigma_1 = ||H_\phi||$ there exists a stable solution p_1 of

$$np_1 = \lambda d^* p_1^* + d\pi_1.$$

Dividing this equation by dp_1 we get

$$\frac{n}{d} - \frac{\pi_1}{p_1} = \lambda_1 \frac{d^* p_1^*}{dp_1}.$$

So, with

$$q = \frac{\pi_1}{p_1} = \frac{n}{d} - \lambda_1 \frac{d^* p_1^*}{dp_1} \in H^\infty$$

we get

$$||\phi - q||_\infty = \sigma_1 = ||H_\phi||. \qquad \square$$

11.5 Nevanlinna–Pick Interpolation

We now discuss briefly the connection between Nehari's theorem in the rational case and the finite Nevanlinna–Pick interpolation problem, which is described next.

Definition 11.5.1 *Given points $\lambda_1, \ldots, \lambda_n$ in the open right half-plane and complex numbers $c_1, \ldots, c_n$, then $\psi \in \mathbf{RH}_+^\infty$ is a **Nevanlinna–Pick** interpolant if it is a function of minimum $\mathbf{RH}_+^\infty$ norm that satisfies*

$$\psi(\lambda_i) = c_i, \quad i = 1, \ldots, n.$$

We define the polynomial d by

$$d(z) = \prod_{i=1}^n (z - \lambda_i).$$

Clearly, d is antistable and d^* stable. We now construct one $\mathbf{RH}_+^\infty$ interpolant.

Let n be the unique polynomial, with $\deg n < \deg d$, that satisfies the following interpolation constraints:

$$n(\lambda_i) = d^*(\lambda_i)c_i, \qquad i = 1, \ldots, n. \tag{11.41}$$

This interpolant can be constructed easily by Lagrange interpolation or any other equivalent method. We note that, as d^* is stable, $d^*(\lambda_i) \neq 0$ for $i = 1, \ldots, n$ and $(n/d^*) \in \mathbf{RH}_+^\infty$. Moreover, Eq. (11.41) implies that

$$\frac{n(\lambda_i)}{d^*(\lambda_i)} = c_i, \qquad i = 1, \ldots, n, \tag{11.42}$$

that is, n/d^* is an $\mathbf{RH}_+^\infty$ interpolant.

Any other interpolant is of the form $(n/d^*) - (d/d^*)\theta$ for some $\theta \in \mathbf{RH}_+^\infty$. To find

$$\inf_{\theta \in \mathbf{RH}_+^\infty} \left\| \frac{n}{d^*} - \frac{d}{d^*}\theta \right\|_\infty$$

is equivalent, d/d^* being inner, to finding

$$\inf_{\theta \in \mathbf{RH}_+^\infty} \left\| \frac{n}{d} - \theta \right\|_\infty.$$

However, this is just the content of Nehari's theorem and

$$\inf_{\theta \in \mathbf{RH}_+^\infty} \left\| \frac{n}{d} - \theta \right\|_\infty = \sigma_1.$$

Moreover, the minimizing function is

$$\theta = \frac{\pi_1}{p_1} = \frac{n}{d} - \lambda_1 \frac{d^* p_1^*}{d p_1}.$$

Going back to the interpolation problem, we get for the Nevanlinna–Pick interpolant ψ

$$\psi = \frac{n}{d^*} - \frac{d}{d^*}\theta = \frac{d}{d^*}\lambda_1 \frac{d^* p_1^*}{d p_1} = \lambda_1 \frac{p_1^*}{p_1}.$$

Therefore, we have proved the following:

Theorem 11.5.1 *Given the Nevanlinna–Pick interpolation problem of Definition 11.5.1, let $d(z) = \prod_{i=1}^{n}(z - \lambda_i)$, and let n be the minimal degree polynomial satisfying the interpolation constraints*

$$n(\lambda_i) = d^*(\lambda_i)c_i.$$

Let p_1 be the minimal degree solution of

$$np_1 = \lambda_1 d^* p_1^* + d\pi_1,$$

corresponding to the largest singular value σ_1. Then the Nevanlinna–Pick interpolant is given by

$$\psi = \lambda_1 \frac{p_1^*}{p_1}. \qquad \square$$

11.6   Hankel Approximant Singular Values

Our aim in this section is, given a Hankel operator with a rational symbol ϕ, to study the singular values and singular vectors corresponding to the Hankel operators with symbols equal to the best Hankel norm approximant and the Nehari complement. For the simplicity of exposition, in the rest of this chapter we will make the genericity assumption that, for $\phi = (n/d) \in \mathbf{RH}_-^\infty$, all of the singular values of $H_{n/d}$ are simple.

Theorem 11.6.1 *Let $\phi = (n/d) \in \mathbf{RH}_-^\infty$, and let p_i be the minimal degree solutions of*

$$np_i = \lambda_i d^* p_i^* + d\pi_i.$$

Consider the best Hankel norm approximant

$$\frac{\pi_n}{p_n} = \frac{n}{d} - \lambda_n \frac{d^* p_n^*}{dp_n},$$

which corresponds to the smallest nonzero singular value. Then:

1. *$(\pi_n/p_n) \in \mathbf{RH}_-^\infty$ and H_{π_n/p_n} has the singuar values $\sigma_i = |\lambda_i|$, $i = 1, \ldots, n-1$, and the σ_i-Schmidt pairs of H_{π_n/p_n} are given by $\{\alpha_i/p_n^*, \alpha_i^*/p_n\}$, where the α_i are given by*

$$\lambda_i p_n^* p_i - \lambda_n p_n p_i^* = \lambda_i d^* \alpha_i. \tag{11.43}$$

2. *Moreover, we have*

$$\frac{\alpha_i}{p_n^*} = P_{X^{p_n^*}} \frac{p_i}{d^*},$$

that is, the singular vectors of H_{π_n/p_n} are projections of the singular vectors of $H_{n/d}$ onto $X^{p_n^*}$, the orthogonal complement of $\mathrm{Ker}\, H_{\pi_n/p_n} = (p_n/p_n^*)\mathbf{RH}_+^2$.

3. *We have*

$$\|\frac{\alpha_i}{p_n^*}\|^2 = (1-|\frac{\lambda_n}{\lambda_i}|^2)\,\|\frac{p_i}{d^*}\|^2. \tag{11.44}$$

Proof:

1. Rewrite Eq. (11.40) as

$$\lambda_i \frac{p_i}{d^*} - \lambda_n \frac{p_n p_i^*}{d^* p_n^*} = \lambda_i \frac{\alpha_i}{p_n^*}. \tag{11.45}$$

So

$$\lambda_i \frac{\pi_n}{p_n} \frac{p_i}{d^*} - \lambda_n \frac{\pi_n}{p_n} \frac{p_n p_i^*}{d^* p_n^*} = \lambda_i \frac{\pi_n}{p_n} \frac{\alpha_i}{p_n^*}.$$

Projecting on $\mathbf{RH}_-^2$, and recalling that p_n is antistable, we get

$$P_- \frac{\pi_n}{p_n} \frac{p_i}{d^*} = P_- \frac{\pi_n}{p_n} \frac{\alpha_i}{p_n^*}.$$

So

$$\frac{p_i}{d^*} - \frac{\alpha_i}{p_n^*} \in \operatorname{Ker} H_{\pi_n/p_n}.$$

This is also clear from

$$\frac{p_i}{d^*} - \frac{\alpha_i}{p_n^*} = \frac{\lambda_n}{\lambda_i} \frac{p_n}{p_n^*} \frac{p_i}{d^*} \in \frac{p_n}{p_n^*} \mathbf{RH}_+^2 = \operatorname{Ker} H_{\pi_n/p_n}.$$

Now

$$\frac{\pi_n}{p_n} = \frac{n}{d} - \lambda_n \frac{d^* p_n^*}{dp_n},$$

so

$$
\begin{aligned}
H_{\pi_n/p_n} \frac{p_i}{d^*} &= H_{((n/d)-\lambda_n (d^* p_n^*/dp_n))} \frac{p_i}{d^*} \\
&= H_{n/d} \frac{p_i}{d^*} - \lambda_n H_{(d^* p_n^*/dp_n)} \frac{p_i}{d^*} \\
&= \lambda_i \frac{p_i^*}{d} - \lambda_n P_- \frac{d^* p_n^*}{dp_n} \frac{p_i}{d^*} = \lambda_i \frac{p_i^*}{d} - P_- \frac{\lambda_n p_n^* p_i}{dp_n} \\
&= \lambda_i \frac{p_i^*}{d} - \frac{\lambda_n p_n^* p_i}{dp_n} = \lambda_i \frac{p_i^*}{d} - \frac{\{\lambda_i p_n p_i^* - \lambda_i d^* \alpha_i\}}{dp_n} \\
&= \lambda_i \frac{d\alpha_i^*}{dp_n} = \lambda_i \frac{\alpha_i^*}{p_n}.
\end{aligned}
$$

So finally we get

$$H_{\pi_n/p_n} \frac{\alpha_i}{p_n^*} = \lambda_i \frac{\alpha_i^*}{p_n}.$$

2. Note that Eq. (11.45) can be written as

$$\frac{p_i}{d^*} = \frac{\alpha_i}{p_n^*} + \frac{\lambda_n}{\lambda_i} \frac{p_n}{p_n^*} \frac{p_i^*}{d^*}. \tag{11.46}$$

Since $(p_i^*/d^*) \in \mathbf{RH}_+^2$, this yields, projecting on $X^{p_n^*} = \{(p_n/p_n^*) \times \mathbf{RH}_+^2\}^\perp$,

$$P_{X^{p_n^*}} \frac{p_i}{d^*} = \frac{\alpha_i}{p_n^*}.$$

3. Follows from Eq. (11.46), using orthogonality and computing norms. $\qquad \square$

Corollary 11.6.1 *There exist polynomials ζ_i of degree $\leq n-3$ such that*

$$\pi_n \alpha_i - \lambda_i p_n^* \alpha_i^* = p_n \zeta_i, \quad i = 1, \ldots, n-1.$$

Orthogonality Relations

We now present the derivation of some polynomial identities out of the singular value/singular vector equations. We proceed to interpret these relations as orthogonality relations between singular vectors associated with different singular values. Furthermore, the same equations provide useful orthogonal decompositions of singular vectors.

Equation (11.43), and, in fact, more general relations, could be derived directly by simple computations, and this we proceed to explain. Starting from the singular value equations, that is,

$$\begin{cases} np_i = \lambda_i d^* p_i^* + d\pi_i \\[2mm] np_j = \lambda_j d^* p_j^* + d\pi_j, \end{cases}$$

we get

$$0 = d^* \{\lambda_i p_i^* p_j - \lambda_j p_i p_j^*\} + d\{\pi_i p_j - \pi_j p_i\}. \tag{11.47}$$

Since d and d^* are coprime, there exist polynomials α_{ij}, of degree $\leq n-2$, for which

$$\lambda_i p_i^* p_j - \lambda_j p_i p_j^* = d\alpha_{ij}, \tag{11.48}$$

and $\pi_i p_j - \pi_j p_i = -d^* \alpha_{ij}$.

These equations have a very nice interpretation as orthogonality relations. In fact, we know that, for any self-adjoint operator, eigenvectors corresponding to different eigenvalues are orthogonal. In particular, this applies to singular vectors. Thus, under our genericity assumption, we have for $i \neq j$,

$$\left(\frac{p_i}{d^*}, \frac{p_j}{d^*} \right)_{\mathbf{RH}_+^2} = 0. \tag{11.49}$$

This orthogonality relation could be derived directly from the polynomial equations by contour integration in the complex plane. Indeed, Eq. (11.48) could be rewritten as

$$\frac{p_i^*}{d} \frac{p_j}{d^*} - \frac{\lambda_j}{\lambda_i} \frac{p_i}{d^*} \frac{p_j^*}{d} = \frac{\alpha_{ij}}{d^*}.$$

This equation can be integrated over the boundary of a half-disk of radius R and centered at origin, and which lies in the right half-plane. Since d^* is stable, the integral on the right-hand side is zero. A standard estimate, using the fact that $\deg p_i^* p_j \leq 2n-2$, leads in the limit, as $R \to \infty$, to

$$\left(\frac{p_j}{d^*}, \frac{p_i}{d^*} \right)_{\mathbf{RH}_+^2} = \frac{\lambda_j}{\lambda_i} \left(\frac{p_i}{d^*}, \frac{p_j}{d^*} \right)_{\mathbf{RH}_+^2}.$$

This indeed implies the orthogonality relation (11.49).

Equation (11.48) can be rewritten as

$$p_i^* p_j = \frac{\lambda_j}{\lambda_i} p_i p_j^* + \frac{1}{\lambda_i} d\alpha_{ij}.$$

However, if we rewrite Eq. (11.48) as $\lambda_j p_i p_j^* = \lambda_i p_i^* p_j - d\alpha_{ij}$, then, after conjugation, we get $p_i^* p_j = (\lambda_i/\lambda_j)p_i p_j^* - (1/\lambda_j)d^* \alpha_{ij}^*$. Equating the two expressions leads to

$$\left(\frac{\lambda_i}{\lambda_j} - \frac{\lambda_j}{\lambda_i} \right) p_i p_j^* = \frac{1}{\lambda_i} d\alpha_{ij} + \frac{1}{\lambda_j} d^* \alpha_{ij}^*.$$

Putting $j = i$, we get $\alpha_{ii} = 0$. Otherwise, we have

$$p_i p_j^* = \frac{1}{\lambda_i^2 - \lambda_j^2} \{\lambda_j d\alpha_{ij} - \lambda_i d^* \alpha_{ij}^*\}. \tag{11.50}$$

Conjugating this last equation and interchanging indices leads to

$$p_i p_j^* = \frac{1}{\lambda_j^2 - \lambda_i^2} \{\lambda_j d\alpha_{ji} - \lambda_i d^* \alpha_{ji}^*\}. \tag{11.51}$$

Comparing the two expressions leads to

$$\alpha_{ji} = -\alpha_{ij}. \tag{11.52}$$

We continue by studying two special cases. For the case $j = n$, we put $\alpha_i = \alpha_{in}^*/\lambda_i$ to obtain

$$\lambda_i p_n^* p_i - \lambda_n p_n p_i^* = \lambda_i d^* \alpha_i, \quad i = 1, \ldots, n-1, \tag{11.53}$$

or

$$\lambda_i p_i^* p_n - \lambda_n p_i p_n^* = \lambda_i d\alpha_i^*, \quad i = 1, \ldots, n-1. \tag{11.54}$$

From Eq. (11.47) it follows that $\pi_n p_i - \pi_i p_n = \lambda_i d^* \alpha_i^*$, or, equivalently, $\pi_n^* p_i^* - \pi_i^* p_n^* = \lambda_i d\alpha_i$, If we specialize now to the case $i = 1$, we obtain $\pi_n p_1 - \pi_1 p_n = \lambda_1 d^* \alpha_1^*$, which, after dividing through by $p_1 p_n$ and conjugating, yields

$$\frac{\pi_n^*}{p_n^*} - \frac{\pi_1^*}{p_1^*} = \lambda_1 \frac{d\alpha_1}{p_1^* p_n^*}. \tag{11.55}$$

Similarly, starting from Eq. (11.48) and putting $i = 1$, we get $\lambda_1 p_1^* p_i - \lambda_i p_1 p_i^* = d\alpha_{1i}$; by also putting $\beta_i = \lambda_1^{-1} \alpha_{1i}$, we get

$$\lambda_1 p_1^* p_i - \lambda_i p_1 p_i^* = \lambda_1 d\beta_i, \tag{11.56}$$

and of course $\beta_1 = 0$. This can be rewritten as

$$p_1^* p_i - \frac{\lambda_i}{\lambda_1} p_1 p_i^* = d\beta_i, \tag{11.57}$$

or

$$p_1 p_i^* - \frac{\lambda_i}{\lambda_1} p_1^* p_i = d^* \beta_i^*. \tag{11.58}$$

This is equivalent to

$$\frac{p_i^*}{d^*} = \frac{\beta_i^*}{p_1} + \frac{\lambda_i}{\lambda_1} \frac{p_1^*}{p_1} \frac{p_i}{d^*}. \tag{11.59}$$

We note that Eq. (11.59) is nothing else but the orthogonal decomposition of p_i^*/d^* relative to $\mathbf{RH}_+^2 = X^{p_1} \oplus (p_1^*/p_1)\mathbf{RH}_+^2$. Therefore we have

$$||\frac{\beta_i^*}{p_1}||^2 = ||\frac{p_i^*}{d^*}||^2 - \frac{|\lambda_i|^2}{|\lambda_1|^2}||\frac{p_i}{d^*}||^2 = ||\frac{p_i}{d^*}||^2 \left(1 - \frac{|\lambda_i|^2}{|\lambda_1|^2}\right). \tag{11.60}$$

Notice that if we specialize Eq. (11.54) to the case of $i = 1$ and Eq. (11.56) to the case $i = n$, we obtain the relation $\beta_n = \alpha_1^*$.

In the rest of this section we will shed some light on intrinsic duality properties of problems of Hankel norm approximation and extensions. Results strongly suggesting an underlying duality appeared before. This in fact turns out to be the case, although the duality analysis is far from being obvious. In the process we will prove a result dual to Theorem 11.6.1. While this analysis, after leading to the form of the Schmidt pairs in Theorem 11.6.3, is not necessary for the proof, it is felt that it is of independent interest and its omission would leave all intuition out of the exposition.

The analysis of duality can be summed up in the following scheme, which exhibits the relevant Hankel operators, their singular values, and the corresponding Schmidt pairs.

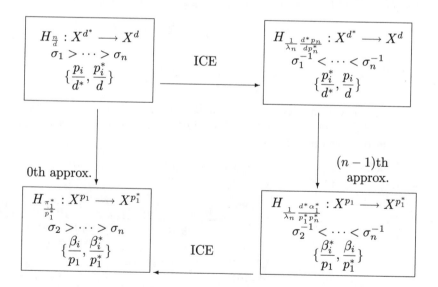

We would like to analyze the truncation that corresponds to the largest singular value. To this end we invert (I) the Hankel operator $H_{n/d}$ and conjugate (C) it, that is, take its adjoint, as in Theorem 11.4.2. This operation preserves Schmidt pairs and inverts singular values; however, the operator so obtained is not a Hankel operator. This we correct by replacing it with (E) an equivalent Hankel operator. This preserves singular values but changes the Schmidt pairs. Thus ICE in the previous diagram stands for a sequence of these three operations. To the Hankel operator so obtained, that is, to $H_{(1/\lambda_n)\,(d^*p_n/dp_n^*)}$, we apply Theorem 11.6.1, which leads to the Hankel operator $H_{(1/\lambda_n)\,(d^*\alpha_1^*/p_1^*p_n^*)}$. This is done in Theorem 11.6.2. To this Hankel operator we apply again the sequence of three operations ICE, and this leads to Theorem 11.6.3.

We proceed to study this Hankel map.

Theorem 11.6.2 *For the Hankel operator* $H_{(1/\lambda_n)\,(d^*p_n/dp_n^*)}$, *the Hankel norm approximant corresponding to the least singular value, that is, to* σ_1^{-1}, *is* $(1/\lambda_n)\,(d^*\alpha_1^*/p_1^*p_n^*)$.

For the Hankel operator $H_{(1/\lambda_n)\,(d^*\alpha_1^*/p_1^*p_n^*)}$, *we have*

1.
$$\operatorname{Ker} H_{(1/\lambda_n)\,(d^*\alpha_1^*/p_1^*p_n^*)} = \frac{p_1^*}{p_1}\mathbf{RH}_+^2.$$

2.
$$X^{p_1} = \left\{\frac{p_1^*}{p_1}\mathbf{RH}_+^2\right\}^\perp.$$

3. *The singular values of* $H_{(1/\lambda_n)\,(d^*\alpha_1^*/p_1^*p_n^*)}$ *are* $\sigma_2^{-1} < \cdots < \sigma_n^{-1}$.

4. *The Schmidt pairs of* $H_{(1/\lambda_n)\,(d^*\alpha_1^*/p_1^*p_n^*)}$ *are* $\{\beta_i^*/p_1,\ \beta_i/p_1^*\}$, *where*
$$\frac{\beta_i^*}{p_1} = P_{X^{p_1}}\frac{p_i^*}{d^*}$$

or

$$\frac{\beta_i^*}{p_1} = \frac{p_i^*}{d^*} - \frac{\lambda_i}{\lambda_1}\frac{p_1^*}{p_1}\frac{p_i}{d^*}.$$

Proof: By Theorem 11.4.3 the Schmidt pairs for $H_{(1/\lambda_n)\,(d^*p_n/dp_n^*)}$ are $\{p_i^*/d^*,\ p_i/d\}$. Therefore, the best Hankel norm approximant associated to σ_1^{-1} is, also using Eq. (11.53),

$$\frac{1}{\lambda_n}\frac{d^*p_n}{dp_n^*} - \frac{1}{\lambda_1}\frac{p_1}{d}\bigg/\frac{p_1^*}{d^*} = \frac{1}{\lambda_n}\frac{d^*p_n}{dp_n^*} - \frac{1}{\lambda_1}\frac{d^*p_1}{dp_1^*}$$

$$= \frac{1}{\lambda_1\lambda_n}\frac{d^*\{\lambda_1 p_n p_1^* - \lambda_n p_1 p_n^*\}}{dp_1^*p_n^*} = \frac{1}{\lambda_1\lambda_n}\frac{d^*\{\lambda_1 d\alpha_1^*\}}{dp_1^*p_n^*}$$

$$= \frac{1}{\lambda_n}\frac{d^*\alpha_1^*}{p_1^*p_n^*}.$$

1. Let $f \in (p_1^*/p_1)\mathbf{RH}_+^2$, that is, $f = (p_1^*/p_1)g$, for some $g \in \mathbf{RH}_+^2$. Then
$$P_-\frac{1}{\lambda_n}\frac{d^*\alpha_1^*}{p_1^*p_n^*}\frac{p_1^*}{p_1}g = \frac{1}{\lambda_n}P_-\frac{d^*\alpha_1^*}{p_1 p_n^*}g = 0,$$
as $(d^*\alpha_1^*/p_1 p_n^*) \in \mathbf{RH}_+^\infty$ and $g \in \mathbf{RH}_+^2$.

Conversely, let $f \in \text{Ker } H_{(1/\lambda_n)\,(d^*\alpha_1^*/p_1^*p_n^*)}$, that is,

$$P_-\frac{d^*\alpha_1^*}{p_1^*p_n^*}f = 0.$$

This implies that $p_1^* \mid d^*\alpha_1 f$. Now p_1 and d are coprime as the first polynomial is stable, whereas the second is antistable. This naturally implies the coprimeness of p_1^* and d^*. We have also

$$\lambda_1 p_1^* p_n - \lambda_n p_1 p_n^* = \lambda_1 d\alpha_1^*.$$

If p_1^* and α_1^* are not coprime, then, by the previous equation, p_1^* has a common factor with $p_1 p_n^*$. However, p_1 and p_n are coprime as the first is stable and the second antistable. So are p_1 and p_1^*, and for the same reason. Therefore we must have that f/p_1^* is analytic in the right half-plane. So $(p_1/p_1^*)f \in \mathbf{RH}_+^2$, that is, $f \in (p_1^*/p_1)\mathbf{RH}_+^2$.

2. Follows from the previous part.

3. This is a consequence of Theorem 11.6.1.

4. Also follows from Theorem 11.6.1, as the singular vectors of $H_{(1/\lambda_n)\,(d^*\alpha_1^*/p_1^*p_n^*)}$ are given by

$$P_{X^{p_1}}\frac{p_i^*}{d^*}.$$

This can be computed. Indeed, starting from Eq. (11.56), we have

$$\beta_i = \frac{\lambda_1 p_1^* p_i - \lambda_i p_1 p_i^*}{\lambda_1 d}.$$

We compute

$$
\begin{aligned}
\lambda_i \beta_n p_i^* - \lambda_n \beta_i p_n^* &= \lambda_i \{\frac{\lambda_1 p_1^* p_n - \lambda_n p_1 p_n^*}{\lambda_1 d}\}p_i^* \\
&\quad - \lambda_n \{\frac{\lambda_1 p_1^* p_i - \lambda_i p_1 p_i^*}{\lambda_1 d}\}p_n^* \\
&= \frac{\lambda_1 p_1^*}{\lambda_1 d}\{\lambda_i p_i^* p_n - \lambda_n p_i p_n^*\} \\
&= \frac{p_1^*}{d}\{\lambda_i d\alpha_i^*\}\{\lambda_i p_i^* p_n - \lambda_n p_i p_n^*\} \\
&= \lambda_i p_1^* \alpha_i^*.
\end{aligned}
$$

Recalling that

$$\frac{\beta_i^*}{p_1} = \frac{p_i^*}{d^*} - \frac{\lambda_i}{\lambda_1}\frac{p_1^*}{p_1}\frac{p_i}{d^*} \tag{11.61}$$

and $\beta_n = \alpha_1^*$, we have

$$H_{(1/\lambda_n)\,(d^*\alpha_1^*/p_1^*p_n^*)}\frac{\beta_i^*}{p_1} - \frac{1}{\lambda_i}\frac{\beta_i}{p_1^*} = P_-\frac{1}{p_1^*p_n^*p_1}\left\{\frac{1}{\lambda_n}d^*\beta_n\beta_i^* - \frac{1}{\lambda_i}p_1p_n^*\beta_i\right\}.$$

Thus it suffices to show that the last term is zero. Now, from Eq. (11.58),

$$d^*\beta_i^* = p_1p_i^* - \frac{\lambda_i}{\lambda_1}p_1^*p_i.$$

Hence

$$P_-\frac{1}{p_1^*p_n^*p_1}\left\{\frac{1}{\lambda_n}d^*\beta_n\beta_i^* - \frac{1}{\lambda_i}p_1p_n^*\beta_i\right\}$$

$$= P_-\frac{1}{p_1^*p_n^*p_1}\left\{\frac{1}{\lambda_n}p_1p_i^*\beta_n - \frac{\lambda_i}{\lambda_1}p_ip_1^*\beta_n - \frac{1}{\lambda_i}p_1p_n^*\beta_i\right\}$$

$$= P_-\frac{1}{p_1^*p_n^*p_1}\left\{\frac{p_1}{\lambda_i\lambda_n}(\lambda_ip_i^*\beta_n - \lambda_np_n^*\beta_i) - \frac{\lambda_i}{\lambda_1}p_ip_1^*\beta_n\right\}$$

$$= P_-\frac{1}{p_1^*p_n^*p_1}\left\{\frac{p_1}{\lambda_i\lambda_n}\lambda_1\lambda_ip_1^*\alpha_i^* - \frac{\lambda_i}{\lambda_1}p_ip_1^*\beta_n\right\}$$

$$= P_-\frac{1}{p_n^*p_1}\left\{\frac{\lambda_1}{\lambda_n}p_1\alpha_i^* - \frac{\lambda_i}{\lambda_1}p_i\beta_n\right\} = 0,$$

as $p_1p_n^*$ is stable. □

Theorem 11.6.3 Let $\phi = (n/d) \in \mathbf{RH}_-^\infty$ and $n \wedge d = 1$, that is, d is antistable. Let π_1/p_1 be the optimal causal, that is, $\mathbf{RH}_+^\infty$, approximant to n/d. Then:

1. $(n/d) - (\pi_1/p_1)$ is all-pass.

2. The singular values of the Hankel operator $H_{\pi_1^*/p_1^*}$ are $\sigma_2 > \cdots > \sigma_n$, and the corresponding Schmidt pairs of $H_{\pi_1^*/p_1^*}$ are $\{\beta_i/p_1,\ \beta_i^*/p_1^*\}$, where the β_i are defined by

$$\beta_i = \frac{\lambda_1 p_1^* p_i - \lambda_i p_1 p_i^*}{\lambda_1 d}.$$

Proof:

1. Follows from the equality

$$\frac{n}{d} - \frac{\pi_1}{p_1} = \lambda_1\frac{d^*p_1^*}{dp_1}.$$

2. We saw, in Eq. (11.55), that $\pi_1^*/p_1^* = (\pi_n^*/p_n^*) - \lambda_1(d\alpha_1/p_1^*p_n^*)$. Since $(\pi_n^*/p_n^*) \in \mathbf{RH}^\infty$, the associated Hankel operator is zero. Hence, $H_{\pi_1^*/p_1^*} = H_{-\lambda_1(d\alpha_1/p_1^*p_n^*)}$. Thus we have to show that

$$H_{-\lambda_1(d\alpha_1/p_1^*p_n^*)} \frac{\beta_i}{p_1} = \lambda_i \frac{\beta_i^*}{p_1^*}.$$

To this end, we start from Eq. (11.57), which, multiplied by α_1, yields

$$d\alpha_1\beta_i = p_1^*p_i\alpha_1 - \frac{\lambda_i}{\lambda_1}p_1p_i^*\alpha_1. \tag{11.62}$$

This in turn implies that

$$\frac{d\alpha_1}{p_1^*p_n^*}\frac{\beta_i}{p_1} = \frac{p_i\alpha_1}{p_1p_n^*} - \frac{\lambda_i}{\lambda_1}\frac{\alpha_1p_i^*}{p_1^*p_n^*}. \tag{11.63}$$

Since $(p_i\alpha_1/p_1p_n^*) \in H_+^2$, we have

$$-\lambda_1 P_- \frac{d\alpha_1}{p_1^*p_n^*}\frac{\beta_i}{p_1} = \lambda_i P_- \frac{\alpha_1p_i^*}{p_1^*p_n^*}. \tag{11.64}$$

All we have to do is to obtain a partial fraction decomposition of the last term. To this end, we go back to Eq. (11.57), from which we get

$$d\beta_i p_n = p_1^*p_ip_n - \frac{\lambda_i}{\lambda_1}p_1p_i^*p_n$$

$$d\beta_n p_i = p_1^*p_np_i - \frac{\lambda_i}{\lambda_1}p_1p_n^*p_i. \tag{11.65}$$

Hence,

$$d(\beta_n p_i - d\beta_i p_n) = \frac{p_1}{\lambda_1}\{\lambda_i p_i^*p_n - \lambda_n p_n^*p_i\} = \frac{p_1}{\lambda_1}\{\lambda_i d\alpha_i^*\} \tag{11.66}$$

and

$$\beta_n p_i - d\beta_i p_n = \frac{\lambda_i}{\lambda_1}p_1\alpha_i^*. \tag{11.67}$$

Now,

$$\beta_n^*p_i^* - \beta_i^*p_n^* = \alpha_1p_i^* - \beta_i^*p_n^* = \frac{\lambda_i}{\lambda_1}p_1^*\alpha_i.$$

Dividing through by $p_1^*p_n^*$, we get

$$\frac{\alpha_1p_i^*}{p_1^*p_n^*} = \frac{\beta_i^*}{p_1^*} + \frac{\lambda_i}{\lambda_1}\frac{\alpha_i}{p_n^*},$$

and from this it follows that

$$P_- \frac{\alpha_1p_i^*}{p_1^*p_n^*} = \frac{\beta_i^*}{p_1^*}.$$

Using Eq. (11.64), we have

$$-\lambda_1 P_-\frac{d\alpha_1}{p_1^*p_n^*}\frac{\beta_i}{p_1} = \lambda_i\frac{\beta_i^*}{p_1^*}, \tag{11.68}$$

and this completes the proof.   □

As an immediate corollary, we obtain the dual of Corollary 11.6.1.

Corollary 11.6.2 *There exist polynomials ω_i of degree $\leq n-2$ such that*

$$\pi_1^*\beta_i - \lambda_i p_1\beta_i^* = p_1^*\omega_i, \quad i = 2,\ldots,n. \tag{11.69}$$

There is another way of looking at duality, and this is summed up in the following diagram.

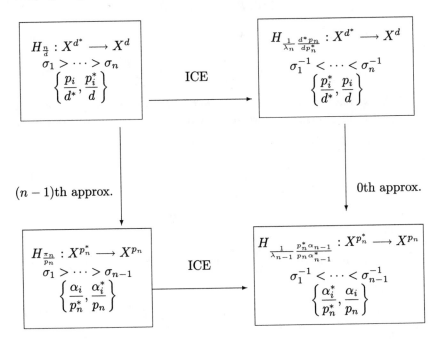

We will not go into the details except for the following.

Theorem 11.6.4 *The Hankel operator $(1/\lambda_{n-1})H_{(p_n^*\alpha_{n-1})/(p_n\alpha_{n-1}^*)} : X^{p_n^*}$ $\longrightarrow X^{p_n}$ has singular values $\sigma_1^{-1} < \cdots < \sigma_{n-1}^{-1}$ and Schmidt pairs $\{\alpha_i^*/p_n^*, \alpha_i/p_n\}$.*

Proof: Starting from

$$\pi_n\alpha_i = \lambda_i p_n^*\alpha_i^* + p_n\zeta_i$$

$$\pi_n\alpha_j = \lambda_j p_n^*\alpha_j^* + p_n\zeta_j,$$

we get $0 = p_n^*\{\lambda_i \alpha_j \alpha_i^* - \lambda_j \alpha_i \alpha_j^*\} + p_n\{\alpha_j \zeta_i - \alpha_i \zeta_j\}$. For $j = n - 1$, we can write $\lambda_i \alpha_{n-1} \alpha_i^* - \lambda_{n-1} \alpha_i \alpha_{n-1}^* = \lambda_i p_n \kappa_i$, or $\alpha_{n-1} \alpha_i^* = p_n \kappa_i + (\lambda_{n-1}/\lambda_i) \alpha_i \alpha_{n-1}^*$, that is,

$$\frac{\alpha_i^*}{p_n} = \frac{\kappa_i}{\alpha_{n-1}} + \frac{\lambda_{n-1}}{\lambda_i} \frac{\alpha_{n-1}^*}{\alpha_{n-1}} \frac{\alpha_i}{p_n}.$$

Thus we have

$$\frac{1}{\lambda_{n-1}} H_{(p_n^* \alpha_{n-1})/(p_n \alpha_{n-1}^*)} \frac{\alpha_i^*}{p_n^*} = \frac{1}{\lambda_{n-1}} P_- \frac{1}{p_n \alpha_{n-1}^*} \{p_n \kappa_i + \frac{\lambda_{n-1}}{\lambda_i} \alpha_i \alpha_{n-1}^*\}$$

$$= \frac{1}{\lambda_{n-1}} P_- \frac{\kappa_i}{\alpha_{n-1}^*} + \frac{1}{\lambda_i} P_- \frac{\alpha_i}{p_n} = \frac{1}{\lambda_i} \frac{\alpha_i}{p_n}. \qquad \square$$

11.7 Exercises

1. Define the map $J : \mathbf{RL}^2 \longrightarrow \mathbf{RL}^2$ by $Jf(z) = f^*(z) = \overline{f(-\bar{z})}$. Clearly this is a unitary map in $\mathbf{RL}^2$ and it satisfies $J\mathbf{RH}_{\pm}^2 = \mathbf{RH}_{\mp}^2$. Let H_ϕ be a Hankel operator.

   (a) Show that $JP_- = P_+ J$ and $JH_\phi = H_\phi^* J \mid \mathbf{RH}_+^2$.

   (b) If $\{f, g\}$ is a Schmidt pair of H_ϕ, then $\{Jf, Jg\}$ is a Schmidt pair of H_ϕ^*.

   (c) Let $\sigma > 0$. Show that

   i. The map $\hat{U} : \mathbf{RH}_+^2 \longrightarrow \mathbf{RH}_+^2$ defined by

   $$\hat{U}f = \frac{1}{\sigma} H^* Jf$$

   is a bounded linear operator in $\mathbf{RH}_+^2$.

   ii. $\mathrm{Ker}\,(H^* H - \sigma^2 I)$ is an invariant subspace for $\hat{U}$.

   iii. The map $U : \mathrm{Ker}\,(H^* H - \sigma^2 I) \longrightarrow \mathrm{Ker}\,(H^* H - \sigma^2 I)$ defined by $U = \hat{U} \mid \mathrm{Ker}\,(H^* H - \sigma^2 I)$ satisfies $U = U^* = U^{-1}$.

   iv. Defining

   $$K_+ = \frac{I + U}{2} \qquad K_- = \frac{I - U}{2},$$

   show that $K_\pm$ are orthogonal projections and

   $$\mathrm{Ker}\,(I - U) = \mathrm{Im}\,K_+,$$

   $$\mathrm{Ker}\,(I + U) = \mathrm{Im}\,K_-,$$

   and

   $$\mathrm{Ker}\,(H^* H - \sigma^2 I) = \mathrm{Im}\,K_+ \oplus \mathrm{Im}\,K_-.$$

   Also, $U = K_+ - K_-$, which is the spectral decomposition of U, that is, U is a signature operator.

2. Let σ be a singular value of the Hankel operator H_ϕ, and let J be defined as before. Let p be the minimal degree solution of Eq. (11.30). Assume that $\deg d = n$ and $\deg p = m$. If $\epsilon = \lambda/\sigma$, show the following:

$$\dim \operatorname{Ker}(H_\phi - \lambda J) = \left[\frac{n - m + 1}{2} \right]$$

$$\dim \operatorname{Ker}(H_\phi + \lambda J) = \left[\frac{n - m}{2} \right]$$

$$\dim \operatorname{Ker}(H_\phi^* H_\phi - \sigma^2 I) = n - m$$

$$\dim \operatorname{Ker}(H_\phi - \lambda J) - \dim \operatorname{Ker}(H_\phi + \lambda J) = \begin{cases} 0 & n - m \text{ even} \\ 1 & n - m \text{ odd.} \end{cases}$$

3. A minimal realization

$$\left(\begin{array}{c|c} A & B \\ \hline C & D \end{array} \right)$$

of an asymptotically stable (antistable) transfer function G is called **balanced** if there exists a diagonal matrix $\operatorname{diag}(\sigma_1, \ldots, \sigma_n)$ such that

$$\begin{cases} A\Sigma + \Sigma\tilde{A} &=& -B\tilde{B} \\ \tilde{A}\Sigma + \Sigma A &=& -\tilde{C}C \end{cases}$$

(with the minus signs removed in the antistable case). The matrix Σ is called the **gramian** of the system and its diagonal entries are called the **system singular values**. Let $\phi = (n/d) \in \mathbf{RH}^\infty$. Assume that all singular values of $H_{n/d}$ are distinct. Let p_i be the minimal degree solutions of the FPE, normalized so that $||p_i/d^*||^2 = \sigma_i$. Show that:

(a) The system singular values are equal to the singular values of $H_{n/d}$.

(b) The function ϕ has a balanced realization of the form

$$\begin{cases} A &=& \left(\dfrac{\epsilon_j b_i b_j}{\lambda_i + \lambda_j} \right) \\ B &=& (b_1, \ldots, b_n)\tilde{} \\ C &=& (\epsilon_1 b_1, \ldots, \epsilon_n b_n) \\ D &=& \phi(\infty), \end{cases}$$

with

$$b_i = (-1)^n \epsilon_i p_{i,n-1}$$
$$c_i = (-1)^{n-1} p_{i,n-1} = -\epsilon_i b_i.$$

(c) The balanced realization is sign-symmetric. Specifically, with $\epsilon_i = \lambda_i/\sigma_i$ and $J = \text{diag}(\epsilon_1, \ldots, \epsilon_n)$, we have

$$JA = \tilde{A}J, \quad JB = \tilde{C}.$$

(d) Relative to a conformal block decomposition

$$\Sigma = \begin{pmatrix} \Sigma_1 & 0 \\ 0 & \Sigma_2 \end{pmatrix}, \quad A = \begin{pmatrix} A_{11} & A_{12} \\ A_{21} & A_{22} \end{pmatrix}$$

we have

$$B\tilde{B} = \begin{pmatrix} A_{11}\Sigma_1 + \Sigma_1\tilde{A}_{11} & A_{12}\Sigma_2 + \Sigma_1\tilde{A}_{21} \\ A_{21}\Sigma_1 + \Sigma_2\tilde{A}_{12} & A_{22}\Sigma_2 + \Sigma_2\tilde{A}_{22} \end{pmatrix}.$$

(e) With respect to the constructed balanced realization, we have the following representation:

$$\frac{p_i^*(z)}{d(z)} = C(zI - A)^{-1}e_i.$$

4. Let $\phi = (n/d) \in \mathbf{RH}^\infty$, and let π_1/p_1 of Theorem 11.4.6 be the Nehari extension of n/d. With respect to the balanced realization of ϕ given in exercise 3, show that π_1/p_1 admits a balanced realization

$$\left(\begin{array}{c|c} A_N & B_N \\ \hline C_N & D_N \end{array} \right)$$

with

$$\begin{cases} A_N &= -\left(\dfrac{\epsilon_j \mu_i \mu_j b_i b_j}{\lambda_i + \lambda_j} \right) \\[2ex] B_N &= (\mu_2 b_2, \ldots, \mu_n b_n) \\[2ex] C_N &= (\mu_2 \epsilon_2 b_2, \ldots, \mu_n \epsilon_n b_n) \\[2ex] D_N &= \lambda_1, \end{cases}$$

where

$$\mu_i = \sqrt{\left(\frac{\lambda_1 - \lambda_i}{\lambda_1 + \lambda_i} \right)} \quad \text{for } i = 2, \ldots, n.$$

11.8   Notes and Remarks

The topics covered in this chapter have their roots early in this century in the work of Caratheodory and Fejer [1911] and Schur [1917]. In a classic paper, Schur [1917], a complete study of contractive analytic functions in the unit disk is given. The problem is attacked by a variety of methods, including what has become known as the Schur algorithm as well as the usage of quadratic forms. One of the early problems considered was the minimum H^∞-norm extension of polynomials. This in turn led to the Nevanlinna–Pick interpolation problem. Many of the interpolation problems can be recast as best approximation problems, which are naturally motivated by computational considerations.

From different considerations, Nehari [1957] was led to proving his celebrated theorem. The connection to modern operator theory and interpolation problems was made in Sarason [1967], leading to the general commutant lifting theorem. The connection between the commutant lifting theorem and the theory of Hankel operators was clarified by Page [1970].

Independently of Sarason's work, Krein and his students started a detailed study of Hankel operators, motivated by classical extension and approximation problems. This was consolidated in a series of articles that became known as the AAK theory.

The relevance of the AAK theory to control problems was recognized by Helton and Dewilde. It was immediately and widely taken up, due to the influence of the work of Zames, which brought a resurgence of frequency domain methods. A most influential contribution to state space aspects of the Hankel norm approximation problems was given in the classic paper by Glover [1984]. The content of this chapter is based mostly on Fuhrmann [1991, 1994a]. Young [1988] is a very readable account of elementary Hilbert space operator theory and contains an infinite-dimensional version of the AAK theory. Nikolskii [1985] is a comprehensive study of the shift operator.

Proposition 11.2.2 has a very simple proof. However, if we remove the assumption of rationality and work in the H^∞ setting, then the equivalence of the coprimeness condition $\sum_{i=1}^{s} |a_i(z)| \geq \delta > 0$ for all z in the open right half-plane and the existence of an H^∞ solution to the Bezout identity is a deep result in analysis, due to Carleson [1962].

What we refer to in this chapter as Kronecker's theorem was not proved in this form. Actually, Kronecker proved that an infinite Hankel matrix has finite rank if and only if its generating function is rational. Hardy spaces were introduced later. The duality theory outlined in Section 6 has far-reaching extensions; see Fuhrmann [1994b].

References

Adamjan, V.M., D.Z. Arov and M.G. Krein [1968a] "Infinite Hankel matrices and generalized problems of Caratheodory-Fejer and F. Riesz," *Funct. Anal. Appl.* 2, 1–18.

Adamjan, V.M., D.Z. Arov and M.G. Krein [1968b] "Infinite Hankel matrices and generalized problems of Caratheodory-Fejer and I. Schur," *Funct. Anal. Appl.* 2, 269–281.

Adamjan, V.M., D.Z. Arov and M.G. Krein [1971] "Analytic properties of Schmidt pairs for a Hankel operator and the generalized Schur-Takagi problem," *Math. USSR Sbornik* 15, 31–73.

Adamjan, V.M., D.Z. Arov and M.G. Krein [1978] "Infinite Hankel block matrices and related extension problems," *Amer. Math. Soc. Transl.* (series 2), 111, 133–156.

Axler, S. [1995] "Down with determinants," *Amer. Math. Monthly*, 102, 139–154.

Beurling, A. [1949] "On two problems concerning linear transformations in Hilbert space," *Acta Math.*, 81, 239–255.

Caratheodory, C., and L. Fejér [1911] "Über den Zusammenhang der Extremen von harmonischen Funktionen mit ihren coeffizienten und über den Picard-Landauschen Satz," *Rend. Circ. Mat. Palermo* 32, 218–239.

Carleson, L. [1962] "Interpolation by bounded analytic functions and the corona problem," *Ann. of Math.* 76, 547–559.

Davis, P.J. [1979] *Circulant Matrices*, J. Wiley, New York.

Douglas, R.G., H.S. Shapiro and A.L. Shields [1971] "Cyclic vectors and invariant subspaces for the backward shift," *Ann. Inst. Fourier, Grenoble* 20(1), 37–76.

Dunford, N., and J.T. Schwartz [1958] *Linear Operators, Part I*, Interscience, New York.

Dunford, N., and J.T. Schwartz [1963] *Linear Operators, Part II*, Interscience, New York.

Duren, P. [1970] *Theory of H^p Spaces*, Academic Press, New York.

Fejér, L. [1915] "Über trigonometrische Polynome," *J. Reine und Angew. Math.* 146, 53–82.

Fuhrmann, P.A. [1968a] "On the corona problem and its application to spectral problems in Hilbert space," *Trans. Amer. Math. Soc.* 132, 55–67.

Fuhrmann, P.A. [1968b] "A functional calculus in Hilbert space based on operator valued analytic functions," *Israel J. Math.* 6, 267–278.

Fuhrmann, P.A. [1975] "On Hankel operator ranges, meromorphic pseudocontinuation and factorization of operator valued analytic functions," *J. Lon. Math. Soc.* (2) 13, 323–327.

Fuhrmann, P.A. [1976] "Algebraic system theory: An analyst's point of view," *J. Franklin Inst.* 301, 521–540.

Fuhrmann, P.A. [1977] "On strict system equivalence and similarity," *Int. J. Contr.* 25, 5–10.

Fuhrmann, P.A. [1981a] *Linear Systems and Operators in Hilbert Space*, McGraw-Hill, New York.

Fuhrmann, P.A. [1981b] "Polynomial models and algebraic stability criteria," *Proceedings of Joint Workshop on Synthesis of Linear and Nonlinear Systems*, Bielefeld, June 1981, 78–90.

Fuhrmann, P.A. [1991] "A polynomial approach to Hankel norm and balanced approximations," *Lin. Alg. Appl.* 146, 133–220.

Fuhrmann, P.A. [1994a] "An algebraic approach to Hankel norm approximation problems," in Differential Equations, Dynamical Systems, and Control Science, the *L. Markus Festschrift*, Edited by K.D. Elworthy, W.N. Everitt, and E.B. Lee, M. Dekker, New York, 523–549.

Fuhrmann, P.A. [1994b] "A duality theory for robust control and model reduction," *Lin. Alg. Appl.*, 203–204, 471–578.

Gantmacher, F.R. [1959] *The Theory of Matrices*, Chelsea, New York.

Garnett, J.B. [1981] *Bounded Analytic Functions*, Academic Press, New York.

Glover, K. [1984] "All optimal Hankel-norm approximations and their L^∞-error bounds," *Int. J. Contr.* 39, 1115–1193.

Glover, K. [1986] "Robust stabilization of linear multivariable systems, relations to approximation," *Int. J. Contr.* 43, 741–766.

Gohberg, I.C., and M.G. Krein [1969] *Introduction to the Theory of Nonselfadjoint Operators*, Amer. Math. Soc., Providence.

Gragg, W.B., and A. Lindquist [1983] "On the partial realization problem," *Lin. Alg. Appl.* 50, 277–319.

Halmos, P.R. [1958] *Finite-Dimensional Vector Spaces*, Van Nostrand, Princeton.

Helmke, U., and P. A. Fuhrmann [1989] "Bezoutians," *Lin. Alg. Appl.* 122–124, 1039–1097.

Hermite, C. [1856] "Sur le nombre des racines d'une equation algebrique comprise entre des limites donnes," *J. Reine Angew. Math.* 52, 39–51.

Hoffman, K. [1962] *Banach Spaces of Analytic Functions*, Prentice-Hall, Englewood Cliffs.

Hoffman, K., and R. Kunze [1961] *Linear Algebra*, Prentice-Hall, Englewood Cliffs.

Hurwitz, A. [1895] "Uber die bedingungen, unter welchen eine Gleichung nur Wurzeln mit negativen reelen Teilen besitzt," *Math. Annal.* 46, 273–284.

Kailath, T. [1980] *Linear Systems*, Prentice-Hall, Englewood Cliffs.

Kalman, R.E. [1968] "Lectures on controllability and observability," in *Proceedings C.I.M.E. Summer School at Pontecchio Marconi*, Bologna, 1–149.

Kalman, R.E. [1969] "Algebraic characterization of polynomials whose zeros lie in algebraic domains," *Proc. Nat. Acad. Sci.* 64, 818–823.

Kalman, R.E. [1970] "New algebraic methods in stability theory," In *Proceeding V. International Conference on Nonlinear Oscillations*, Kiev.

Kalman, R.E., P. Falb and M. Arbib [1969] *Topics in Mathematical System Theory*, McGraw-Hill, New York.

Krein, M.G., and M.A. Naimark [1936] "The method of symmetric and Hermitian forms in the theory of the separation of the roots of algebraic equations," English translation in *Linear and Multilinear Algebra* 10 [1981], 265–308.

Lang, S. [1965] *Algebra*, Addison-Wesley, Reading.

Liapunov, A.M. [1893] "Probleme general de la stabilite de mouvement," *Ann. Fac. Sci. Toulouse* 9 (1907), 203–474. (French translation of the Russian paper published in *Comm. Soc. Math. Kharkow*).

Malcev, A.I. [1963] *Foundations of Linear Algebra*, W.H. Freeman & Co., San Francisco.

Maxwell, J.C. [1868] "On governors," *Proc. Roy. Soc.* Ser. A, 16, 270–283.

Nehari, Z. [1957] "On bounded bilinear forms," *Ann. of Math.* 65, 153–162.

Nikolskii, N.K. [1985] *Treatise on the Shift Operator*, Springer-Verlag, Berlin.

Page, L. [1970] "Applications of the Sz.-Nagy and Foias lifting theorem," *Indiana Univ. Math. J.* 20, 135–145.

Prasolov, V.V. [1994] *Problems and Theorems in Linear Algebra*, Trans. of Math. Monog. 134, Amer. Math. Soc., Providence.

Rosenbrock, H.H. [1970] *State-Space and Multivariable Theory*, J. Wiley, New York.

Rota, G.C. [1960] "On models for linear operators," *Comm. Pure and Appl. Math.* 13, 469–472.

Routh, E.J. [1877] *A Treatise on the Stability of a Given State of Motion*, Macmillan, London.

Sarason, D. [1967] "Generalized interpolation in H^∞," *Trans. Amer. Math. Soc.* 127, 179–203.

Schur, I. [1917] "Über Potenzreihen, die im Innern des Einheitskreiss beschränkt sind," *J. Reine und Angew. Math.* 147, 205–232; and 148, 122–145.

Sz.-Nagy, B., and C. Foias [1970] *Harmonic analysis of Operators on Hilbert Space*, North Holland, Amsterdam.

Vidyasagar, M. [1985] *Control System Synthesis: A Coprime Factorization Approach*, M.I.T. Press, Cambridge.

Van der Waerden, B.L. [1931] *Moderne Algebra*, Springer-Verlag, Berlin.

Willems, J.C., and P.A. Fuhrmann [1992] "Stability theory for high order systems," *Lin. Alg. Appl.* 167, 131–149.

Young, N. [1988] *An Introduction to Hilbert Space*, Cambridge University Press, Cambridge.

Index

Universitext   *(continued)*